Programming Skills
for Data Science

The Pearson Addison-Wesley Data and Analytics Series

livelessons▶
Building Spark Applications
Jonathan Dinu
video

PRACTICAL **DATA SCIENCE**
with
Hadoop and **Spark**
Designing and Building Effective Analytics at Scale
OFER MENDELEVITCH
CASEY STELLA
DOUGLAS EADLINE

R for **Everyone**
Advanced Analytics and Graphics SECOND EDITION
JARED P. LANDER

EXPERT
HADOOP ADMINISTRATION
Managing, Tuning and Securing Spark, YARN, and HDFS
SAM R. ALAPATI

livelessons▶
Data Science Fundamentals
Jonathan Dinu
video

Visit **informit.com/awdataseries** for a complete list of available publications.

The **Pearson Addison-Wesley Data and Analytics Series** provides readers with practical knowledge for solving problems and answering questions with data. Titles in this series primarily focus on three areas:

1. **Infrastructure:** how to store, move, and manage data
2. **Algorithms:** how to mine intelligence or make predictions based on data
3. **Visualizations:** how to represent data and insights in a meaningful and compelling way

The series aims to tie all three of these areas together to help the reader build end-to-end systems for fighting spam; making recommendations; building personalization; detecting trends, patterns, or problems; and gaining insight from the data exhaust of systems and user interactions.

Make sure to connect with us!
informit.com/socialconnect

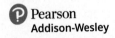 Pearson Addison-Wesley    inform**IT.com** the trusted technology learning source   O'REILLY® Safari

Programming Skills for Data Science

Start Writing Code to Wrangle, Analyze, and Visualize Data with R

Michael Freeman
Joel Ross

✦✦Addison-Wesley

Boston • Columbus • New York • San Francisco • Amsterdam • Cape Town
Dubai • London • Madrid • Milan • Munich • Paris • Montreal • Toronto • Delhi • Mexico City
São Paulo • Sydney • Hong Kong • Seoul • Singapore • Taipei • Tokyo

For information about buying this title in bulk quantities, or for special sales opportunities (which may include electronic versions; custom cover designs; and content particular to your business, training goals, marketing focus, or branding interests), please contact our corporate sales department
at corpsales@pearsoned.com or (800) 382-3419.

For government sales inquiries, please contact governmentsales@pearsoned.com.

For questions about sales outside the U.S., please contact intlcs@pearson.com.

Visit us on the Web: informit.com/aw

Library of Congress Control Number: 2018953978

ISBN-13: 978-0-13-513310-1
ISBN-10: 0-13-513310-6

1 18

❖

To our students who challenged us to develop better resources, and our families who supported us in the process.

❖

Contents

Foreword   **xi**

Preface   **xiii**

Acknowledgments   **xvii**

About the Authors   **xix**

I: Getting Started   1

1   Setting Up Your Computer   3
1.1   Setting up Command Line Tools   4
1.2   Installing `git`   5
1.3   Creating a GitHub Account   6
1.4   Selecting a Text Editor   6
1.5   Downloading the R Language   7
1.6   Downloading RStudio   8

2   Using the Command Line   9
2.1   Accessing the Command Line   9
2.2   Navigating the File System   11
2.3   Managing Files   15
2.4   Dealing with Errors   18
2.5   Directing Output   20
2.6   Networking Commands   20

II: Managing Projects   25

3   Version Control with `git` and GitHub   27
3.1   What Is `git`?   27
3.2   Configuration and Project Setup   30
3.3   Tracking Project Changes   32
3.4   Storing Projects on GitHub   36
3.5   Accessing Project History   40
3.6   Ignoring Files from a Project   42

4   Using Markdown for Documentation   45
4.1   Writing Markdown   45
4.2   Rendering Markdown   48

III: Foundational R Skills   51

5 Introduction to R   53
   5.1   Programming with R   53
   5.2   Running R Code   54
   5.3   Including Comments   58
   5.4   Defining Variables   58
   5.5   Getting Help   63

6 Functions   69
   6.1   What Is a Function?   69
   6.2   Built-in R Functions   71
   6.3   Loading Functions   73
   6.4   Writing Functions   75
   6.5   Using Conditional Statements   79

7 Vectors   81
   7.1   What Is a Vector?   81
   7.2   Vectorized Operations   83
   7.3   Vector Indices   88
   7.4   Vector Filtering   90
   7.5   Modifying Vectors   92

8 Lists   95
   8.1   What Is a List?   95
   8.2   Creating Lists   96
   8.3   Accessing List Elements   97
   8.4   Modifying Lists   100
   8.5   Applying Functions to Lists with `lapply()`   102

IV: Data Wrangling   105

9 Understanding Data   107
   9.1   The Data Generation Process   107
   9.2   Finding Data   108
   9.3   Types of Data   110
   9.4   Interpreting Data   112
   9.5   Using Data to Answer Questions   116

10 Data Frames   119

10.1 What Is a Data Frame?   119

10.2 Working with Data Frames   120

10.3 Working with CSV Data   124

11 Manipulating Data with dplyr   131

11.1 A Grammar of Data Manipulation   131

11.2 Core dplyr Functions   132

11.3 Performing Sequential Operations   139

11.4 Analyzing Data Frames by Group   142

11.5 Joining Data Frames Together   144

11.6 dplyr in Action: Analyzing Flight Data   148

12 Reshaping Data with tidyr   155

12.1 What Is "Tidy" Data?   155

12.2 From Columns to Rows: gather( )   157

12.3 From Rows to Columns: spread( )   158

12.4 tidyr in Action: Exploring Educational Statistics   160

13 Accessing Databases   167

13.1 An Overview of Relational Databases   167

13.2 A Taste of SQL   171

13.3 Accessing a Database from R   175

14 Accessing Web APIs   181

14.1 What Is a Web API?   181

14.2 RESTful Requests   182

14.3 Accessing Web APIs from R   189

14.4 Processing JSON Data   191

14.5 APIs in Action: Finding Cuban Food in Seattle   197

V: Data Visualization   205

15 Designing Data Visualizations   207

15.1 The Purpose of Visualization   207

15.2 Selecting Visual Layouts   209

15.3 Choosing Effective Graphical Encodings   220

15.4 Expressive Data Displays   227

15.5 Enhancing Aesthetics   229

16 Creating Visualizations with `ggplot2`   231

16.1 A Grammar of Graphics   231

16.2 Basic Plotting with `ggplot2`   232

16.3 Complex Layouts and Customization   238

16.4 Building Maps   248

16.5 ggplot2 in Action: Mapping Evictions in San Francisco   252

17 Interactive Visualization in R   257

17.1 The `plotly` Package   258

17.2 The `rbokeh` Package   261

17.3 The `leaflet` Package   263

17.4 Interactive Visualization in Action: Exploring Changes to the City of Seattle   266

VI: Building and Sharing Applications   273

18 Dynamic Reports with R Markdown   275

18.1 Setting Up a Report   275

18.2 Integrating Markdown and R Code   279

18.3 Rendering Data and Visualizations in Reports   281

18.4 Sharing Reports as Websites   284

18.5 R Markdown in Action: Reporting on Life Expectancy   287

19 Building Interactive Web Applications with Shiny   293

19.1 The Shiny Framework   293

19.2 Designing User Interfaces   299

19.3 Developing Application Servers   306

19.4 Publishing Shiny Apps   309

19.5 Shiny in Action: Visualizing Fatal Police Shootings   311

20 Working Collaboratively   319

20.1 Tracking Different Versions of Code with Branches   319

20.2 Developing Projects Using Feature Branches   329

20.3 Collaboration Using the Centralized Workflow   331

20.4 Collaboration Using the Forking Workflow   335

21 Moving Forward   341

21.1 Statistical Learning   341

21.2 Other Programming Languages   342

21.3 Ethical Responsibilities   343

Index   345

Foreword

The data science skill set is ever-expanding to include more and more of the analytics pipeline. In addition to fitting statistical and machine learning models, data scientists are expected to ingest data from different file formats, interact with APIs, work at the command line, manipulate data, create plots, build dashboards, and track all their work in git. By combining all of these components, data scientists can produce amazing results. In this text, Michael Freeman and Joel Ross have created *the* definitive resource for new and aspiring data scientists to learn foundational programming skills.

Michael and Joel are best known for leveraging visualization and front-end interfaces to compose explanations of complex data science topics. In addition to their written work, they have created interactive explanations of statistical methods, including a particularly clarifying and captivating introduction to hierarchical modeling. It is this sensibility and deep commitment to demystifying complicated topics that they bring to their new book, which teaches a plethora of data science skills.

This tour of data science begins by setting up the local computing environment such as text editors, RStudio, the command line, and git. This lays a solid foundation—that is far too often glossed over—making it easier to learn core data skills. After this, those core skills are given attention, including data manipulation, visualization, reporting, and an excellent explanation of APIs. They even show how to use git collaboratively, something data scientists all too often neglect to integrate into their projects.

Programming Skills for Data Science lives up to its name in teaching the foundational skills needed to get started in data science. This book provides valuable insights for both beginners and those with more experience who may be missing some key knowledge. Michael and Joel made full use of their years of teaching experience to craft an engrossing tutorial.

—Jared Lander, series editor

Preface

Transforming data into actionable information requires the ability to clearly and reproducibly wrangle, analyze, and visualize that data. These skills are the foundations of *data science*, a field that has amplified our collective understanding of issues ranging from disease transmission to racial inequities. Moreover, the ability to *programmatically interact with data* enables researchers and professionals to quickly discover and communicate patterns in data that are often difficult to detect. Understanding how to write code to work with data allows people to engage with information in new ways and on larger scales.

The existence of free and open source software has made these tools accessible to anyone with access to a computer. The purpose of this book is to teach people how to leverage programming to ask questions of their data sets.

Focus of the Book

This book revolves around the practical steps needed to *program for data science* using the R programming language. It takes a holistic approach to teaching the topic, recognizing that an entire ecosystem of tools and technologies is needed to do this. While writing code is a core part of being a data scientist (and this book), many more foundational skills must be acquired as part of this journey. Data science requires installing and configuring software to write, execute, and manage code; tracking the version of (and changes to) your projects; leveraging core concepts from computer science to understand how to accomplish a given task; accessing and processing data from a variety of sources; leveraging visual communication to expose patterns in your data; and building applications to share insights with others. The purpose of this text is to help people develop a strong foundation across these areas so that they can enter the data science field (or bring data science to *their field*).

Who Should Read This Book

This book is written for people with no programming or data science experience, though it would still be helpful for people active in the field. This book was originally developed to support a course in the Informatics undergraduate degree program at the University of Washington, so it is (not surprisingly) well suited for college students interested in entering the data science field. We also believe that *anyone* whose job involves working with data can benefit from learning how to reproducibly create analyses, visualizations, and reports.

If you are interested in pursuing a career in data science, or if you use data on a regular basis and want to use programming techniques to gain information from that data, then this text is for you.

Book Structure

The book is divided into six sections, each of which is summarized here.

Part I: Getting Started

This section walks through the steps of downloading and installing necessary software for the rest of the book. More specifically, Chapter 1 details how to install a text editor, Bash terminal, the R interpreter, and the RStudio program. Then, Chapter 2 describes how to use the command line for basic file system navigation.

Part II: Managing Projects

This section walks through the technical basis of project management, including keeping track of the version of your code and producing documentation. Chapter 3 introduces the git software to track line-by-line code changes, as well as the corresponding popular code hosting and collaboration service *GitHub*. Chapter 4 then describes how to use Markdown to produce the well-structured and -styled documentation needed for sharing and presenting data.

Part III: Foundational R Skills

This section introduces the R programming language, the primary language used throughout the book. In doing so, it introduces the basic syntax of the language (Chapter 5), describes fundamental programming concepts such as functions (Chapter 6), and introduces the basic data structures of the language: vectors (Chapter 7), and lists (Chapter 8).

Part IV: Data Wrangling

Because the most time-consuming part of data science is often loading, formatting, exploring, and reshaping data, this section of the book provides a deep dive into the best ways to *wrangle* data in R. After introducing techniques and concepts for understanding the structure of real-world data (Chapter 9), the book presents the data structure most commonly used for managing data in R: the data frame (Chapter 10). To better support working with this data, the book then describes two *packages* for programmatically interacting with the data: dplyr (Chapter 11), and tidyr (Chapter 12). The last two chapters of the section describe how to load data from databases (Chapter 13) and web-based data services with application programming interfaces (APIs) (Chapter 14).

Part V: Data Visualization

This section of the book focuses on the conceptual and technical skills necessary to design and build *visualizations* as part of the data science process. It begins with an overview of data visualization principles (Chapter 15) to guide your choices in designing visualizations. Chapter 16 then describes in granular detail how to use the ggplot2 visualization package in R. Finally, Chapter 17 explores the use of three additional R packages for producing engaging interactive visualizations.

Part VI: Building and Sharing Applications

As in any domain, data science insights are valuable only if they can be shared with and understood by others. The final section of the book focuses on using two different approaches to creating interactive platforms to share your insights (directly from your R program!). Chapter 18 uses the R

Markdown framework to transform analyses into sharable documents and websites. Chapter 19 takes this a step further with the Shiny framework, which allows you to create interactive web applications using R. Chapter 20 then describes approaches for working on collaborative teams of data scientists, and Chapter 21 details how you can further your education beyond this book.

Book Conventions

Throughout the book, you will see computer code appear inline with the text, as well as in distinct blocks. When code appears inline, it will appear in monospace font. A distinct code block looks like this:

```
# This is a comment - it describes the code that follows
# The next line of code prints the text "Hello world!"
print("Hello world!")
```

The text in the code blocks is colored to reflect the syntax of the programming language used (typically the R language). Example code blocks often include values that you need to replace. These replacement values appear in UPPER_CASE_FONT, with words separated by underscores. For example, if you need to work with a folder of your choosing, you would put the name of your folder where it says FOLDER_NAME in the code. Code sections will all include comments: in programming, *comments* are bits of text that are not interpreted as computer instructions—they aren't code, they're just notes about the code! While a computer is able to understand the code, comments are there to help *people* understand it. Tips for writing your own descriptive comments are discussed in Chapter 5.

To guide your reading, we also include five types of special callout notes:

> **Tip**: These boxes provide best practices and shortcuts that can make your life easier.

> **Fun Fact**: These boxes provide interesting background information on a topic.

> **Remember**: These boxes reinforce key points that are important to keep in mind.

> **Caution**: These boxes describe common mistakes and explain how to avoid them.

> **Going Further**: These boxes suggest resources for expanding your knowledge beyond this text.

Throughout the text there are instructions for using specific keyboard keys. These are included in the text in lowercase monospace font. When multiple keys need to be pressed at the same time, they are separated by a plus sign (+). For example, if you needed to press the Command and "c" keys at the same time, it would appear as Cmd+c.

Whenever the cmd key is used, Windows users should instead use the Control (ctrl) key.

How to Read This Book

The individual chapters in this book will walk you through the process of programming for data science. Chapters often build upon earlier examples and concepts (particularly through Part III and Part IV).

This book includes a large number of code examples and demonstrations, with reported output and results. That said, the best way to learn to program is to *do it*, so we highly recommend that as you read, you type out the code examples and try them yourself! Experiment with different options and variations—if you're wondering how something works or if an option is supported, the best thing to do is try it yourself. This will help you not only practice the actual *writing* of code, but also better develop your own mental model of how data science programs work.

Many chapters conclude by applying the described techniques to a real data set in an *In Action* section. These sections take a *data-driven approach* to understanding issues such as gentrification, investment in education, and variation in life expectancy around the world. These sections use a *hands-on* approach to using new skills, and all code is available online.[1]

As you move through each chapter, you may want to complete the accompanying set of online exercises.[2] This will help you practice new techniques and ensure your understanding of the material. Solutions to the exercises are also available online.

Finally, you should know that this text does not aim to be comprehensive. It is both impractical and detrimental to learning to attempt to explain every nuance and option in the R language and ecosystem (particularly to people who are just starting out). While we discuss a large number of popular tools and packages, the book cannot explain all possible options that exist now or will be created in the future. Instead, this text aims to provide a *primer* on each topic—giving you enough details to understand the basics and to get up and running with a particular data science programming task. Beyond those basics, we provide copious links and references to further resources where you can explore more and dive deeper into topics that are relevant or of interest to you. This book will provide the foundations of using R for data science—it is up to each reader to apply and build upon those skills.

Accompanying Code

To guide your learning, a set of online exercises (and their solutions) is available for each chapter. The complete analysis code for all seven *In Action* sections is also provided. See the book website[3] for details.

> Register your copy of *Programming Skills for Data Science* on the InformIT site for convenient access to updates and/or corrections as they become available. To start the registration process, go to informit.com/register and log in or create an account. Enter the product ISBN (9780135133101) and click Submit. Look on the Registered Products tab for an Access Bonus Content link next to this product, and follow that link to access any available bonus materials. If you would like to be notified of exclusive offers on new editions and updates, please check the box to receive email from us.

---

[1] **In-Action Code:** https://github.com/programming-for-data-science/in-action
[2] **Book Exercises:** https://github.com/programming-for-data-science
[3] https://programming-for-data-science.github.io

Acknowledgments

We would like to thank the University of Washington Information School for providing us with an environment in which to collaborate and develop these materials. We had the support of many faculty members—in particular, David Stearns (who contributed to the materials on version control) as well as Jessica Hullman and Ott Toomet (who provided initial feedback on the text). We also thank Kevin Hodges, Jason Baik, and Jared Lander for their comments and insights, as well as Debra Williams Cauley, Julie Nahil, Rachel Paul, Jill Hobbs, and the staff at Pearson for their work bringing this book to press.

Finally, this book would not have been possible without the extraordinary open source community around the R programming language.

About the Authors

Michael Freeman is a Senior Lecturer at the University of Washington Information School, where he teaches courses in data science, interactive data visualization, and web development. Prior to his teaching career, he worked as a data visualization specialist and research fellow at the Institute for Health Metrics and Evaluation. There, he performed quantitative global health research and built a variety of interactive visualization systems to help researchers and the public explore global health trends.

Michael is interested in applications of data science to social justice, and holds a Master's in Public Health from the University of Washington. (His faculty page is at https://faculty.washington.edu/mikefree/.)

Joel Ross is a Senior Lecturer at the University of Washington Information School, where he teaches courses in web development, mobile application development, software architecture, and introductory programming. While his primary focus is on teaching, his research interests include games and gamification, pervasive systems, computer science education, and social computing. He has also done research on crowdsourcing systems, human computation, and encouraging environmental sustainability.

Joel earned his M.S. and Ph.D. in Information and Computer Sciences from the University of California, Irvine. (His faculty page is at https://faculty.washington.edu/joelross/.)

Getting Started

The first part of this book is designed to help you install necessary software for doing data science (Chapter 1), and to introduce you to the syntax needed to provide text-based instructions to your computer using the command line (Chapter 2). Note that all of the software that you will download is free, and instructions are included for both Mac and Windows operating systems.

1

Setting Up Your Computer

In order to write code to work with data, you will need to use a number of different (free) software programs for *writing, executing,* and *managing* your code. This chapter details which software you will need and explains how to install those programs. While there are a variety of options for each task, we discuss software programs that are largely supported within the data science community, and whose popularity continues to grow.

It is an unfortunate reality that one of the most frustrating and confusing barriers to working with code is getting your machine properly set up. This chapter aims to provide sufficient information for setting up your machine and troubleshooting the installation process.

In short, you will need to install the following programs, each of which is described in detail in the following sections.

For Writing Code

There are two different programs that we suggest you use for writing code:

- **RStudio**: An *integrated development environment* (IDE) for writing and executing R code. This will be your primary work environment for doing data science. You will also need to install the R software so that RStudio will be able to execute your code (discussed later in this section).

- **Atom**: A lightweight text editor that supports programming in lots of different languages. (Other text editors will also work effectively; some further suggestions are included in this chapter.)

For Managing Code

To manage your code, you will need to install and set up the following programs:

- **git**: An application used to track changes to your files (namely, your code). This is crucial for maintaining an organized project, and can help facilitate collaboration with other developers. This program is already installed on Macs.

- **GitHub**: A web service for hosting code online. You don't actually need to *install* anything (GitHub uses git), but you will need to create a free account on the GitHub website. The corresponding exercises for this book are hosted on GitHub.

For Executing Code

To provide instructions to your machine (i.e., run code), you will need to have an **environment** in which to provide those instructions, while also ensuring that your machine is able to understand the language in which you're writing your code.

- **Bash shell**: A *command line interface* for controlling your computer. This will provide you with a text-based interface you can use to work with your machine. Macs already have a Bash shell program called *Terminal*, which you can use "out of the box." On Windows, installing git will also install an application called *Git Bash*, which you can use as your Bash shell.

- **R**: A programming language commonly used for data science. This is the primary programming language used throughout this book. "Installing R" actually means downloading and installing tools that will let your computer understand and run R code.

The remainder of this chapter has additional information about the purpose of each software system, how to install it, and alternative configurations or options. The programs are described in the order they are introduced in the book (though in many cases, the software programs are used in tandem).

1.1 Setting up Command Line Tools

The command line provides a text-based interface for giving instructions to your computer (much more on this in Chapter 2). As you are getting started with data science, you will largely use the command line for navigating your computer's file structure and executing commands that allow you to keep track of changes to the code you write (i.e., version control with git).

To use the command line, you will need to use a **command shell** (also called a *command prompt* or *terminal*). This computer program provides the interface in which you type commands. In particular, this book discusses the Bash shell, which provides a particular set of commands common to Mac and Linux machines.

1.1.1 Command Line on a Mac

On a Mac, you will want to use the built-in app called **Terminal** as your Bash shell. This application is part of the Mac operating system, so you don't need to install anything. You can open Terminal by searching via Spotlight (press cmd+spacebar together, type in "terminal", then select the app to open it), or by finding it in the Applications > Utilities folder. This will open your Terminal window, as described in Chapter 2.

1.1.2 Command Line on Windows

On Windows, we recommend using **Git Bash** as your Bash shell, which is installed along with git. Open this program to open the command shell. This works great, since you will primarily be using the command line for performing version control.

Alternatively, the 64-bit Windows 10 Anniversary Update (August 2016) includes a version of an integrated Bash shell. You can access this by enabling the subsystem for Linux[1] and then running `bash` in the command prompt.

> **Caution:** Windows includes its own command shell, called *Command Prompt* (previously *DOS Prompt*), but it has a different set of commands and features. If you try to use the commands described in Chapter 2 with *DOS Prompt*, they will not work. For a more advanced Windows Management Framework, you can look into using *Powershell*.[a] Because Bash is more common in open source programming like in this book, we will focus on that set of commands.
>
> ---
> [a] https://docs.microsoft.com/en-us/powershell/scripting/getting-started/getting-started-with-windows-powershell

1.1.3 Command Line on Linux

Most Linux flavors come with a command shell pre-installed; for example, in Ubuntu you can use the *Terminal* application (use `ctrl+alt+t` to open it, or search for it from the Ubuntu dashboard).

1.2 Installing `git`

One of the most important aspects of doing data science is keeping track of the changes that you (and others) make to your code. `git` is a *version control system* that provides a set of commands that you can use to manage changes to written code, particularly when collaborating with other programmers (version control is described in more detail in Chapter 3).

`git` comes pre-installed on Macs, though it is possible that the first time you try to use the tool you will be prompted to install the *Xcode command line developer tools* via a dialog box. You can choose to install these tools, or download the latest version of `git` online.

On Windows, you will need to download[2] the `git` software. Once you have downloaded the installer, double-click on your downloaded file, and follow the instructions to complete installation. This will also install a program called *Git Bash*, which provides a command line (text-based) interface for executing commands on your computer. See Section 1.1.2 for alternative and additional Windows command line tools.

On Linux, you can install `git` using `apt-get` or a similar command. For more information, see the download page for Linux.[3]

---
[1] **Install the Windows subsystem for Linux:** https://msdn.microsoft.com/en-us/commandline/wsl/install_guide
[2] **`git` downloads:** https://git-scm.com/downloads
[3] **`git` download for Linux and Unix:** https://git-scm.com/download/linux

1.3   Creating a GitHub Account

GitHub[4] is a website that is used to store copies of computer code that are being managed with `git`. To use GitHub, you will need to create a free GitHub account.[5] When you register, remember that your profile is public, and future collaborators or employers may review your GitHub account to assess your background and ongoing projects. Because GitHub leverages the `git` software package, you don't need to install anything else on your machine to use GitHub.

1.4   Selecting a Text Editor

While you will be using RStudio to write R code, you will sometimes want to use another text editor that is more lightweight (e.g., runs faster), more robust, or supports a different programming language than R. A coding-focused text editor provides features such as automatic formatting and coloring for easier interpretation of the code, auto-completion, and integration with version control (features that are also available in RStudio).

Many different text editors are available, all of which have slightly different appearances and features. You only need to download and use one of the following programs (we recommend Atom as a default), but feel free to try out different ones until you find something you like (and then evangelize about it to your friends!).

> **Tip:** Programming involves working with many different file types, each of which is indicated by its **extension** (the letters after the `.` in the file name, such as `.pdf`). It is useful to specify that your computer should show these extensions in File Explorer or Finder; see instructions for Windows[a] or for Mac[b] to enable this.
>
> ─────────────
> [a] https://helpx.adobe.com/x-productkb/global/show-hidden-files-folders-extensions.html
> [b] https://support.apple.com/kb/PH25381?locale=en_US

1.4.1   Atom

Atom[6] is a text editor built by the folks at GitHub. As it is an open source project, people are continually building (and making available) interesting and useful extensions to Atom. Atom's built-in spell-check is a great feature, especially for documents that require lots of written text. It also has excellent support for *Markdown*, a markup language used regularly in this book (see Chapter 4). In fact, much of this text was written using Atom!

To download Atom, visit the application's webpage and click the "Download" button to download the program. On Windows, you will download the installer `AtomSetup.exe` file; double-click on that icon to install the application. On a Mac, you will download a zip file; open that file and drag the `Atom.app` file to your "Applications" folder.

─────────────

[4] **GitHub:** https://github.com
[5] **Join GitHub:** https://github.com/join
[6] **Atom:** https://atom.io

Once you've installed Atom, you can open the program and create a new text file (just like you would create a new file with a word processor such as Microsoft Word). When you save a document that is a particular file type (e.g., FILE_NAME.R or FILE_NAME.md), Atom (or any other modern text editor) will apply a language specific color scheme to your text, making it easier to read.

The trick to using Atom more efficiently is to get comfortable with the Command Palette.[7] If you press cmd+shift+p (Mac) or ctrl+shift+p (Windows), Atom will open a small window where you can search for whatever you want the editor to do. For example, if you type in markdown, you can get a list of commands related to Markdown files (including the ability to open up a preview right in Atom).

For more information about using Atom, see the manual.[8]

1.4.2 Visual Studio Code

Visual Studio Code[9] (or VS Code; not to be confused with Visual Studio) is a free, open source editor developed by Microsoft—yes, really. While it focuses on web programming and JavaScript, it readily supports lots of languages, including Markdown and R, and provides a number of extensions for adding even more features. It has a similar command palette to Atom, but isn't quite as nice for editing Markdown specifically. Although fairly new, it is updated regularly and has become one of the authors' main editors for programming.

1.4.3 Sublime Text

Sublime Text[10] is a very popular text editor with excellent defaults and a variety of available extensions (though you will need to manage and install extensions to achieve the functionality offered by other editors out of the box). While the software can be used for free, every 20 or so saves it will prompt you to purchase the full version (an offer that you can decline without loss of functionality).

1.5 Downloading the R Language

The primary programming language used throughout this book is called R.[11] It is a very powerful statistical programming language that is built to work well with large and diverse data sets. Chapter 5 provides a more in-depth introduction to the language.

To program with R, you will need to install the R *Interpreter* on your machine. This software is able to "read" code written in R and use that code to control your computer, thereby "programming" it.

The easiest way to install R is to download it from the Comprehensive R Archive Network (CRAN).[12] Click on the appropriate link for your operating system to find a link to the installer. On a Mac,

---

[7] **Atom Command Palette**: http://flight-manual.atom.io/getting-started/sections/atom-basics/#command-palette
[8] **Atom Flight Manual**: http://flight-manual.atom.io
[9] **Visual Studio Code**: https://code.visualstudio.com
[10] **Sublime Text**: https://www.sublimetext.com/3
[11] **The R Project for Statistical Computing**: https://www.r-project.org
[12] **The Comprehensive R Archive Network (CRAN)**: https://cran.rstudio.com

click the link to download the `.pkg` file for the latest version supported by your computer. Double-click on the `.pkg` file and follow the prompts to install the software. On Windows, follow the Windows link to "*install* R *for the first time*," then click the link to download the latest version of R for Windows. You will need to double-click on the `.exe` file and follow the prompts to install the software.

1.6   Downloading RStudio

While you are able to execute R scripts without a dedicated application, the RStudio program provides a wonderful way to engage with the R language by providing a single interface to write and execute code, search documentation, and view results such as charts and maps. RStudio is described in more detail in Chapter 5. This book assumes you are using RStudio to write R code.

To install the RStudio program, visit the download page,[13] select to "Download" the *free* version of *RStudio Desktop*, and then select the installer for your operating system to download it.

After the download is complete, double-click on the `.exe` or `.dmg` file to run the installer. Follow the steps of the installer, and you should be prepared to use RStudio.

This chapter has walked you through setting up the necessary software for basic data science, including the following programs:

- Bash for controlling your computer

- R for programmatically analyzing and working with data

- RStudio as an IDE for writing and executing R code

- `git` for version control

- Atom as a general text editor for creating and editing documents

With this software installed, you are ready to get started programming for data science!

---

[13] **Download RStudio:** https://www.rstudio.com/products/rstudio/download/

Using the Command Line

The **command line** is an *interface* to a computer—a way for you (the human) to communicate with the machine. Unlike common graphical interfaces that use "windows, icons, menus, and pointers" (i.e., *WIMP*), the command line is *text-based*, meaning you type commands instead of clicking on icons. The command line lets you do everything you would normally do by clicking with a mouse, but by typing in a manner similar to programming! As a data scientist, you will mostly use the command line to manage your files and keep track of your code using a version control system (see Chapter 3).

While the command line is not as friendly or intuitive as a graphical interface, it has the advantage of being both more powerful and more efficient (it's faster to type than to move a mouse, and you can do *lots* of "clicks" with a single command). The command line is also necessary when working on remote servers (other computers that often do not have graphical interfaces enabled). Thus, the command line is an essential tool for data scientists, particularly when working with large amounts of data or files.

This chapter provides a brief introduction to basic tasks using the command line—enough to get you comfortable navigating the interface and to enable you to interpret commands.

2.1 Accessing the Command Line

To use the command line, you will need to open a **command shell** (also known as a *command prompt* or *terminal*). This program provides the interface you type commands into. You should have installed a command shell, here also referred to as "the terminal" or the "command line," as detailed in Chapter 1.

Once you open up the command shell (the Terminal program on Mac, or Git Bash on Windows), you should see something like the screen shown in Figure 2.1.

A command shell is the textual equivalent of having opened up Finder or File Explorer and having it display the user's "Home" folder. While every command shell program has a slightly different interface, most will display at least the following information:

- The **machine** you are currently interfacing with (you can use the command line to control different computers across a network or the internet). In Figure 2.1 the Mac machine (top) is work-laptop1, and the Windows machine (bottom) is is-joelrossm13.

Figure 2.1   Newly opened command shells: Terminal on a Mac (top) and Git Bash on Windows (bottom). Red notes are added.

- The **directory** (folder) you are currently looking at. In Figure 2.1 the Mac directory is ~/Documents, while the Windows directory is ~/Desktop. The ~ is a shorthand for the "home directory": /Users/CURRENT_USER/ on a Mac, or C:/Users/CURRENT_USER/ on Windows.

- The **user** you are logged in as. In Figure 2.1 the users are mikefree (Mac) and joelross (Windows).

- The command **prompt** (typically denoted as the $ symbol), which is where you will type in your commands.

> **Remember:** Lines of code that begin with a pound symbol (#) are *comments*: They are included to explain the code to human readers (they will be ignored by your computer!).

2.2 Navigating the File System

Although the command prompt gives you the name of the folder you are in, you might like more detail about where that folder is. Time to send your first command! At the prompt, type the **pwd** command:

```
# Print the working directory (which folder the shell is currently inside)
pwd
```

This command stands for **p**rint **w**orking **d**irectory (shell commands are highly abbreviated to make them faster to type), and will tell the computer to print the folder you are currently "in." You can see the results of the pwd command (among others) in Figure 2.2.

> **Fun Fact**: Command line functions like pwd actually start a tiny program (app) that does exactly one thing. In this case, the app prints the working directory. When you run a command, you're actually executing a tiny program!

Folders on computers are stored in a hierarchy: each folder has more folders inside it, which have more folders inside them. This produces a **tree** structure similar to the one shown in Figure 2.3.

You describe what folder you are in by putting a slash / between each folder in the tree. Thus /Users/mikefree means "the mikefree folder, which is inside the Users folder." You can optionally include a trailing / at the end of a directory: /Users/mikefree and /Users/mikefree/ are identical. The final / can be useful for indicating that something is a folder, rather than just a file that lacks an extension.

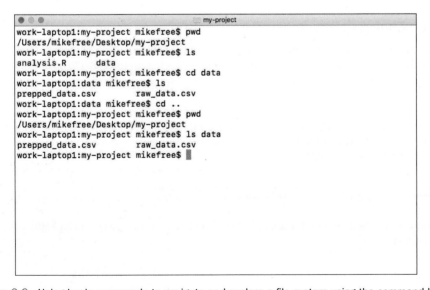

Figure 2.2   Using basic commands to navigate and explore a file system using the command line.

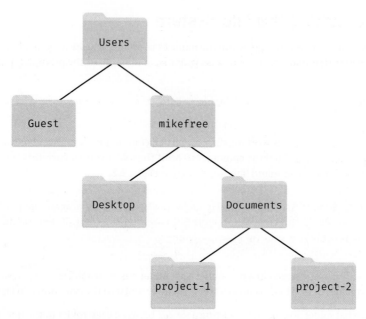

Figure 2.3   The tree structure of directories (folders) on a Mac.

At the very top (or bottom, depending on your point of view) is the **root** / directory—which has no name, and so is just indicated with that single slash. Thus /Users/mikefree really means "the mikefree folder, which is inside the Users folder, which is inside the *root* folder."

2.2.1   Changing Directories

To interact with your files of interest, you will need to change the directory in the command shell. In a graphical system like Finder, you would simply double-click on the folder to open it. On the command line, you perform this type of navigation by typing in commands for what you want to do.

> **Caution:** There is *no clicking* with the mouse on the command line (at all!). This includes clicking to move the cursor to an earlier part of the command you have typed, which can be frustrating. You will need to use your left and right arrow keys to move the cursor instead. However, you can make the cursor jump over segments of your syntax if you hold down the alt (or option) key when you press the left and right arrow keys.

The command to change your directory is called **cd** (for change **d**irectory). You type this command as:

```
# Change the working directory to the child folder with the name "FOLDER_NAME"
cd FOLDER_NAME
```

The first word in this example is the **command,** or what you want the computer to do. In this case, you're issuing the cd command.

The second word is an example of an **argument,** which is a programming term that means "more details about what to do." In this case, you're providing a *required* argument of what folder you want to change to! You will, of course, need to replace FOLDER_NAME with the name of the folder to change to (which need not be in all caps).

For practice, you can try changing to the Desktop folder and printing your current location to confirm that you have moved locations.

> **Tip:** The up and down arrow keys will let you cycle though your previous commands so you don't need to retype them!

2.2.2   Listing Files

In a graphical system, File Explorer or Finder will show you the contents of a folder. The command line doesn't do this automatically; instead, you need another command:

```
# List the contents of the current folder
ls
```

The **ls** command says to **list** the folder contents. If you just issue this command without an argument (as shown in the example), it will list the contents of the current folder. If you include an *optional* argument (e.g., ls FOLDER_NAME), you can "peek" at the contents of a folder you are not currently in (as in Figure 2.2).

> **Caution:** The command line often gives limited or no feedback for your actions. For example, if there are no files in the folder, then ls will show nothing, so it may seem as if it "didn't work." Additionally, when you're typing a password, the letters you type won't be displayed (not even as *) as a security measure.
>
> Just because you don't see any results from your command/typing, that doesn't mean it didn't work! Trust in yourself, and use basic commands like ls and pwd to confirm any changes if you're unsure. Take it slow, one step at a time.

> **Caution:** The ls command is specific to Bash shells, such as Terminal or Git Bash. Other command shells such as the Windows Command Prompt use different commands. This book focuses on the syntax for Bash shells, which are available across all operating systems and are more common on remote servers where the command line becomes a necessity (see Section 2.6).

2.2.3 Paths

Both the cd and ls commands work even for folders that are not "immediately inside" the current directory! You can refer to *any* file or folder on the computer by specifying its **path**. A file's path is "how you get to that file": the list of folders you would need to click through to get to the file, with each folder separated by a slash (/). For example, user mikefree could navigate to his Desktop by describing the *path* to that location in his file system:

```
# Change the directory to the Desktop using an absolute path (from the root)
cd /Users/mikefree/Desktop/
```

This code says to start at the root directory (that initial /), then go to Users, then go to mikefree, then to Desktop. Because this path starts with a specific directory (the root directory), it is called an **absolute path**. No matter what folder you currently happen to be in, that path will refer to the correct directory because it always starts on its journey from the root.

Contrast that with the following example:

```
# Change the directory to `mikefree/Desktop`, relative to the current location
cd mikefree/Desktop/
```

Because this path doesn't have the leading slash, it just says to "go to the mikefree/Desktop/ folder *from the current location.*" This is an example of a **relative path**: it gives you directions to a file *relative to the current folder.* As such, the relative path mikefree/Desktop/ will refer to the correct location only if you happen to be in the /Users folder; if you start somewhere else, who knows where you will end up!

> **Remember**: You should *always* use relative paths, particularly when programming! Because you will almost always be managing multiples files in a project, you should refer to the files *relatively* within your project. That way, your program can easily work across computers. For example, if your code refers to /Users/YOUR_USER_NAME/PROJECT_NAME/data, it can run only on the YOUR_USER_NAME account. However, if you use a relative path within your code (i.e., PROJECT_NAME/data), the program will run on multiple computers—which is crucial for collaborative projects.

You can refer to the "current folder" by using a single dot (.). So the command

```
# List the contents of the current directory
ls .
```

means "list the contents of the current folder" (the same thing you get if you leave off the argument entirely).

If you want to go *up* a directory, you use *two* dots (..) to refer to the **parent** folder (that is, the one that contains this one). So the command

```
# List the contents of the parent directory
ls ..
```

means "list the contents of the folder that contains the current folder."

Note that . and .. act just like folder names, so you can include them anywhere in paths: ../../my_folder says to "go up two directories, and then into my_folder."

> **Tip:** Most command shells like Terminal and Git Bash support **tab-completion**. If you type out just the first few letters of a file or folder name and then press the tab key, it will automatically fill in the rest of the name! If the name is ambiguous (e.g., you type Do and there is both a Documents and a Downloads folder), you can press Tab *twice* to see the list of matching folders. Then add enough letters to distinguish them and press Tab to complete the name. This shortcut will make your life easier.

Additionally, you can use a tilde ~ as shorthand for the *absolute path* to the home directory of the current user. Just as dot (.) refers to "current folder," ~ refers to the user's home directory (usually /Users/USERNAME). And of course, you can use the tilde as part of a path as well (e.g., ~/Desktop is an absolute path to the desktop for the current user).

You can specify a path (relative or absolute) to a *file* as well as to a folder by including the full filename at the end of the folder path—like the "destination":

```
# Use the `cat` command to conCATenate and print the contents of a file
cat ~/Desktop/my_file.txt
```

Files are sometimes discussed as if they were part of the folder that contains them. For example, telling someone to "go up a directory from ~/Desktop/my_file.txt" is just shorthand for saying "go up a directory from the folder that contains ~/Desktop/my_file.txt" (e.g., from ~/Desktop/ to the ~ home directory).

2.3 Managing Files

Once you're comfortable navigating folders using the command line, you can start to use it to do all the same things you would do with Finder or File Explorer, simply by using the correct command. Table 2.1 provides some commonly used commands to get you started using the command line, though there are many more.[1]

Table 2.1 **Basic command line commands**

Command	Behavior
mkdir	**m**a**k**e a **dir**ectory
rm	**r**e**m**ove a file or folder
cp	**c**o**p**y a file from one location to another
open	open a file or folder (Mac only)
start	open a file or folder (Windows only)
cat	con**cat**enate (combine) file contents and display the results
history	show previous commands executed
!!	repeat the previous command

---

[1] An example list of Unix commands can be found here: http://www.lagmonster.org/docs/unix/intro-137.html

> **Caution**: The command line makes it dangerously easy to *permanently delete* multiple files
> or folders and *will not* ask you to confirm that you want to delete them (or move them to the
> "recycling bin"). Be very careful when using the terminal to manage your files, as it is very
> powerful.

Be aware that many of these commands won't print anything when you run them. This often
means that they worked; they just did so quietly. If it *doesn't* work, you will know because you will
see a message telling you so (and why, if you read the message). So just because you didn't get any
output, that doesn't mean you did something wrong—you can use another command (such as ls)
to confirm that the files or folders changed in the way you wanted!

2.3.1 Learning New Commands

Given the evolving nature of the data science field, you will frequently have to learn new things.
One way to do this is to consult the official written descriptions (generically called the
documentation) that explain how the syntax works. This information is available online, but
many command shells (though *not* Git Bash, unfortunately) also include their own manual you
can use to look up commands. On the command line, you can use the **man** command to look up a
specific command in the manual:

```
# View the manual for the `mkdir` command (not available in Git Bash)
man mkdir
```

This command will display the **man**ual for the **mkdir** command (shown in Figure 2.4). Because
manuals are often long, they are opened up in a command line viewer called **less**. You can "scroll"
up and down by using the arrow keys. Press the q key to **q**uit and return to the command prompt.

If you look under "Synopsis," you can see a summary of all the different arguments this command
understands. A few notes about reading this syntax:

- Anything written in brackets [ ] is optional. Arguments that are not in brackets (e.g.,
  directory_name) are required.

- Underlined arguments are ones you choose: You don't actually type the word
  directory_name, but instead insert your own directory name. Contrast this with the
  options: if you want to use the -p option, you need to type -p exactly.

- "Options" (or "*flags*") for command line programs are often marked with a leading hyphen
  - to distinguish them from file or folder names. Options may change the way a command
  line program behaves—just as you might set "easy" or "hard" as the mode in a game. You can
  either write out each option individually or combine them: mkdir -p -v and mkdir -pv
  are equivalent.

  Some options may require an additional argument beyond just indicating a particular
  operation style. In Figure 2.4 you can see that the -m option requires you to specify an
  additional mode argument; check the details in the "Description" for exactly what that
  argument should be.

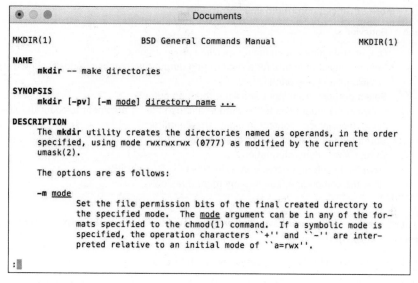

Figure 2.4   The manual ("man") page for the mkdir command, as shown on a Mac Terminal.

Command line manuals ("man pages") are often very difficult to read and understand. Start by looking at just the required arguments (which are usually straightforward), and then search for and use a particular option if you want to change a command's behavior. For practice, read the man page for rm and try to figure out how to delete a folder and not just a single file. Be careful, as this is a good way to unintentionally *permanently* delete files.

> **Tip**: Manual pages are a good example of the kind of syntax explanations you will find when learning about a particular command, but are not necessarily the best way to actually learn to *use* a command. To do that, we recommend more focused resources, such as Michael Hartle's excellent online tutorial *Learn Enough Command Line to Be Dangerous*.[a] Try searching online for a particular command to find many different tutorials and examples!
>
> ---
> [a]https://www.learnenough.com/command-line-tutorial

Some other useful commands you could explore are listed in Table 2.2.

2.3.2   Wildcards

One last note about working with files: since you will often work with multiple files, command shells offer some shortcuts for talking about files with similar names. In particular, you can use an asterisk * as a **wildcard** when referring to files. This symbol acts like a "wild" or "blank" tile in the board game Scrabble—it can be "replaced" by any character (or any set of characters) when determining which file(s) you're talking about.

Table 2.2 **More advanced command line commands**

Command	Behavior
head	Output first *n* lines of an input (specified as an argument)
grep	Search the list of inputs for a pattern and output the matches (**g**lobally search **r**egular **e**xpression and **p**rint)
cut	Select portions from input and write them as output
uniq	Copy unique input lines to the output (and use the -c argument to count the lines!)
sed	"Find and replace" content in input (**s**tream **ed**itor)
sort	Sort input lines (ascending or descending)
wc	Output word count information
curl	Download content/webpage at a URL ("see URL"—get it?)
say	Have the computer speak the argument (Mac only)

- *.txt refers to all files that have .txt at the end. cat *.txt would output the contents of every .txt file in the folder.

- hello* refers to all files whose names start with hello.

- hello*.txt refers to all files that start with hello and end with .txt, no matter how many characters are in the middle (including no characters!).

- *.* refers to all files that have an extension (usually all files).

As an example, you could remove all files that have the extension .txt by using the following syntax (again, be careful!):

```
# Remove all files with the extension `.txt` (careful!)
rm *.txt
```

2.4 Dealing with Errors

The syntax of the command line commands (how you write them out) is rather inflexible. Computers aren't good at figuring out what you meant if you aren't really specific; forgetting a space may result in an entirely different action.

Consider another command: **echo** lets you "echo" (print out) some text. For example, you can echo "Hello World", which is the traditional first computer program written for a new language or environment:

```
# Echo (print) "Hello world" to the terminal
echo "Hello world"
```

What happens if you forget the closing quotation mark (")? You keep pressing enter but the shell just shows a > each time!

What's going on? Because you didn't "close" the quote, the shell thinks you are still typing the message you want to echo! When you press enter, it adds a *line break* instead of ending the command, and the > indicates that you're still going. If you finally close the quote, you will see your multi-line message printed.

Tip: If you ever get stuck in the command line, press ctrl+c (the control and c keys together). This almost always means "cancel" and will "stop" whatever program or command is currently running in the shell so that you can try again. Just remember: "ctrl+c to flee."

If that doesn't work, try pressing the esc key, or typing exit, q, or quit. Those commands will cover *most* command line programs.

This book discusses a variety of approaches to handling errors in computer programs. Many programs do provide **error messages** that explain what went wrong, though the density of these messages may make it tempting to disregard them. If you enter an unrecognized command, the shell will inform you of your mistake, as shown in Figure 2.5. In that example, a simple typo (lx instead of ls) is invalid syntax, yielding a fairly helpful error message (*command not found*—the computer can't find the lx command you are trying to use).

However, forgetting arguments yields different results. In some cases, there will be a default behavior (consider what happens if you enter cd without any arguments). If some arguments are *required* to run a command, the shell may provide you with a brief summary of the command's usage, as shown in Figure 2.6.

Remember: Whenever the command line (or any other code interpreter, for that matter) provides you with feedback, take the time to read the message and think about what the problem might be before you try again.

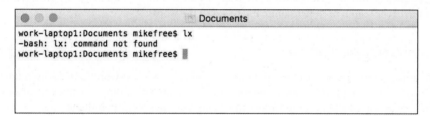

```
work-laptop1:Documents mikefree$ lx
-bash: lx: command not found
work-laptop1:Documents mikefree$ 
```

Figure 2.5   An error on the command line due to a typo in the command name.

```
work-laptop1:Documents mikefree$ mkdir
usage: mkdir [-pv] [-m mode] directory ...
work-laptop1:Documents mikefree$ 
```

Figure 2.6   Executing a command without the required arguments may provide information about how to use the command.

2.5 Directing Output

All commands discussed so far have either modified the file system or printed some output to the terminal. But you can also specify that you want the output to go somewhere else (e.g., to save it to a file for later). This is done using **redirects**. Redirect commands are usually single punctuation marks, because the commands are supposed to be as quick to type (but hard to read!) as possible.

- **>** says "take the output of the command and put it in this file." For example, `echo "Hello World" > hello.txt` will put the outputted text `"Hello World"` into a file called `hello.txt`. Note that this will replace any previous content in the file, or create the file if it doesn't exist. This is a great way to save the output of your command line work!

- **>>** says "take the output of the command and *append* it to the end of this file." This will keep you from overwriting previous content.

- **|** (the *pipe*) says "take the output of this command and send it to the next command." For example, `cat hello.txt | less` would take the output of the `hello.txt` file and send it to the `less` program, which provides the arrow-based "scrolling" interface that man pages use. This is primarily used when you need to "chain" multiple commands together—that is, take the result of one command and send it to the next, and then send the result of that to the *next* command. This type of sequencing is used in R, as described in Chapter 11.

You might not use this syntax on a regular basis, but it is useful to be familiar with the symbols and concepts. Indeed, you can use them to quickly perform some complex data tasks, such as determining how often a word appears in a set of files. For example, the text of this book was written across a number of different files, all with the extension `.Rmd` (more on this in Chapter 18). To see how frequently the word "data" appears in these `.Rmd` files, you could first *search* for the word using the `grep` command (using a wildcard to specify *all* files with that extension), then redirect the output of the search to the `wc` command to count the words:

```
# Search .Rmd files for "data", then perform a word count on the results
grep -io data *.Rmd | wc -w
```

This command shows the value of interest on the command line: The word *"data"* is used 1897 times! While this example is somewhat dense and requires understanding the different *options* each command makes available, it demonstrates the potential power of the command line.

2.6 Networking Commands

One of the most common uses of the command line is for accessing and controlling **remote computers**—that is, machines to which you can connect over the internet. This includes web servers that may host data or reports you wish to share, or cloud-based clusters (such as Microsoft Azure) that may process data much more quickly than your own machine. Because these computers are located somewhere else, you often can't use a mouse, keyboard, and monitor to control them. The command line is the most effective way to control these machines as if you were actually there.

To access a remote computer, you will most commonly use the **ssh** (secure shell) command. ssh is a command utility and protocol for securely transferring information over a network. In this case, the information being transferred will be the commands you run on the machine and the output they produce. At the most basic level, you can use the ssh command to connect to a remote machine by specifying the *host URL* of that machine. For example, if you wanted to connect to a computer at ovid.washington.edu, you would use the command:

```
# Use the secure shell (ssh) utility to connect to a remote computer
ssh ovid.washington.edu
```

However, most remote machines don't let just anyone connect to them for security reasons. Instead, you need to specify your *username* for that machine. You do this by putting the username followed by an @ symbol at the beginning of the host URL:

```
# Use the secure shell (ssh) to connect to a remote computer as mikefree
ssh mikefree@ovid.washington.edu
```

When you give this command, the remote server will prompt you for your password to that machine. Remember that the command line won't show anything (even *) as you type in the password, but it is being entered nonetheless!

> **Tip:** If you connect to a remote server repeatedly, it can become tedious to constantly retype your password. Instead, you can create and use an **ssh key**,[a] which "saves" your authentication information on the server so you don't need to put in a password each time. Check with the administrator of the remote machine for specific instructions.
>
> ---
> [a]https://help.github.com/articles/generating-a- new-ssh-key-and-adding-it-to-the-ssh-agent/

Once you connect to a remote server, you will see the command prompt change to that of the remote server, as shown in Figure 2.7.

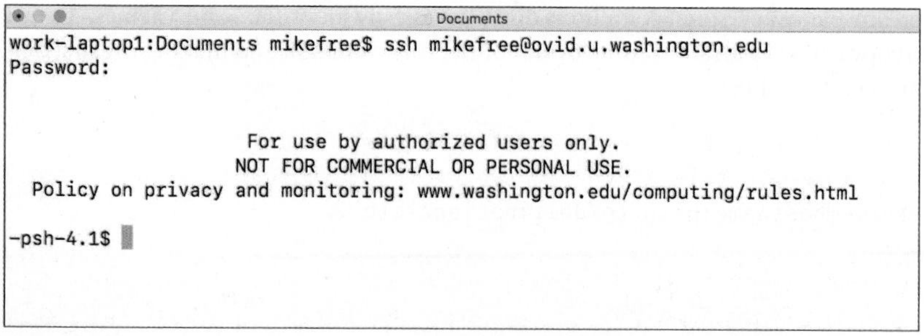

```
● ● ●                          Documents
work-laptop1:Documents mikefree$ ssh mikefree@ovid.u.washington.edu
Password:


                    For use by authorized users only.
                    NOT FOR COMMERCIAL OR PERSONAL USE.
      Policy on privacy and monitoring: www.washington.edu/computing/rules.html

-psh-4.1$ ▮
```

Figure 2.7   Connecting to a remote server using the ssh command on a Mac Terminal.

At that point, you can use commands like pwd and ls to see where you are on that remote computer, cd to navigate to another folder, and any other command line command you wish—just as if you had opened a terminal on that machine!

Once you are finished working on the remote machine, you can disconnect by using the exit command. Closing the command shell will also usually end your connection, but using exit will more explicitly stop any ongoing processes on a remote machine.

The ssh utility will let you connect to a remote machine and control it as if it were right in front of you. But if you want to move files between your local machine and the remote one, you will need to use the **scp** (secure **c**opy) command. This command works exactly like the cp command mentioned earlier, but copies files over the secure SSH protocol.

To copy a local file *to* a location on a remote machine, you need to specify the username and host URL of that machine, similar to what you would use to connect via ssh. In addition, you will need to specify the destination *path* (which folder to copy the file to) on that remote machine. You can specify a path on a remote machine by including it after a colon (:) following the host URL. For example, to refer to the ~/projects folder on the ovid.washington.edu machine (for user mikefree), you would use

mikefree@ovid.washington.edu:~/projects

Thus to copy a local file to a folder on a remote machine, user mikefree would use a command like this:

```
# Securely copy the local file data.csv into the projects folder on the
# remote machine
scp data.csv mikefree@ovid.washington.edu:~/projects


# Or more generically:
scp MY_LOCAL_FILE username@hostname:path/to/destination
```

It is important to note that file paths are *relative to the currently connected machine*—that is why you need to specify the host URL. For example, if you had connected to a remote server via ssh and wanted to copy a file *back to* your local machine, you would need to specify the remote path to your computer! Since most personal computers don't have easily identifiable hostnames, it's usually easiest to copy a file *to* a local machine by disconnecting from ssh and making the first scp argument the remote host:

```
# Run from local machine (not connected through SSH)
# Copies the remote file to the current folder (indicated with the dot .)
scp username@hostname:path/to/destination/file .
```

> **Going Further**: Other utilities can also be used to copy files between machines. For example, the **rsync** command will copy only *changes* to a file or folder, which helps avoid the need to frequently transfer large amounts of data.

Overall, being able to use basic terminal commands will allow you to navigate to and interact with a wide variety of machines, and provides you with a quick and powerful interface to your computer. For practice using the command line, see the set of accompanying book exercises.[2]

---

[2]**Command line exercises:** https://github.com/programming-for-data-science/chapter-02-exercises

II

Managing Projects

This section of the book teaches you the necessary skills for managing data science projects. The two core skills involved are keeping track of the version of your code (Chapter 3), and producing documentation for your code using a language called Markdown (Chapter 4).

Version Control with git and GitHub

One of the most important parts of writing code to work with data is keeping track of changes to your code. Maintaining a clear and well-documented history of your work is crucial for transparency and collaboration. Even if you are working independently, tracking your changes will enable you to revert to earlier versions of your project and more easily identify errors.

Alternatives to proper version control systems—such as emailing code to others, or having dozens of versions of the same file—lack any structured way of backing up work, and are time-consuming and error-prone. This is why you should be using a version control system like git.

This chapter introduces the git command line program and the GitHub cloud storage service, two wonderful tools that track changes to your code (git) and facilitate collaboration (GitHub). git and GitHub are the industry standards for the family of tasks known as **version control**. Being able to manage changes to your code and share that code with others is one of the most important technical skills a data scientist can learn, and is the focus of this chapter as well as Chapter 20.

> **Tip**: Because this chapter revolves around using new interfaces and commands to track file changes—which can be difficult to understand abstractly—we suggest that you follow along with the instructions as they are introduced throughout the chapter. The best way to learn is by doing!

3.1 What Is git?

git[1] is an example of a **version control system**. Open source software guru Eric Raymond defines version control as follows:

---

[1] **Git** homepage: http://git-scm.com/

> *A version control system (VCS) is a tool for managing a collection of program code that provides you with three important capabilities:* **reversibility, concurrency, and annotation.**[2]

Version control systems work a lot like Dropbox or Google Docs: they allow multiple people to work on the same files at the same time, and to view or "roll back" to previous versions. However, systems like `git` differ from Dropbox in a couple of key ways:

- Each new version or "checkpoint" of your files must be explicitly created (*committed*). `git` doesn't save a new version of your entire project each time you save a file to disk. Instead, after making progress on your project (which may involve editing multiple files), you take a *snapshot* of your work, along with a description of what you've changed.

- For text files (which almost all programming files are), `git` tracks changes *line by line*. This means it can easily and automatically combine changes from multiple people, and give you very precise information about which lines of code have changed.

Like Dropbox and Google Docs, `git` can show you all previous versions of a file and can quickly roll back to one of those previous versions. This is often helpful in programming, especially if you embark on making a massive set of changes, only to discover partway through that those changes were a bad idea (we speak from experience here).

But where `git` really comes in handy is in team development. Almost all professional development work is done in teams, which involves multiple people working on the same set of files at the same time. `git` helps teams coordinate all these changes, and provides a record so that anyone can see how a given file ended up the way it did.

There are a number of different version control systems that offer these features, but `git` is the de facto standard—particularly when used in combination with the cloud-based service GitHub.

3.1.1   `git` Core Concepts

To understand how `git` works, you need to understand its core concepts and terms:

- **repository (repo):** A database of your file history, containing all the checkpoints of all your files, along with some additional meta-data. This database is stored in a hidden subdirectory named `.git` within your project directory. If you want to sound cool and in-the-know, call the project folder itself a "repo" (even though the repository is technically the database inside the project folder).

- **commit:** A snapshot or checkpoint of your work at a given time that has been added to the repository (saved in the database). Each commit will also maintain additional information, including the name of the person who did the commit, a message describing the commit, and a timestamp. This extra tracking information allows you to see when, why, and by whom changes were made to a given file. Committing a set of changes creates a snapshot of what that work looks like at the time, which you can return to in the future.

---

[2]Raymond, E. S. (2009). Understanding version-control systems. http://www.catb.org/esr/writings/version-control/version-control.html

- **remote:** A link to a copy of your repository on a different machine. This link points to a location on the web where the copy is stored. Typically this will be a central ("master") version of the project that all local copies point to. This chapter generally deals with copies stored on GitHub as remote repositories. You can push (upload) commits to, and pull (download) commits from, a remote repository to keep everything in sync.

- **merging:** `git` supports having multiple different versions of your work that all live side by side (in what are called **branches**), which may be created by one person or by many collaborators. `git` allows the commits (checkpoints) saved in different versions of the code to be easily *merged* (combined) back together without any need to manually copy and paste different pieces of the code. This makes it easy to separate and then recombine work from different developers.

3.1.2   What Is GitHub?

`git` was created to support completely decentralized development, in which developers pull commits (sets of changes) from one another's machines directly. But in practice, most professional teams take the approach of creating one central repository on a server that all developers push to and pull from. This repository contains the authoritative version of the source code, and all deployments to the "rest of the world" are done by downloading from this centralized repository.

Teams can set up their own servers to host these centralized repositories, but many choose to use a server maintained by someone else. The most popular of these in the open source world is **GitHub**,[3] which as of 2017 had more than 24 million developers using the site.[4] In addition to hosting centralized repositories, GitHub offers other team development features such as issue tracking, wiki pages, and notifications. Public repositories on GitHub are free, but you have to pay for private ones.

In short, GitHub is a site that will host a copy of your project in the cloud, enabling multiple people to collaborate (using `git`). `git` is what you use to do version control; GitHub is one possible place where repositories of code can be stored.

Going Further: Although GitHub is the most popular service that hosts "git" repositories, it is not the only such site. *BitBucket*[a] offers a similar set of features to GitHub, though it has a different pricing model (you get unlimited free private repos, but are limited in the number of collaborators). *GitLab*[b] offers a hosting system that incorporates more operations and deployment services for software projects.

[a]https://bitbucket.org
[b]https://gitlab.com

---

[3]**GitHub:** https://github.com
[4]**The State of the Octoverse 2017:** https://octoverse.github.com

> **Caution**: The interface and functionality of websites such as GitHub are constantly evolving and may change. Additional features may become available, and the current structure may be reorganized to better support common usage.

3.2 Configuration and Project Setup

This section walks you through all the commands needed to set up version control for a project using `git`. It focuses on using `git` from the command line, which is the most effective way to *learn* (if not use) the program, and is how most professional developers interact with the software. That said, it is also possible to use `git` directly through code editors and IDEs such as Atom or RStudio—as well as through dedicated graphical software such as GitHub Desktop[5] or Sourcetree.[6]

The first time you use `git` on your machine after having installed it, you will need to configure[7] the installation, telling `git` who you are so you can commit changes to a repository. You can do this by using the `git` command line command with the `config` option (i.e., running the `git config` command):

```
# Configure `git` on your machine (only needs to be done once)

# Set your name to appear alongside your commits
# This *does not* need to be your GitHub username
git config --global user.name "YOUR FULLNAME"


# Set your email address
# This *does* need to be the email associated with your GitHub account
git config --global user.email "YOUR_EMAIL_ADDRESS"
```

Even after `git` knows who you are, it will still prompt you for your password before pushing your code up to GitHub. One way to save some time is by setting up an SSH key for GitHub. This will allow GitHub to recognize and permit interactions coming from *your machine*. If you don't set up the key, you will need to enter your GitHub password each time you want to push changes up to GitHub (which may be multiple times a day). Instructions for setting up an SSH key are available from GitHub Help.[8] Make sure you set up your key on a machine that you control and trust!

3.2.1 Creating a Repo

To work with `git`, you will need to create a **repository**. A repository acts as a "database" of changes that you make to files in a directory.

[5]**GitHub Desktop:** https://desktop.github.com
[6]**Sourcetree:** https://www.sourcetreeapp.com
[7]**GitHub: Set Up Git:** https://help.github.com/articles/set-up-git/
[8]**GitHub: Authenticating to GitHub:** https://help.github.com/articles/generating-a-new-ssh-key-and-adding-it-to-the-ssh-agent/

A repository is always created in an existing directory (folder) on your computer. For example, you could create a new folder called learning_git on your computer's Desktop. You can turn this directory into a repository by telling the git program to run the init action (running the git init command) *inside* that directory:

```
# Create a new folder in your current location called `learning_git`
mkdir learning_git

# Change your current directory to the new folder you just created
cd learning_git

# Initialize a new repository inside your `learning_git` folder
git init
```

The **git init** command creates a new *hidden* folder called .git inside the current directory. Because it's hidden, you won't see this folder in Finder, but if you use ls -a (the "list" command with the all option) you can see it listed. This folder is the "database" of changes that you will make—git will store all changes you commit in this folder. The inclusion of the .git folder causes a directory to become a repository; you refer to the whole directory as the "repo." However, you won't ever have to directly interact with this hidden folder; instead, you will use a short set of terminal commands to interact with the database.

> **Caution:** *Do not put one repo inside of another!* Because a git repository tracks *all* of the content inside of a single folder (including the content in subfolders), this will turn one repo into a "sub-repo" of another. Managing changes to both the repo and sub-repo will be difficult and should be avoided.
>
> Instead, you should create a lot of different repos on your computer (one for each project), making sure that they are in *separate* folders.
>
> Note that it is also not a good idea to have a git repository inside of a shared folder, such as one managed with Dropbox or Google Drive. Those systems' built-in file tracking will interfere with how git manages changes to files.

3.2.2 Checking Status

Once you have a repo, the next thing you should do is check its **status**:

```
# Check the status of your repository
# (this and other commands will only work inside git project folders)
git status
```

The **git status** command will give you information about the current "state" of the repo. Running this command on a new repo tells you a few things (as shown in Figure 3.1):

- That you're actually in a repo (otherwise you will get an error)
- That you're on the master branch (think: line of development)

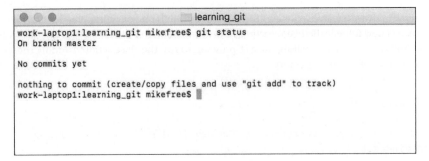

Figure 3.1 Checking the status of a new (empty) repository with the `git status` command.

- That you're at the initial commit (you haven't committed anything yet)

- That currently there are no changes to files that you need to commit (save) to the database

- What to do next! (namely, *create/copy files and use "git add" to track*)

That last point is important. `git status` messages are verbose and somewhat awkward to read (this is the command line after all). Nevertheless, if you look at them carefully, they will almost always tell you which command to use next.

> **Tip:** If you ever get stuck, use `git status` to figure out what to do next!

This makes `git status` the most useful command in the entire process. As you are learning the basics of `git`, you will likely find it useful to run the command *before and after* each other command to see how the status of your project changes. Learn it, use it, love it.

3.3 Tracking Project Changes

Running `git status` in a new repository will tell you to create a file—which we suggest you do now to practice the steps of using version control. For example, open up your favorite text editor (e.g., Atom) and create a plain text file with a list of your favorite books. Save this file in your `learning_git` folder as `favorite_books.txt`. `git` will be able to detect and manage changes to your file as long as it was saved inside the repo (project directory).

> **Remember:** After editing a file, always save it to your computer's hard drive (e.g., with `File > Save`). `git` can track only changes that have been saved!

3.3.1 Adding Files

After making a change to your repository (such as creating *and saving* the `favorite_books.txt` file), run `git status` again. As shown in Figure 3.2, `git` now gives a list of changed and "untracked" files, as well as instructions about what to do next to save those changes to the repo's database.

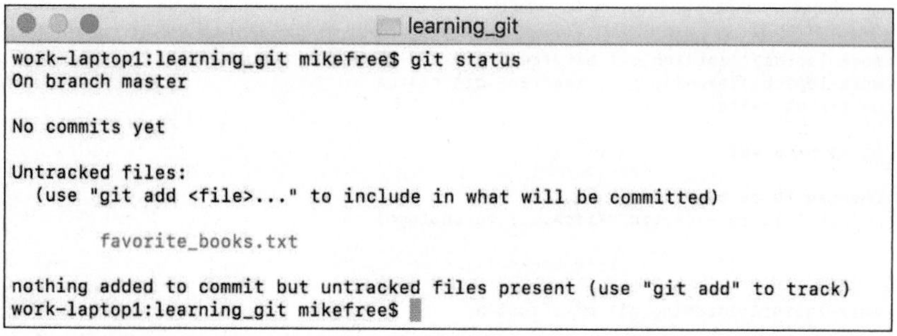

```
●  ●  ●                          learning_git
work-laptop1:learning_git mikefree$ git status
On branch master

No commits yet

Untracked files:
  (use "git add <file>..." to include in what will be committed)

        favorite_books.txt

nothing added to commit but untracked files present (use "git add" to track)
work-laptop1:learning_git mikefree$ ▊
```

Figure 3.2   The status of a repository with changes that have not (yet) been added and are therefore shown in red.

The first step is to add those changes to the **staging area**. The staging area is like a shopping cart in an online store: you put changes in temporary storage before you commit to recording them in the database (e.g., before clicking "purchase").

You add files to the staging area using the **git add** command (replacing FILENAME in the following example with the name/path of the file or folder you want to add):

```
# Add changes to a file with the name FILENAME to the staging area
# Replace FILENAME with the name of your file (e.g., favorite_books.txt)
git add FILENAME
```

This will add a single file *in its current saved state* to the staging area. For example, git add favorite_books.txt would add that file to the staging area. If you change the file later, you will need to add the updated version by running the git add command again.

You can also add all of the contents of the current directory (tracked or untracked) to the staging area with the following command:

```
# Add all saved contents of the directory to the staging area
git add .
```

This command is the most common way to add files to the staging area, unless you've made changes to specific files that you aren't ready to commit yet. Once you've added files to the staging area, you've "changed" the repo and so can run git status again to see what it says to do next. As you can see in Figure 3.3, git will tell you which files are in the staging area, as well as the command to *unstage* those files (i.e., remove them from the "cart").

3.3.2   Committing

When you're happy with the contents of your staging area (i.e., you're ready to purchase), it's time to **commit** those changes, saving that snapshot of the files in the repository database. You do this with the **git commit** command:

```
# Create a commit (checkpoint) of the changes in the staging area
# Replace "Your message here" with a more informative message
git commit -m "Your message here"
```

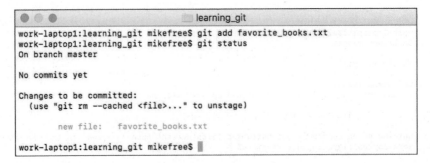

Figure 3.3 The status of a repository after adding changes (added files are displayed in green).

You should replace "`Your message here`" with a short message saying what changes that commit makes to the repo. For example, you could type `git commit -m "Create favorite_books.txt file"`.

> **Caution:** If you forget the -m option, `git` will put you into a command line *text editor* so that you can compose a message (then save and exit to finish the commit). If you haven't done any other configuration, you might be dropped into the **vim** editor. Type **:q** (*colon* then *q*) and press `enter` to flee from this place and try again, remembering the -m option! Don't panic: getting stuck in *vim* happens to everyone.[a]
>
> ───────────────
>
> [a]https://stackoverflow.blog/2017/05/23/stack-overflow-helping-one-million-developers-exit-vim/

3.3.2.1 Commit Message Etiquette

Your commit messages should be informative[9] about which changes the commit is making to the repo. "`stuff`" is not a good commit message. In contrast, "`Fix critical authorization error`" is a good commit message.

Commit messages should use the **imperative mood** ("`Add feature`", not "`Added feature`"). They should complete the following sentence:

>   *If applied, this commit will {**your message**}*

Other advice suggests that you limit your message to 50 characters (like an email subject line), at least for the first line—this helps when you are going back and looking at previous commits. If you want to include more detail, do so after a blank line. (For more detailed commit messages, we recommend you learn to use *vim* or another command line text editor.)

A specific commit message format may also be required by your company or project team. Further consideration of good commit messages can be found in this blog post.[10]

───────────────

[9]Do not do this: https://xkcd.com/1296/

[10]**Chris Beams: How to Write a Git Commit Message** blog post: http://chris.beams.io/posts/git-commit/

As you make commits, remember that they are a public part of your project history, and will be read by your professors, bosses, coworkers, and other developers on the internet.[11]

After you've committed your changes, be sure and check `git status`, which should now say that there is nothing to commit!

3.3.3   Reviewing the local `git` Process

This cycle of edit files–add files–commit changes is the standard "development loop" when working with `git`, and is illustrated in Figure 3.4.

In general, you will make lots of changes to your code (editing lots of files, running and testing your code, and so on). Once you're at a good "break point"—you've got a feature working, you're stuck and need some coffee, you're about to embark on some radical changes—be sure to `add` and `commit` your changes to make sure you don't lose any work and you can always get back to that point.

> **Remember:** Each commit represents a *set of changes*, which can and usually does include multiple files. Do not think about each commit being a change to a file; instead, think about each commit as being a snapshot of your entire project!

> **Tip:** If you accidentally add files that you want to "unadd," you can use the **`git reset`** command (with no additional arguments) to remove all added files from the staging area.
>
> If you accidentally `commit` files when you didn't want to, you can "undo" the commit using the command `git reset --soft HEAD~1`. This command makes it so the commit you *just made* never occurred, leaving the changed files in your working directory. You can then edit which files you wish to commit before running the `git commit` command again. Note that this works only on the most recent commit, and you cannot (easily) undo commits that have been pushed to a remote repository.

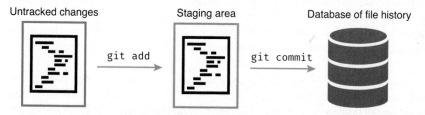

Figure 3.4   The local `git` process: add changes to the staging area, then create a checkpoint of your project by making a `commit`. The commit saves a version of the project at this point in time to the database of file history.

---

[11] Don't join this group: https://twitter.com/gitlost

3.4   Storing Projects on GitHub

Once you are able to track your changes locally with `git`, you will often want to access your project from a different computer, or share your project with other people. You can do this using GitHub, an online service that stores copies of repositories in the cloud. These repositories can be *linked* to your **local** repositories (the ones on your machine, like those you've been working with so far) so that you can synchronize changes between them. The relationship between `git` and GitHub is the same as that between a *Photos* application on your computer and a photo hosting service such as *Flickr*: `git` is the program you use to (locally) create and manage repositories (like Photos); GitHub is simply a website that stores these repositories (like Flickr). Thus you use `git`, but upload to and download from GitHub.

Repositories stored on GitHub are examples of **remotes**: other repos that are linked to your local one. Each repo can have multiple remotes, and you can synchronize commits between them. Each remote has a URL associated with it (indicating where on the internet the remote copy of the repo can be found), but they are given "alias" names—similar to browser bookmarks. By convention, the remote repo stored on GitHub's servers is named **origin**, since it tends to be the "origin" of any code you've started working on.

To use GitHub, you will need to create a free GitHub account, which is discussed in Chapter 1.

Next, you will need to "link" your local repository to the remote one on GitHub. There are two common processes for doing this:

1. *If you already have a project tracked with `git` on your computer*, you can create a new repository on GitHub by clicking the green "New Repository" button on the GitHub homepage (you will need to be logged in). This will create a new empty repo on GitHub's servers under your account. Follow the provided instructions on how to link a repository on your machine to the new one on GitHub.

2. *If there is a project on GitHub that you want to edit on your computer*, you can *clone* (download) a copy of a repo that already exists on GitHub, allowing you to work with and modify that code. This process is more common, so it is described in more detail here.

Each repository on GitHub has a *web portal* at a unique location. For example, https://github.com/ programming-for-data-science/book-exercises is the webpage for the programming exercises that accompany this book. You can click on the files and folders on this page to view their source and contents online, but you won't change them through the browser.

> **Remember:** You should *always* create a local copy of the repository when working with code. Although GitHub's web interface supports it, you should *never make changes or commit directly to GitHub*. All development work is done locally, and changes you make are then uploaded and *merged* into the remote. This allows you to *test* your work and to be more flexible with your development.

3.4.1 Forking and Cloning

Just like with Flickr or other image-hosting sites, all GitHub users have their own account under which repos are stored. The repo mentioned earlier is under this book's account `programming-for-data-science`. Because it's under the book's user account, you won't be able to modify it—just as you can't change someone else's picture on Flickr. So the first thing you will need to do is copy the repo over to *your own account on GitHub's servers*. This process is called **forking** the repo (you're creating a "fork" in the development, splitting off to your own version).

To fork a repo, click the "Fork" button in the upper right of the screen (shown in Figure 3.5). This will copy the repo over to your own account; you will be able to download and upload changes to that copy but not to the original. Once you have a copy of the repo under your own account, you need to download the entire project (files and their history) to your local machine to make changes. You do this by using the **git clone** command:

```
# Change to the folder that will contain the downloaded repository folder
cd ~/Desktop

# Download the repository folder into the current directory
git clone REPO_URL
```

This command creates a new repo (directory) *in the current folder*, and downloads a copy of the code and all the commits from the URL you specify into that new folder.

> **Caution:** Make sure that you are in the desired location in the command line before running any `git` commands. For example, you would want to `cd` out of the `learning_git` directory described earlier; you don't want to `clone` into a folder that is already a repo!

You can get the URL for the `git clone` command from the address bar of your browser, or by clicking the green "Clone or Download" button. If you click that button, you will see a pop-up that contains a small clipboard icon that will copy the URL to your clipboard, as shown in Figure 3.6. This allows you to use your terminal to clone the repository. If you click *"Open in Desktop,"* it will prompt you to use a program called GitHub Desktop[12] to manage your version control (a technology not discussed in this book). But do not click the *"Download Zip"* option, as it contains code without the previous version history (the code, but not the repository itself).

Figure 3.5 The Fork button for a repository on GitHub's web portal. Click this button to create your own copy of a repository on GitHub.

---

[12]https://desktop.github.com

Figure 3.6   The Clone button for a repository on GitHub's web portal. Click this button to open the dialog box, then click the clipboard icon to copy the GitHub URL needed to clone the repository to your machine. Red notes are added.

> **Remember**: Make sure you clone from the *forked* version (the one under your account!) so that the repo downloads with a proper link back to the `origin` remote.

Note that you will only need to `clone` once per machine. `clone` is like `init` for repos that are on GitHub; in fact, the `clone` command *includes* the `init` command (so you do not need to init a cloned repo). After cloning, you will have a full copy of the repository—which includes the full project history—on your machine.

3.4.2   Pushing and Pulling

Once you have a copy of the repo code, you can make changes to that code on your machine and then *push* those changes up to GitHub. You can edit the files (e.g., the README.md) in an editor as if you had created them locally. After making changes, you will, of course, need to add the changed files to the staging area and `commit` the changes to the repo (don't forget the -m message!).

Committing will save your changes locally, but it does not push those changes to GitHub. If you refresh the web portal page (make sure you're looking at the one under your account), you shouldn't see your changes yet.

To get the changes to GitHub (and share your code with others), you will need to **push** (upload) them to GitHub's computers. You can do this with the following command:

```
# Push commits from your computer up to a remove server (e.g., GitHub)
git push
```

By default, this command will push the current code to the `origin` remote (specifically, to its `master` branch of development). When you cloned the repo, it came with an `origin` "bookmark" link to the original repo's location on GitHub. To check where the remote is, you can use the following command:

```
# Print out (verbosely) the remote location(s)
git remote -v
```

Once you've pushed your code, you should be able to refresh the GitHub webpage and see your changes on the web portal.

If you want to download the changes (commits) that someone else has made, you can do that using the **pull** command. This command will download the changes from GitHub and *merge* them into the code on your local machine:

```
# Pull changes down from a remove server (e.g., GitHub)
git pull
```

> **Caution:** Because pulling down code involves merging versions of code together, you will need to keep an eye out for merge conflicts! Merge conflicts are discussed in more detail in Chapter 20.

> **Going Further:** The commands git pull and git push have the default behavior of interacting with the master branch at the origin remote location. git push is thus equivalent to the more explicit command git push origin master. As discussed in Chapter 20, you will adjust these arguments when engaging in more complex and collaborative development processes.

The overall process of using git and GitHub together is illustrated in Figure 3.7.

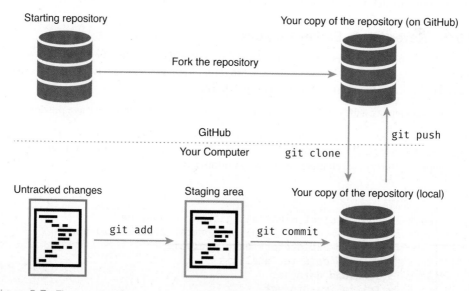

Figure 3.7 The remote git process: *fork* a repository to create a copy on GitHub, then clone it to your machine. Then add and commit changes, and push them up to GitHub to share.

> **Tip:** If you are working with others (or just on different computers), always `pull` in the latest changes *before* you start working. This will get you the most up-to-date changes, and reduce the chances that you will encounter an issue when you try to `push` your code.

3.5   Accessing Project History

The benefit of making each commit (checkpoint) is that you can easily view your project or revert to that checkpoint at any point in the future. This section details the introductory approaches for viewing files at an earlier point in time, and reverting to those checkpoints.

3.5.1   Commit History

You can view the history of commits you've made by using the **`git log`** command while inside of your repo on the command line:

```
# Print out a repository's commit history
git log
```

This will give you a list of the *sequence* of commits you've made: you can see who made which changes and when. (The term **HEAD** refers to the most recent commit made.) The optional `--oneline` argument gives you a nice compact printout, though it displays less information (as shown in Figure 3.8). Note that each commit is listed with its **SHA-1 hash** (the sequence of random-looking numbers and letters), which you can use to identify that commit.

3.5.2   Reverting to Earlier Versions

One of the key benefits of version control systems is **reversibility**, meaning the ability to "undo" a mistake (and we all make lots of mistakes when programming!). `git` provides two basic ways that you can go back and fix a mistake you've made previously:

1. You can replace a file (or the entire project directory!) with a version saved as a previous commit.

2. You can have `git` "reverse" the changes that you made with a previous commit, effectively applying the *opposite* changes and thereby undoing it.

```
● ● ●                            project
work-laptop1:project mikefree$ git log --oneline
e4894a0 (HEAD -> master) Implement first statistical model
2abd8f3 Perform exploratory data analysis
086850f Wrangle data for analysis
6fc0078 Download data set
e6cfd89 Initialize project
```

Figure 3.8 A project's commit history displayed using the `git log --oneline` command in the terminal. Each commit is identified by a six-digit *hash* (e.g., e4894a0), the most recent of which is referred to as the HEAD.

Note that both of these approaches require you to have committed a working version of the code that you want to go back to. `git` only knows about changes that have been committed: if you don't commit, `git` can't help you!

> **Tip**: Commit early; commit often.

For both forms of undoing, you first need to decide which version of the file to revert to. Use the `git log --oneline` command described earlier, and note the *SHA-1 hash* for the commit that saved the version you want to revert to. The first six characters of each hash is a unique ID and acts as the "name" for the commit.

To go back to an older version of the file (to "revert" it to the version of a previous *commit*), you can use the **git checkout** command:

```
# Print a list of commit hashes
git log --oneline

# Checkout (load) the version of the file from the given commit
git checkout COMMIT_HASH FILENAME
```

Replace `COMMIT_HASH` and `FILENAME` with the commit ID hash and the file you want to revert, respectively. This will replace the current version *of that single file* with the version saved in `COMMIT_HASH`. You can also use **--** as the commit hash to refer to the *most recent commit* (called the HEAD), such as if you want to discard current changes:

```
# Checkout the file from the HEAD (the most recent commit)
git checkout -- FILENAME
```

This will *change the file* in your working directory, so that it appears just as it did when you made the earlier commit.

> **Caution**: You can use the `git checkout` command to *view* project files at the time of a particular commit by leaving off the filename (i.e., `git checkout COMMIT_HASH`). However, you can't actually commit any changes to these files when you do this. Thus you should use this command only to explore the files at a previous point in time.
>
> If you do this (or if you forget the filename when checking out), you can return to your most recent version of the code with the following command:
>
> ```
> # Checkout the most recent version of the master branch
> git checkout master
> ```

If you just had one bad commit but don't want to throw out other valuable changes you made to your project later, you can use the **git revert** command:

```
# Apply the opposite changes made by the given commit
git revert COMMIT_HASH --no-edit
```

This command will determine which changes the specified commit made to the files, and then apply the *opposite* changes to effectively "back out" the commit. Note that this does not go back to the given commit number (that's what `git checkout` is for!), but rather *reverses only the commit you specify*.

The `git revert` command does create a new commit (the `--no-edit` option tells `git` that you don't want to include a custom commit message). This is great from an archival point of view: you never "destroy history" and lose the record of which changes were made and then reverted. History is important; don't mess with it!

> **Caution:** The `git reset` command can destroy your commit history. Be very careful when using it. We recommend you never reset beyond the most recent commit—that is, use it only to unstage files (git reset) or undo the most recent commit (git reset --soft HEAD~1).

3.6   Ignoring Files from a Project

Sometimes you want `git` to always ignore particular directories or files in your project. For example, if you use a Mac and you tend to organize your files in Finder, the operating system will create a hidden file in that folder named .DS_Store (the leading dot makes it "hidden") to track the positions of icons, which folders have been "expanded," and so on. This file will likely be different from machine to machine, and has no meaningful information for your project. If it is added to your repository and you work from multiple machines (or as part of a team), it could lead to a lot of merge conflicts (not to mention cluttering up the folders for Windows users).

You can tell `git` to ignore files like these by creating a special *hidden* file in your project directory called .gitignore (note the leading dot). This text file contains a list of files or folders that `git` should "ignore" and therefore not "see" as one of the files in the folder. The file uses a very simple format: each line contains the path to a directory or file to ignore; multiple files are placed on multiple lines. For example:

```
# This is an example .gitignore file
# The leading "#" marks a comment describing the code

# Ignore Mac system files;
.DS_Store

# Don't check in passwords stored in this file
secret/my_password.txt

# Don't include large files or libraries
movies/my_four_hour_epic.mov

# Ignore everything in a particular folder; note the slash
raw-data/
```

The easiest way to create the `.gitignore` file is to use your preferred text editor (e.g., Atom). Select `File > New` from the menu and choose to make the `.gitignore` file *directly inside* your repo (in the **root folder** of that repo, not in a subfolder).

If you are on a Mac, we strongly suggest *globally ignoring* your `.DS_Store` file. There's no need to ever share or track this file. To always ignore this file on your machine, you can create a "global" `.gitignore` file (e.g., in your ~ home directory), and then tell `git` to always exclude files listed there through the `core.excludesfile` configuration option:

```
# Append `.DS_Store` to your `.gitignore` file in your home directory
echo ".DS_Store" >> ~/.gitignore

# Always ignore files listed in that central file
git config --global core.excludesfile ~/.gitignore
```

Note that you may still want to list `.DS_Store` in a repo's local `.gitignore` file in case you are collaborating with others.

Additionally, GitHub provides a number of suggested `.gitignore` files for different languages,[13] including R.[14] These are good places to start when creating a local `.gitignore` file for a project.

Whew! You made it through! This chapter has a lot to take in, but really you just need to understand and use the following half-dozen commands:

- `git status`: Check the status of a repo.

- `git add`: Add files to the staging area.

- `git commit -m "Message"`: Commit changes.

- `git clone`: Copy a repo to the local machine.

- `git push`: Upload commits to GitHub.

- `git pull`: Download commits from GitHub.

While it's tempting to ignore version control systems, they will save you time in the long run. `git` is a particularly complex and difficult-to-understand system given its usefulness and popularity. As such, a wide variety of tutorials and explanations are available online if you need further clarification. Here are a few recommendations to get started:

- *Atlassian's Git Tutorial*[15] is an excellent introduction to all of the major `git` commands.

- GitHub's cheatsheet[16] and supplemental resources[17,18] provide clearly documented "how-to" guides for performing specific actions.

---

[13] **.gitignore templates:** https://github.com/github/gitignore
[14] **.gitignore template for R:** https://github.com/github/gitignore/blob/master/R.gitignore
[15] https://www.atlassian.com/git/tutorials/what-is-version-control
[16] https://education.github.com/git-cheat-sheet-education.pdf
[17] https://help.github.com/articles/git-and-github-learning-resources/
[18] https://try.github.io

- Jenny Bryan's free online book *Happy Git and GitHub for the useR*[19] provides an in-depth approach to using version control for R users.

- DataCamp's online course *Introduction to Git for Data Science*[20] will also cover the basics of `git`.

- The *Pro Git Book*[21] is the official reference for full (if not necessarily clear) details on any and all `git` commands.

For practice working with `git` and GitHub, see the set of accompanying book exercises.[22]

---

[19] http://happygitwithr.com
[20] https://www.datacamp.com/courses/introduction-to-git-for-data-science/
[21] https://git-scm.com/book/en/v2
[22] **Version control exercises:** https://github.com/programming-for-data-science/chapter-03-exercises

4

Using Markdown for Documentation

As a data scientist, you will often encounter the somewhat trivial task of adding formatting to plain text (e.g., making it **bold** or *italic*) without the use of a program like Microsoft Word. This chapter introduces Markdown, a simple programming syntax that can be used to describe text formatting and structure by adding special characters to the text. Being comfortable with this simple syntax to describe text rendering will help you document your code, and post well-formatted messages to question forums (such as StackOverflow[1]) or chat applications (such as Slack[2]), as well as create clear documentation that describes your code's purpose when hosted on GitHub (called the "README" file). In this chapter, you will learn the basics of Markdown syntax, and how to leverage it to produce readable code documents.

4.1   Writing Markdown

Markdown[3] is a lightweight syntax that is used to describe the format and structure of text documents. With only a small handful of options, Markdown allows you to apply formatting to your text (such as making text **bold** or *italic*), as well as to provide structure to a document (such as headers or bullet points). Mastering the basics of writing Markdown will allow you to quickly and easily create well-formatted documents.

> **Fun Fact**: Markdown belongs to a family of programming languages used to describe document formatting known as *markup* languages (confusing, right?). For example, HTML (*HyperText **Markup** Language*) is used to describe the content and format of websites.
>
> **Additional Fun Fact**: This book was written in Markdown!

---

[1] **StackOverflow**: https://stackoverflow.com
[2] **Slack**: https://slack.com
[3] **Markdown: Syntax** original specification by John Gruber: https://daringfireball.net/projects/markdown/syntax

4.1.1   Text Formatting

At its most basic, Markdown is used to declare text formatting options. You do this by adding special symbols (punctuation) *around* the text you wish to "mark." For example, if you want text to be rendered (displayed) in *italics*, you would surround that text with underscores (_): you would type _italics_, and a program would know to render that text as *italics*. You can see how this looks in Figure 4.1.

There are a few different ways you can format text, as summarized in Table 4.1.

While there are further variations and syntax options, these are the most common.

4.1.2   Text Blocks

Markdown isn't just about adding bold and italics in the middle of text; it also enables you to create distinct blocks of formatted content (such as a header or a chunk of code). You do this by adding a symbol in front of the text. For example, in Figure 4.2, the document (shown on the right) is produced using the Markdown syntax (shown on the left) described in Table 4.2.

4.1.3   Hyperlinks

Providing hyperlinks in documentation is a great way to reference other resources on the web. You turn text into a hyperlink in Markdown by surrounding the text in square brackets [], and placing the URL to link to immediately after that in parentheses (). Here's an example:

```
[text to display](https://some/url/or/path)
```

The text between the brackets (*"text to display"*) will be displayed in your document with hyperlink formatting. Clicking on the hyperlink will direct a web browser to the URL in the parentheses (https://some/url/or/path). Note that hyperlinks can be included *inline* in the middle of a paragraph or list item; the text to display can also be formatted with Markdown to make it bold or italic.

`This is a paragraph in which we'll add **bold text**, _italicized text_, and` `code` `into the middle of a sentence`	This is a paragraph in which we'll add **bold text**, *italicized text*, and `code` into the middle of a sentence

Figure 4.1   Markdown text formatting. The code version is on the left; the rendered version is on the right.

Table 4.1   **Markdown text formatting syntax**

Syntax	Formatting
text	emphasize (*italicize*) using underscores (_)
text	strongly emphasize (**bold**) using two asterisks (*)
`text`	code style using backticks (`)
~~text~~	~~strike-through~~ using two tildes (~)

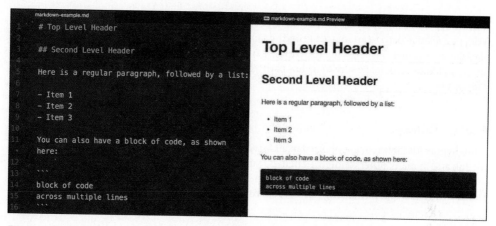

Figure 4.2   Markdown block formatting. Code (left) and rendered output (right).

Table 4.2   **Markdown block formatting syntax**

Syntax	Formatting
#	Header (use ## for second level, ### for third level, etc.)
``	Code section (three backticks) that encapsulate the code
-	Bulleted/unordered lists (hyphens)
>	Block quote

While the URL is most commonly an *absolute path* to a resource on the web, it can also be a *relative path* to another file on the same machine (the file path is relative to the Markdown document that contains the link). This is particularly useful for linking from one Markdown file to another (e.g., if the documentation for a project is spread across multiple pages).

4.1.4   Images

Markdown also supports the rendering of images in your documents, which allows you to include diagrams, charts, and pictures in your documentation. The syntax for including images is similar to that for hyperlinks, except with an exclamation point **!** before the link to indicate that it should be shown as an image:

```
![description of the image](path/to/image)
```

When shown as an image, the "text to display" becomes an *alternate text description* for the image, which will be shown if the image cannot be shown (e.g., if it fails to load). This is particularly important for the **accessibility** of the documents you create, as anyone using a *screenreader* can be read the description provided in place of the image.

As with hyperlinks, the path to an image can be an absolute path (for referencing images on the web), or a relative path to an image file on the same machine (the file path is relative to the Markdown document). Specifying the correct path is the most common problem when rendering images in Markdown; make sure to review paths (Section 2.2.3) if you have any trouble rendering your image.

4.1.5 Tables

While syntax for tables isn't supported in all Markdown environments, tables can be shown on GitHub and in many other rendering engines. Tables are useful for organizing content, though they are somewhat verbose to express in markup syntax. For example, Table 4.2 describing Markdown syntax and formatting was written using the following Markdown syntax:

```
| Syntax        | Formatting
| :------------ | :---------------------------------------------------------
| `#`           | Header (use `##` for second level, `###` for third level, etc.)
| ``` ``` ```` | Code section (3 backticks) that encapsulate the code
| `-`           | Bulleted/unordered lists (hyphens)
| `>`           | Block quote
```

This is known as a *pipe table*, as columns are separated with the pipe symbol (|). The first line contains the column headers, followed by a line of hyphens (-), followed by each row of the table on a new line. The colon (:) next to the hyphens indicates that the content in that column should be aligned to the left. The outer pipe characters and additional spaces in each row are optional, but they help keep the code easy to read; it isn't required to have the pipes line up.

(Note that in the table the triple backticks used for a code section are surrounded by *quadruple* backticks to make sure that they are rendered as the ` symbol, and not interpreted as a Markdown command!)

For other Markdown options—including *blockquotes* and *syntax-colored code blocks*—see, for example, this *GitHub Markdown Cheatsheet*.[4]

4.2 Rendering Markdown

To view the *rendered* version of your Markdown-formatted syntax, you need to use a program that converts from Markdown into a formatted document. Luckily, this often happens automatically with systems that leverage Markdown. For example, GitHub's web portal will automatically render Markdown files (which have the extension **.md**), and Slack and StackOverflow will automatically format your messages.

Indeed, the web portal page for each GitHub repository will automatically format and display as project documentation the Markdown file called **README.md** (it *must* have this name) stored in the root directory of the project repo. The **README** file contains important instructions and details about the program—it asks you to "read me!" Most public GitHub repositories include a README

---

[4] **Markdown Cheatsheet:** https://github.com/adam-p/markdown-here/wiki/Markdown-Cheatsheet

that explains the context and usage of the code in the repo. For example, the documentation describing this book's exercises is written in this README.md file,[5] with individual folders also having their own README files to explain the code in those specific directories.

> **Caution**: The syntax may vary slightly across programs and services that render Markdown. For example, Slack doesn't *technically* support Markdown syntax (though it is very similar to Markdown). GitHub in particular has special limitations and extensions to the language; see the documentation[a] for details or if you run into problems.
>
> ───────────────
> [a]https://help.github.com/categories/writing-on-github/

However, it can be helpful to preview your rendered Markdown before posting code to GitHub or StackOverflow. One of the best ways to do this is to write your marked code in a text editor that supports *preview rendering*, such as Atom.

To preview what your rendered content will look like, simply open a Markdown file (.md) in Atom. Then use the *command palette*[6] (or the shortcut ctrl+shift+m) to toggle the **Markdown Preview**. Once this preview is open, it will automatically update to reflect any changes to the Markdown code as you type.

> **Tip**: You can use the command palette to **Toggle Github Style** for the Markdown preview; this will make the rendered preview look (mostly) the same as it will when uploaded to GitHub, though some syntactical differences may still apply.

Other options for previewing rendered Markdown include the following:

- Many editors (such as *Visual Studio Code*[7]) include automatic Markdown rendering, or have extensions to provide that functionality.

- Stand-alone programs such as *MacDown*[8] (Mac only) will also do the same work, often providing nicer-looking editor windows.

- There are a variety of online Markdown editors that you can use for practice or quick tests. Dillinger[9] is one of the nicer ones, but there are plenty of others if you're looking for something more specific.

- A number of Google Chrome Extensions will render Markdown files for you. For example, Markdown Reader[10] provides a simple rendering of a Markdown file (note it may differ slightly from the way GitHub would render the document). Once you've installed the extension, you can drag-and-drop a .md file into a blank Chrome tab to view the formatted document. Double-click to view the raw code.

───────────────

[5]See https://github.com/programming-for-data-science/book-exercises/blob/master/README.md
[6]**Atom Command Palette**: http://flight-manual.atom.io/getting-started/sections/atom-basics/#command-palette
[7]**Visual Studio Code**: https://code.visualstudio.com
[8]**MacDown**: Markdown editor (Mac Only): http://macdown.uranusjr.com
[9]**Dillinger**: online Markdown editor: http://dillinger.io
[10]**Markdown Reader** extension for Google Chrome: https://chrome.google.com/webstore/detail/markdown-reader/gpoigdifkoadgajcincpilkjmejcaanc?hl=en

- If you want to **render** (compile) your markdown to a `.pdf` file, you can use an Atom extension[11] or a variety of other programs to do so.

This chapter introduced Markdown syntax as a helpful tool for formatting documentation about your code. You will use this syntax to provide information about your code (e.g., in `git` repository `README.md` files), to ask questions about your code (e.g., on StackOverflow), and to present the results of your code analysis (e.g., using R Markdown, described in Chapter 18). For practice writing Markdown syntax, see the set of accompanying book exercises.[12]

---

[11]**Markdown to PDF** extension for Atom: https://atom.io/packages/markdown-pdf
[12]**Markdown exercises:** https://github.com/programming-for-data-science/chapter-04-exercises

III

Foundational R Skills

This section of the book introduces the fundamentals of the R programming language. In doing so, it both explains the syntax of the language and describes the core concepts in computer programming you will need to begin writing code to work with data.

Introduction to R

R is an extraordinarily powerful open source software program built for working with data. It is one of the most popular data science tools because of its ability to efficiently perform statistical analysis, implement machine learning algorithms, and create data visualizations. R is the primary programming language used throughout this book, and understanding its foundational operations is key to being able to perform more complex tasks.

5.1 Programming with R

R is a **statistical programming language** that allows you to write code to work with data. It is an **open source** programming language, which means that it is free and continually improved upon by the R community. The R language has a number of capabilities that allow you to read, analyze, and visualize data sets.

> **Fun Fact**: R is called "R" in part because it was inspired by the language "S," a language for Statistics developed by AT&T, and because it was developed by Ross Ihaka and Robert Gentleman.

In previous chapters, you leveraged formal language to give instructions to your computer, such as by writing syntactically precise instructions at the command line. Programming in R works in a similar manner: you write instructions using R's special language and syntax, which the computer **interprets** as instructions for how to work with data.

However, as projects grow in complexity, it becomes useful if you can write down all the instructions in a single place, and then order the computer to *execute* all of those instructions at once. This list of instructions is called a **script**. Executing or "running" a script will cause each instruction (line of code) to be run *in order, one after the other*, just as if you had typed them in one by one. Writing scripts allows you to save, share, and reuse your work. By saving instructions in a file (or set of files), you can easily check, change, and re-execute the list of instructions as you figure out how to use data to answer questions. And, because R is an *interpreted* language, rather than a *compiled* language like C or Java, R programming environments give you the ability to separately execute each individual line of code in your script if you desire.

As you begin working with data in R, you will be writing multiple instructions (lines of code) and saving them in files with the **.R** extension, representing R scripts. You can write this R code in any text editor (such as Atom), but we recommend you usually use **RStudio**, a program that is specialized for writing and running R scripts.

5.2   Running R Code

There are a few different ways in which you can have your computer execute code that you write in the R language. The most user-friendly approach is to use RStudio.

5.2.1   Using RStudio

RStudio is an open source **integrated development environment (IDE)** that provides an informative user interface for interacting with the R interpreter. Generally speaking, IDEs provide a platform for writing *and* executing code, including viewing the results of the code you have run. This is distinct from a code *editor* (like Atom), which is used just to *write* code.

When you open the RStudio program, you will see an interface similar to that in Figure 5.1. An RStudio session usually involves four sections ("panes"), though you can customize this layout if you wish:

- **Script**: The top-left pane is a simple text editor for writing your R code as different script files. While it is not as robust as a text editing program like Atom, it will colorize code, auto-complete text, and allow you to easily execute your code. Note that this pane is hidden if there are no open scripts; select File > New File > R Script from the menu to create a new script file.

Figure 5.1   RStudio's user interface, showing a script file. Red notes are added.

To execute (run) the code you write, you have two options:

1. You can execute a section of your script by selecting (highlighting) the desired code and clicking the "Run" button (or use the keyboard shortcut[1]: cmd+enter on Mac, or ctrl+enter on Windows). If no lines are selected, this will run the line currently containing the cursor. This is the most common way to execute code in RStudio.

> **Tip:** Use cmd+a (Mac) or ctrl+a (Windows) to select the entire script!

2. You can execute an entire script by clicking the "Source" button (at the top right of the Script pane, or via shift+cmd+enter) to execute all lines of code in the script file, one at a time, from top to bottom. This command will treat the current script file as the "source" of code to run. If you check the "Source on Save" option, your entire script will be executed every time you save the file (which may or may not be appropriate, depending on the complexity of your script and its output). You can also hover your mouse over this or any other button to see keyboard shortcuts.

> **Fun Fact:** The Source button actually calls an R function called source(), described in Chapter 14.

- **Console**: The bottom-left pane is a console for entering R commands. This is identical to an interactive session you would run on the command line, in which you can type and execute one line of code at a time. The console will also show the printed results of executing the code from the Script pane. If you want to perform a task *once*, but don't want to save that task in your script, simply type it in the console and press enter.

> **Tip:** Just as with the command line, you can use the *up arrow* to easily access previously executed lines of code.

- **Environment**: The top-right pane displays information about the current R environment—specifically, information that you have stored inside of *variables*. In Figure 5.1 the value 3 is stored in a variable called num_cups_coffee. You will often create dozens of variables within a script, and the Environment pane helps you keep track of which values you have stored in which variables. *This is incredibly useful for "debugging" (identifying and fixing errors)!*

- **Plots, packages, help, etc.**: The bottom-right pane contains multiple tabs for accessing a variety of information about your program. When you create visualizations, those plots will be rendered in this section. You can also see which packages you have loaded or look up information about files. Most importantly, you can access the official documentation for the R language in this pane. If you ever have a question about how something in R works, this is a good place to start!

---

[1]**RStudio Keyboard Shortcuts**: https://support.rstudio.com/hc/en-us/articles/200711853-Keyboard-Shortcuts

Note that you can use the small spaces between the quadrants to adjust the size of each area to your liking. You can also use menu options to reorganize the panes.

> **Tip**: RStudio provides a built-in link to a "Cheatsheet" for the IDE—as well as for other packages described in this text—through the `Help > Cheatsheets` menu.

5.2.2   Running **R** from the Command Line

While RStudio is the interface that we suggest for running R code, you may find that in certain situations you need to execute some code without the IDE. It is possible to issue R instructions (run lines of code) one by one at the command line by starting an **interactive R session** within your command shell. This will allow you to type R code directly into the terminal, and your computer will interpret and execute each line of code (if you just typed R syntax directly into the terminal, your computer wouldn't understand it).

With the R software installed, you can start an interactive R session on a Mac by typing R (or lowercase r) into the Terminal to run the R program. This will start the session and provide you with some information about the R language, as shown in Figure 5.2.

Notice that this description also includes *instructions on what to do next*—most importantly, `"Type 'q()' to quit R."`

> **Remember**: Always read the output carefully when working on the command line!

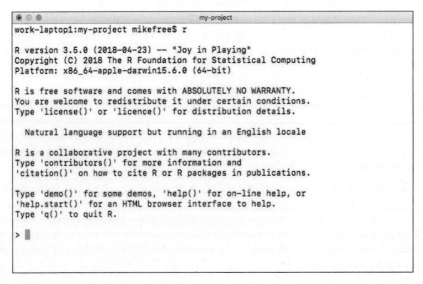

Figure 5.2   An interactive R session running in a command shell.

Once you've started running an interactive R session, you can begin entering one line of code at a time at the prompt (>). This is a nice way to experiment with the R language or to quickly run some code. For example, you can try doing some math at the command prompt (e.g., enter 1 + 1 and see the output).

It is also possible to run entire scripts from the command line by using the RScript program, specifying the .R file you wish to execute, as shown in Figure 5.3. Entering the command shown in Figure 5.3 in the terminal would execute each line of R code written in the analysis.R file, performing all of the instructions that you had saved there. This is helpful if your data has changed, and you want to recalculate the results of your analysis using the same instructions.

On Windows (and some other operating systems), you may need to tell the computer where to find the R and RScript programs to execute—that is, the **path** to these programs. You can do this by specifying the *absolute path* to the R.exe program when you execute it, as in Figure 5.3.

> **Going Further**: If you use Windows and plan to run R from the command line regularly (which is *not* required or even suggested in this book), a better solution is to add the folder containing these programs to your computer's **PATH variable**. This *system-level* variable contains a list of folders that the computer searches when finding programs to execute. The reason the computer knows where to find the git.exe program when you type git in the command line is because that program is "on the PATH."
>
> In Windows, you can add the R.exe and RScript.exe programs to your computer's PATH by editing your machine's **environment variables** through the *Control Panel*.[a] Overall, using R from the command line can be tricky; we recommend you just use RStudio instead as you're starting out.
>
> ---
> [a]https://helpdeskgeek.com/windows-10/add-windows-path-environment-variable/

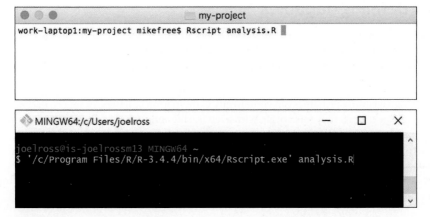

Figure 5.3 Using the RScript command to run an R script from a command shell: Mac (top) and Windows (bottom).

> **Caution**: On Windows, the R interpreter download also installs an "RGui" application (e.g., "R x64 3.4.4"), which will likely be the default program for opening .R scripts. Make sure to use the RStudio IDE for working in R!

5.3   Including Comments

Before discussing how to write programs with R, it's important to understand the syntax that lets you add comments your code. Since computer code can be opaque and difficult to understand, developers use comments to help write down the meaning and purpose of their code. This is particularly important when someone else will be looking at your work—whether that person is a collaborator or simply a future version of you (e.g., when you need to come back and fix something and so need to remember what you were trying to do).

Comments should be clear, concise, and helpful. They should provide information that is not otherwise present or "obvious" in the code itself.

In R, you mark text as a comment by putting it after the pound symbol (**#**). Everything from the # until the end of the line is a comment. You put descriptive comments *immediately above* the code they describe, but you can also put very short notes at the end of the line of code, as in the following example (note that the R code syntax used is described in the following section):

```
# Calculate the number of minutes in a year
minutes_in_a_year <- 365 * 24 * 60 # 525,600 minutes!
```

(You may recognize this # syntax and commenting behavior from command line examples in previous chapters—because the same syntax is used in a Bash shell!)

5.4   Defining Variables

Since computer programs involve working with lots of information, you need a way to store and refer to that information. You do this with **variables**. Variables are labels for information; in R, you can think of them as "boxes" or "name tags" for data. After putting data in a variable box, you can then refer to that data by the label on the box.

In the R language, variable names can contain any combination of letters, numbers, periods (.), or underscores (_), though they must begin with a letter. Like almost everything in programming, variable names are **case sensitive**. It is best practice to make variable names descriptive and informative about what data they contain. For example, x is not a good variable name, whereas num_cups_coffee is a good variable name. Throughout this book, we use the formatting suggested in *the tidyverse style guide*.[2] As such, variable names should be all lowercase letters, separated by underscores (_). This is also known as snake_case.

---

[2] **Tidyverse style guide**: http://style.tidyverse.org

> **Remember**: There is an important distinction between *syntax* and *style*. The syntax of a language describes the rules for writing the code so that a computer can interpret it. Certain operations are permitted, and others are not. Conversely, styles are optional conventions that make it easier for other humans to interpret your code. The use of a *style guide* allows you to describe the conventions you will follow in your code to help keep things like variable names consistent.

Storing information in a variable is referred to as **assigning** a **value** to the variable. You assign a value to a variable using the *assignment operator* **<-**. For example:

```
# Assign the value 3 to a variable named `num_cups_coffee`
num_cups_coffee <- 3
```

Notice that the variable name goes on the left, and the value goes on the right.

You can see which value (data) is "inside" a variable by either executing that variable name as a line of code or by using R's built-in **print( )** function (functions are detailed in Chapter 6):

```
# Print the value assigned to the variable `num_cups_coffee`
print(num_cups_coffee)
# [1] 3
```

The `print( )` function prints out the value (3) stored in the variable (num_cups_coffee). The `[1]` in that output indicates that the first element stored in the variable is the number 3—this is discussed in detail in Chapter 7.

You can also use **mathematical operators** (e.g., +, -, /, *) when assigning values to variables. For example, you could create a variable that is the sum of two numbers as follows:

```
# Use the plus (+) operator to add numbers, assigning the result to a variable
too_much_coffee <- 3 + 4
```

Once a value (like a number) is in a variable, you can use that variable in place of any other value. So all of the following statements are valid:

```
# Calculate the money spent on coffee using values stored in variables
num_cups_coffee <- 3 # store 3 in `num_cups_coffee`
coffee_price <- 3.5 # store 3.5 in `coffee_price`
money_spent_on_coffee <- num_cups_coffee * coffee_price # total spent on coffee
print(money_spent_on_coffee)
# [1] 10.5

# Alternatively, you can use a mixture of numeric values and variables
# Calculate the money spent on 4 cups of coffee
money_spent_on_four_cups <- coffee_price * 4 # total spent on 4 cups of coffee
print(money_spent_on_four_cups)
# [1] 14
```

In many ways, script files are just note pads where you've jotted down the R code you wish to run. Lines of code can be (and often are) executed out of order, particularly when you want to change or fix a previous statement. When you do change a previous line of code, you will need to **re-execute** that line of code to have it take effect, as well as re-execute any subsequent lines if you want them to use the updated value.

As an example, if you had the following code in your script file:

```
# Calculate the amount of caffeine consumed using values stored in variables
num_cups_coffee <- 3 # line 1
cups_of_tea <- 2 # line 2
caffeine_level <- num_cups_coffee + cups_of_tea # line 3
print(caffeine_level) # line 4
# [1] 5
```

Executing all of the lines of code one after another would assign the variables and print a value 5. If you edited line 1 to say num_cups_coffee <- 4, the computer wouldn't do anything different until you re-executed the line (by selecting it and pressing cmd+enter). And re-executing line 1 wouldn't cause another new value to be printed, since that command occurs at line 4! If you then re-executed line 4 (by selecting that line and pressing cmd+enter), it would still print out 5—because you haven't told R to recalculate the value of caffeine_level! You would need to re-execute *all* of the lines of code (e.g., select them *all* and pressing cmd+enter) to have your script print out the desired (new) value of 6. This kind of behavior is common for computer programming languages (though different from environments like Excel, where values are automatically updated when you change other referenced cells).

5.4.1 Basic Data Types

The preceding examples show the storage of numeric values in variables. R is a **dynamically typed language**, which means that you do not need to explicitly state which type of information will be stored in each variable you create. R is intelligent enough to understand that if you have code num_cups_coffee <- 3, then num_cups_coffee will contain a numeric value (and thus you can do math with it).

> **Going Further**: In **statically typed languages,** you need to declare the *type* of variable you want to create. For example, in the Java programming language (which is not used in this text), you have to indicate the type of variable you want to create: if you want the *integer* 10 to be stored in the variable my_num, you would have to write int my_num = 10 (where int indicates that my_num will be an integer).

There are a few "basic types" (or *modes*) for data in R:

- **Numeric**: The default computational data type in R is numeric data, which consists of the set of real numbers (including decimals). You can use **mathematical operators** on numeric data (such as +, -, *, -, etc.). There are also numerous functions that work on numeric data (such as for calculating sums or averages).

Note that you can use multiple operators in a single **expression**. As in algebra, parentheses can be used to enforce order of operations:

```
# Calculate the number of minutes in a year
minutes_in_a_year <- 365 * 24 * 60

# Enforcing order of operations with parentheses
# Calculate the number of minutes in a leap year
minutes_in_a_leap_year <- (365 + 1) * 24 * 60
```

- **Character**: Character data stores *strings* of characters (e.g., letters, special characters, numbers) in a variable. You specify that information is character data by surrounding it with either single quotes (') or double quotes ("); the tidyverse style guide suggests always using double quotes.

```
# Create character variable `famous_writer` with the value "Octavia Butler"
famous_writer <- "Octavia Butler"
```

Note that character data is still data, so it can be assigned to a variable just like numeric data.

There are no special operators for character data, though there are a many built-in functions for working with strings.

> **Caution**: If you see a plus sign (+) in the terminal as opposed to the typical greater than symbol (>)—as in Figure 5.4—you have probably forgotten to close a quotation mark. If you find yourself in this situation, you can press the esc key to cancel the line of code and start over. This will also work if you forget to close a set of parentheses (( )) or brackets ([ ]).

- **Logical**: Logical (boolean) data types store "yes-or-no" data. A logical value can be one of two values: TRUE or FALSE. Importantly, these *are not* the strings "TRUE" or "FALSE"; logical values are a different type! If you prefer, you can use the shorthand T or F in lieu of TRUE and FALSE in variable assignment.

> **Fun Fact**: Logical values are called "booleans" after mathematician and logician George Boole.

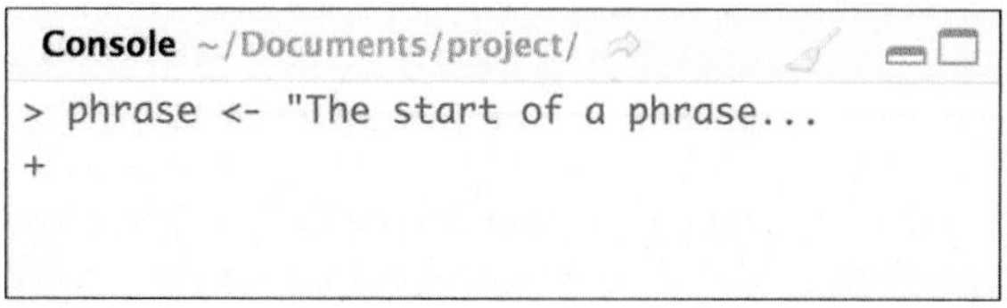

```
Console  ~/Documents/project/
> phrase <- "The start of a phrase...
+
```

Figure 5.4   An unclosed statement in the RStudio console: press the esc key to cancel the statement and return to the command prompt.

Logical values are most commonly produced by applying a **relational operator** (also called a **comparison operator**) to some other data. Comparison operators are used to compare values and include < (less than), > (greater than), <= (less than or equal), >= (greater than or equal), == (equal), and != (not equal). Here are a few examples:

```
# Store values in variables (number of strings on an instrument)
num_guitar_strings <- 6
num_mandolin_strings <- 8

# Compare the number of strings on each instrument
num_guitar_strings > num_mandolin_strings # returns logical value FALSE
num_guitar_strings != num_mandolin_strings # returns logical value TRUE

# Equivalently, you can compare values that are not stored in variables
6 == 8 # returns logical value FALSE

# Use relational operators to compare two strings
"mandolin" > "guitar" # returns TRUE (m comes after g alphabetically)
```

If you want to write a more complex logical expression (i.e., for when something is true **and** something else is false), you can do so using **logical operators** (also called **boolean operators**). These include & (and), | (or), and ! (not).

```
# Store the number of instrument players in a hypothetical band
num_guitar_players <- 3
num_mandolin_players <- 2

# Calculate the number of band members
total_band_members <- num_guitar_players + num_mandolin_players # 5

# Calculate the total number of strings in the band
# Shown on two lines for readability, which is still valid R code
total_strings <- num_guitar_players * num_guitar_strings +
  num_mandolin_strings * num_mandolin_players # 34

# Are there fewer than 30 total strings AND fewer than 6 band members?
total_strings < 30 & total_band_members < 6 # FALSE

# Are there fewer than 30 total strings OR fewer than 6 band members?
total_strings < 30 | total_band_members < 6 # TRUE

# Are there 3 guitar players AND NOT 3 mandolin players?
# Each expression is wrapped in parentheses for increased clarity
(num_guitar_players == 3) & !(num_mandolin_players == 3) # TRUE
```

It's easy to write complex—even overly complex—expressions with logical operators. If you find yourself getting lost in your logic, we recommend rethinking your question to see if there is a simpler way to express it!

- **Integer**: Integer (whole-number) values are technically a different data type than numeric values because of how they are stored and manipulated by the R interpreter. This is something that you will rarely encounter, but it's good to know that you can specify that a number is of the integer type rather than the general numeric type by placing a capital L (for "long integer") after a value in variable assignment (`my_integer <- 10L`). You will rarely do this intentionally, but this is helpful for answering the question, *Why is there an L after my number...?*

- **Complex**: Complex (imaginary) numbers have their own data storage type in R, and are created by placing an i after the number: `complex_variable <- 2i`. We will not be using complex numbers in this book, as they rarely are important for data science.

5.5 Getting Help

As with any programming language, you will inevitably run into problems, confusing situations, or just general questions when working in R. Here are a few ways to start getting help.

1. **Read the error messages**: If there is an issue with the way you have written or executed your code, R will often print out an error message in your console (in red in RStudio). Do your best to decipher the message—read it carefully, and think about what is meant by each word in the message—or you can put that message directly into Google to search for more information. You will soon get the hang of interpreting these messages if you put the time into trying to understand them. For example, Figure 5.5 shows the result of accidentally mistyping a variable name. In that error message, R indicated that the object `cty` was not found. This makes sense, because the code never defined a variable `cty` (the variable was called `city`).

2. **Google**: When you're trying to figure out how to do something, it should come as no surprise that search engines such as Google are often the best resource. Try searching for queries like `"how to DO_THING in R"`. More frequently than not, your question will lead you to a Q&A forum called StackOverflow (discussed next), which is a great place to find potential answers.

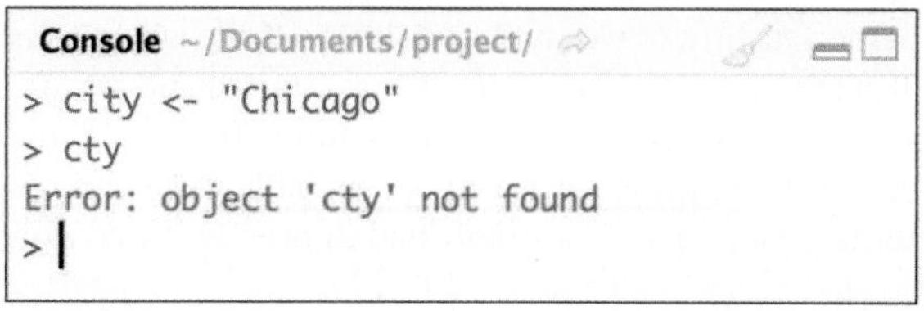

```
Console  ~/Documents/project/
> city <- "Chicago"
> cty
Error: object 'cty' not found
> |
```

Figure 5.5   RStudio showing an error message due to a typo (there is no variable cty).

3. **StackOverflow**: StackOverflow is an amazing Q&A forum for asking/answering programming questions. Indeed, most basic questions have already been asked and answered there. However, don't hesitate to post your own questions to StackOverflow. Be sure to hone in on the specific question you're trying to answer, and provide error messages and sample code. You will often find that by the time you can articulate the question clearly enough to post it, you will have figured out your problem anyway.

> **Tip:** There is a classical method of fixing errors called *rubber duck debugging*, which involves trying to explain your code/problem to an inanimate object (talking to pets works too). You will usually be able to fix the problem if you just step back and think about how you would explain it to someone else!

4. **Built-in documentation**: R's documentation is actually pretty good. Functions and behaviors are all described in the same format, and often contain helpful examples. To search the documentation within R (or in RStudio), type a question mark (?) followed by the function name you're using (e.g, `?sum`). You can perform a broader search of available documentation by typing two questions marks (??) followed by your search term (e.g., `??sum`).

   You can also look up help by using the `help()` function (e.g., `help(print)` will look up information on the `print()` function, just as `?print` does). There is also an `example()` function you can call to see examples of a function in action (e.g., `example(print)`). This will be more applicable starting in Chapter 6.

   In addition, *RDocumentation.org*[3] has a lovely searchable and readable interface to the R documentation.

5. **RStudio Community**: RStudio recently launched an online community[4] for R users. The intention is to build a more positive online community for getting programming help with R and engaging with the open source community using the software.

5.5.1 Learning to Learn R

This chapter has demonstrated the basics of the R programming language, and further features are detailed through the rest of the book. However, it's not possible to cover *all* features of a particular programming language—not to mention its surrounding ecosystem, such as the other frameworks used in data science—especially in a way that is accessible to those who are just getting started. While we will cover all of the material that you need to get started and ask questions of data using code, you will most certainly encounter problems in the future that aren't discussed in this text. *Doing* data science will require continuously learning new skills and techniques that are more advanced, more specific to your problem, or simply hadn't been invented when this book was written!

---

[3]**RDocumentation.org**: https://www.rdocumentation.org
[4]**RStudio Community**: https://community.rstudio.com

Luckily, you're not alone in this process! There is a huge number of resources that you can use to help you learn R or any other topic in programming or data science. This section provides an overview and examples of the types of resources you might use.

- **Books:** Many excellent text resources are available both in print and for free online. Books can provide a comprehensive overview of a topic, usually with a large number of examples and links to even more resources. We typically recommend them for beginners, as they help to cover all of the myriad steps involved in programming and their extensive examples help inform good programming habits. Free online books are easily accessible (and allow you to copy-and-paste code examples), but physical print can provide a useful point of reference (and typing out examples is a great way to practice).

  For learning R in particular, *R for Data Science*[5] is one of the best free online textbooks, covering the programming language through the lens of the `tidyverse` collection of packages (which are used in this book as well). Excellent print books include *R for Everyone*[6] and *The Art of R Programming*.[7]

- **Tutorials and videos:** The internet is also host to a large number of more informal explanations of programming concepts. These range from mini-books (such as the opinionated but clear introduction *aRrgh: a newcomer's (angry) guide to R*[8]), to tutorial series (such as those provided by *R Tutor*[9] or *Quick-R*[10]), to focused articles and guides (e.g., posts on *R-bloggers*[11]), to particularly informative StackOverflow responses. These smaller guides are particularly useful when you're trying to answer a specific question or clarify a single concept—when you want to know how to do one thing, not necessarily understand the entire language. In addition, many people have created and shared online video tutorials (such as Pearson's *LiveLessons*[12]), often in support of a course or textbook. Video code blogging is even more common in other programming languages such as JavaScript. Video demonstrations are great at showing you how to actually use a programming concept in practice—you can see all the steps that go into a program (though there is no substitute for doing it yourself).

  Because such guides can be created and hosted by anyone, the quality and accuracy may vary. It's always a good idea to confirm your understanding of a concept with multiple sources (do multiple tutorials agree?), with your own experience (does the solution actually work for your code?), and your own intuition (does that seem like a sensible explanation?). In general, we encourage you to start with more popular or official guides, as they are more likely to encourage best practices.

- **Interactive tutorials and courses:** The best way to learn any skill is by doing it, and there are multiple interactive websites that will let you learn and practice programming right in your web browser. These are great for seeing topics in action or for experimenting with different

---

[5]Wickham, H., & Grolemund, G. (2016). *R for Data Science*. O'Reilly Media, Inc. http://r4ds.had.co.nz

[6]Lander, J. P. (2017). *R for Everyone: Advanced Analytics and Graphics* (2nd ed.). Boston, MA: Addison-Wesley.

[7]Matloff, N. (2011). *The Art of R Programming: A Tour of Statistical Software Design*. San Francisco, CA: No Starch Press.

[8]**aRrgh: a newcomer's (angry) guide to R**: http://arrgh.tim-smith.us

[9]**R Tutor**: http://www.r-tutor.com/; start with the introduction at http://www.r-tutor.com/r-introduction

[10]**Quick-R**: https://www.statmethods.net/index.html; be sure and follow the hyperlinks.

[11]**R-Bloggers**: https://www.r-bloggers.com

[12]**LiveLessons** video tutorials: https://www.youtube.com/user/livelessons

options (though it is simple enough to experiment inside of RStudio—an approach taken by the *swirl*[13] package).

The most popular set of interactive tutorials for R programming are provided by *DataCamp*[14] and are presented as online courses (a sequence of explanations and exercises that you can learn to use a skill) on different topics. DataCamp tutorials provide videos and interactive tutorials for a wide range of different data science topics. While most of the introductory courses (e.g., *Introduction to R*[15]) are free, more advanced courses require you to sign up and pay for the service. Nevertheless, even at the free level, this is an effective set of resources for picking up new skills.

In addition to these informal interactive courses, it is possible to find more formal online courses in R and data science through massive open online course (MOOC) services such as *Coursera*[16] or *Udacity*.[17] For example, the *Data Science at Scale*[18] course from the University of Washington offers a deep introduction to data science (though it assumes some programming experience, so it may be more appropriate for *after* you've finished this book!). Note that these online courses almost always require a paid fee, though you can sometimes earn university credit or certifications from them.

- **Documentation:** One of the best places to start out when learning a programming concept is the official documentation. In addition to the base R documentation described in the previous section, many system creators will produce useful "getting started" guides and references—called "vignettes" in the R community—that you can use (to encourage adoption of their tool). For example, the `dplyr` package (described in great detail in Chapter 11) has an official "getting started" summary on its homepage[19] as well as a complete reference.[20] Further detail on a package may also often be found linked from that package's homepage on GitHub (where the documentation can be kept under version control); checking the GitHub page for a package or library is often an effective way to gain more information about it. Additionally, many R packages host their documentation in `.pdf` format on CRAN's website; to learn to use a package, you will need to read its explanation carefully and try out its examples!

- **Community resources:** As R is an open source language, many of the R resources described here are created by the community of programmers—and this community can be one of the best resources for learning to program. In addition to community-generated tutorials and answers to questions, in-person meet-ups can be an excellent source for getting help (particularly in larger urban areas). Check whether your city or town has a local "useR" group that may host events or training sessions.

---

[13] swirl interactive tutorial: http://swirlstats.com
[14] DataCamp: https://www.datacamp.com/home
[15] **DataCamp: Introduction to R:** https://www.datacamp.com/courses/free-introduction-to-r
[16] Coursera: https://www.coursera.org
[17] Udacity: https://www.udacity.com
[18] **Data Science at Scale:** online course from the University of Washington: https://www.coursera.org/specializations/data-science
[19] **dplyr** homepage: https://dplyr.tidyverse.org
[20] **dplyr** reference: https://dplyr.tidyverse.org/reference/index.html

This section lists only a few of the many, many resources for learning R. You can find many more online resources on similar topics by searching for "TOPIC tutorial" or "how to DO_SOMETHING in R." You may also find other compilations of resources. For example, RStudio has put together a list[21] of its recommended tutorials and resources.

In the end, remember that the best way to learn about anything—whether about programming or from a set of data—is to *ask questions*. For practice writing code in R and familiarizing yourself with RStudio, see the set of accompanying book exercises.[22]

[21] **RStudio: Online Learning** resource collection: https://www.rstudio.com/online-learning/
[22] **Introductory R exercises**: https://github.com/programming-for-data-science/chapter-05-exercises

6

Functions

As you begin to take on data science projects, you will find that the tasks you perform will involve multiple different instructions (lines of code). Moreover, you will often want to be able to repeat these tasks (both within and across projects). For example, there are many steps involved in computing summary statistics for some data, and you may want to repeat this analysis for different variables in a data set or perform the same type of analysis across two different data sets. Planning out and writing your code will be notably easier if can you group together the lines of code associated with each overarching task into a single step.

Functions represent a way for you to add a label to a group of instructions. Thinking about the tasks you need to perform (rather than the individual lines of code you need to write) provides a useful *abstraction* in the way you think about your programming. It will help you hide the details and generalize your work, allowing you to better reason about it. Instead of thinking about the many lines of code involved in each task, you can think about the task itself (e.g., `compute_summary_stats()`). In addition to helping you better reason about your code, labeling groups of instructions will allow you to save time by reusing your code in different contexts—repeating the task without rewriting the individual instructions.

This chapter explores how to use functions in R to perform advanced capabilities and create code that is flexible for analyzing multiple data sets. After considering a function in a general sense, it discusses using built-in R functions, accessing additional functions by loading R packages, and writing your own functions.

6.1   What Is a Function?

In a broad sense, a **function** is a named sequence of instructions (lines of code) that you may want to perform one or more times throughout a program. Functions provide a way of *encapsulating* multiple instructions into a single "unit" that can be used in a variety of contexts. So, rather than needing to repeatedly write down all the individual instructions for drawing a chart for every one of your variables, you can define a `make_chart()` function once and then just **call** (execute) that function when you want to perform those steps.

In addition to grouping instructions, functions in programming languages like R tend to follow the mathematical definition of functions, which is a set of operations (instructions!) that are performed on some **inputs** and lead to some **outputs**. Function inputs are called **arguments** (also referred to as **parameters**); specifying an argument for a function is called **passing** the argument into the function (like passing a football). A function then **returns** an output to use. For example, imagine a function that can determine the largest number in a set of numbers—that function's input would be the set of numbers, and the output would be the largest number in the set.

Grouping instructions into reusable functions is helpful throughout the data science process, including areas such as the following:

- *Data management*: You can group instructions for loading and organizing data so they can be applied to multiple data sets.

- *Data analysis*: You can store the steps for calculating a metric of interest so that you can repeat your analysis for multiple variables.

- *Data visualization*: You can define a process for creating graphics with a particular structure and style so that you can generate consistent reports.

6.1.1   R Function Syntax

R functions are referred to by name (technically, they are values like any other variable). As in many programming languages, you call a function by writing the name of the function followed immediately (no space) by parentheses **( )**. Inside the parentheses, you put the arguments (inputs) to the function separated by commas (**,**). Thus, computer functions look just like multi-variable mathematical functions, but with names longer than f( ). Here are a few examples of using functions that are included in the R language:

```
# Call the print() function, passing it "Hello world" as an argument
print("Hello world")
# [1] "Hello world"

# Call the sqrt() function, passing it 25 as an argument
sqrt(25) # returns 5 (square root of 25)

# Call the min() function, passing it 1, 6/8, and 4/3 as arguments
# This is an example of a function that takes multiple arguments
min(1, 6 / 8, 4 / 3) # returns 0.75 (6/8 is the smallest value)
```

> **Remember:** In this text, we always include empty parentheses ( ) when referring to a function by name to help distinguish between variables that hold functions and variables that hold values (e.g., add_values( ) versus my_value). This does not mean that the function takes no arguments; instead, it is just a useful shorthand for indicating that a variable holds a function (*not* a value).

If you call any of these functions interactively, R will display the returned value (the output) in the console. However, the computer is not able to "read" what is written in the console—that's for humans to view! If you want the computer to be able to *use* a returned value, you will need to give that value a name so that the computer can refer to it. That is, you need to store the returned value in a variable:

```
# Store the minimum value of a vector in the variable `smallest_number`
smallest_number <- min(1, 6 / 8, 4 / 3)

# You can then use the variable as usual, such as for a comparison
min_is_greater_than_one <- smallest_number > 1 # returns FALSE

# You can also use functions inline with other operations
phi <- .5 + sqrt(5) / 2 # returns 1.618034

# You can pass the result of a function as an argument to another function
# Watch out for where the parentheses close!
print(min(1.5, sqrt(3)))
# [1] 1.5
```

In the last example, the resulting value of the "inner" function function—sqrt( )—is immediately used as an argument. Because that value is used immediately, you don't have to assign it a separate variable name. Consequently, it is known as an **anonymous variable**.

6.2   Built-in **R** Functions

As you have likely noticed, R comes with a variety of functions that are built into the language (also referred to as "*base*" R functions). The preceding example used the print( ) function to print a value to the console, the min( ) function to find the smallest number among the arguments, and the sqrt( ) function to take the square root of a number. Table 6.1 provides a *very* limited list of functions you might experiment with (or see a few more from *Quick-R*[1]).

To learn more about any individual function, you can look it up in the R documentation by using ?FUNCTION_NAME as described in Chapter 5.

> **Tip:** Part of learning any programming language is identifying which functions are available in that language and understanding how to use them. Thus, you should look around and become familiar with these functions—but do not feel that you need to memorize them! It's enough to be aware that they exist, and then be able to look up the name and arguments for that function. As you can imagine, Google also comes in handy here (i.e., "*how to DO_TASK in R*").

This is just a tiny taste of the many different functions available in R. More functions will be introduced throughout the text, and you can also see a nice list of options in the *R Reference Card*[2] cheatsheet.

---

[1]**Quick-R: Built-in Functions:** http://www.statmethods.net/management/functions.html
[2]**R Reference Card:** cheatsheet summarizing built-in R functions: https://cran.r-project.org/doc/contrib/Short-refcard.pdf

Table 6.1 **Examples and descriptions of frequently used R functions**

Function Name	Description	Example
sum(a, b, ...)	Calculates the sum of all input values	sum(1, 5) # *returns* 6
round(x, digits)	Rounds the first argument to the given number of digits	round(3.1415, 3) # *returns* 3.142
toupper(str)	Returns the characters in uppercase	toupper("hi mom") # *returns* "HI MOM"
paste(a, b, ...)	*Concatenates* (combines) characters into one value	paste("hi", "mom") # *returns* "hi mom"
nchar(str)	Counts the number of characters in a string (including spaces and punctuation)	nchar("hi mom") # *returns* 6
c(a, b, ...)	*Concatenates* (combines) multiple items into a *vector* (see Chapter 7)	c(1, 2) # *returns* 1, 2
seq(a, b)	Returns a sequence of numbers from a to b	seq(1, 5) # *returns* 1, 2, 3, 4, 5

6.2.1 Named Arguments

Many functions have both *required arguments* (values that you must provide) and *optional arguments* (arguments that have a "default" value, unless you specify otherwise). Optional arguments are usually specified using **named arguments**, in which you specify that an argument value has a particular name. As a result, you don't need to remember the order of optional arguments, but can instead simply reference them by name.

Named arguments are written by putting the name of the argument (which is like a variable name), followed by the equals symbol (=), followed by the value to pass to that argument. For example:

```
# Use the `sep` named argument to specify the separator is '+++'
paste("Hi", "Mom", sep = "+++") # returns "Hi++Mom"
```

Named arguments are almost always optional (since they have default values), and can be included in any order. Indeed, many functions allow you to specify arguments either as **positional arguments** (called such because they are determined by their position in the argument list) or with a name. For example, the second positional argument to the round() function can also be specified as the named argument digits:

```
# These function calls are all equivalent, though the 2nd is most clear/common
round(3.1415, 3) # 3.142
round(3.1415, digits = 3) # 3.142
round(digits = 3, 3.1415) # 3.142
```

To see a list of arguments—required or optional, positional or named—available to a function, look it up in the documentation (e.g., using ?FUNCTION_NAME). For example, if you look up the paste()

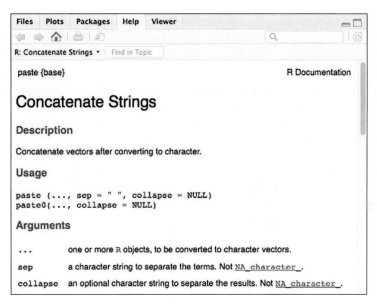

Figure 6.1   Documentation for the paste() function, as shown in RStudio.

function (using ?paste in RStudio), you will see the documentation shown in Figure 6.1. The *usage* displayed —paste (..., sep = " ", collapse = NULL)— specifies that the function takes any number of positional arguments (represented by the ...), as well as two additional named arguments: sep (whose default value is " ", making pasted words default to having a space between them) and collapse (used when pasting *vectors*, described in Chapter 7).

> **Tip**: In R's documentation, functions that require a limited number of unnamed arguments will often refer to them as x. For example, the documentation for round() is listed as follows: round(x, digits = 0). The x just means "the data value to run this function on."

> **Fun Fact**: The mathematical operators (e.g., +) are actually functions in R that take two arguments (the operands). The familiar mathematical notation is just a shortcut.
>
> ```
> # These two lines of code are the same:
> x <- 2 + 3 # add 2 and 3
> x <- '+'(2, 3) # add 2 and 3
> ```

6.3   Loading Functions

Although R comes with lots of built-in functions, you can always use more functions! **Packages** (also broadly, if inaccurately, referred to as *libraries*) are additional sets of R functions that are written and published by the R community. Because many R users encounter the same data management and analysis challenges, programmers are able to use these packages and thereby benefit from the work of others. (This is the amazing thing about the open source community—people solve problems and then make those solutions available to others.) Popular R packages exist for manipulating data (dplyr), making beautiful graphics (ggplot2), and implementing machine learning algorithms (randomForest).

R packages do not ship with the R software by default, but rather need to be downloaded (once) and then loaded into your interpreter's environment (each time you wish to use them). While this may seem cumbersome, the R software would be huge and slow if you had to install and load *all* available packages to do anything with it.

Luckily, it is possible to install and load R packages from within R. The base R software provides **install.packages()** function for installing packages, and the **library()** function for loading them. The following example illustrates installing and loading the stringr package (which contains handy functions for working with character strings):

```
# Install the `stringr` package. Only needs to be done once per computer
install.packages("stringr")

# Load the package (make `stringr` functions available in this `R` session)
library("stringr") # quotes optional here, but best to include them
```

> **Caution**: When you install a package, you may receive a warning message about the package being built under a previous version of R. In all likelihood, this shouldn't cause a problem, but you should pay attention to the details of the messages and keep them in mind (especially if you start getting unexpected errors).
>
> Errors installing packages are some of the trickiest to solve, since they depend on machine-specific configuration details. Read any error messages carefully to determine what the problem may be.

The install.packages() function downloads the necessary set of R code for a given package (which explains why you need to do it only once per machine), while the library() function loads those scripts into your current R session (you connect to the "library" where the package has been installed). If you're curious *where* the library of packages is located on your computer, you can run the R function .libPaths() to see where the files are stored.

> **Caution**: Loading a package sometimes overrides a function of the same name that is already in your environment. This may cause a warning to appear in your R terminal, but it does not necessarily mean you made a mistake. Make sure to read warning messages carefully and attempt to decipher their meaning. If the warning doesn't refer to something that seems to be a problem (such as overriding existing functions you weren't going to use), you can ignore it and move on.

After loading a package with the library() function, you have access to functions that were written as part of that package. For example, stringr provides a function **str_count()** that returns how many times a "substring" appears in a word (see the `stringr` documentation[3] for a complete list of functions included in that package):

```
# How many i's are in Mississippi?
str_count("Mississippi", "i") # 4
```

Because there are so many packages, many of them will provide functions with the same names. You thus might need to distinguish between the `str_count()` function from `stringr` and the `str_count()` function from somewhere else. You can do this by using the full package name of the function (called **namespacing** the function)—written as the package name, followed by a double colon (`::`), followed by the name of the function:

```
# Explicitly call the namespaced `str_count` function. Not very common.
stringr::str_count("Mississippi", "i") # 4

# Equivalently, call the function without namespacing
str_count("Mississippi", "i") # 4
```

Much of the work involved in programming for data science involves finding, understanding, and using these external packages (no need to reinvent the wheel!). A number of such packages will be discussed and introduced in this text, but you must also be willing to extrapolate what you learn (and research further examples) to new situations.

> **Tip**: There are packages available to help you improve the style of your R code. The `lintr`[a] package detects code that violates the tidyverse style guide, and the `styler`[b] package applies suggested formatting to your code. After loading those packages, you can run `lint("MY_FILENAME.R")` and `style_file("MY_FILENAME.R")` (using the appropriate filename) to help ensure you have used good code style.
>
> ---
> [a] https://github.com/jimhester/lintr
> [b] http://styler.r-lib.org

6.4 Writing Functions

Even more exciting than loading other people's functions is writing your own. Anytime that you have a task that you may repeat throughout a script—or if you just want to organize your thinking—it's good practice to write a function to perform that task. This will limit repetition and reduce the likelihood of errors, as well as make things easier to read and understand (and identify flaws in your analysis).

---

[3] https://cran.r-project.org/web/packages/stringr/stringr.pdf

The best way to understand the syntax for defining a function is to look at an example:

```
# A function named `make_full_name` that takes two arguments
# and returns the "full name" made from them
make_full_name <- function(first_name, last_name) {
  # Function body: perform tasks in here
  full_name <- paste(first_name, last_name)

  # Functions will *return* the value of the last line
  full_name
}

# Call the `make_full_name()` function with the values "Alice" and "Kim"
my_name <- make_full_name("Alice", "Kim") # returns "Alice Kim" into `my_name`
```

Functions are in many ways like variables: they have a **name** to which you *assign* a value (using the same assignment operator: <-). One difference is that they are written using the **function** keyword to indicate that you are creating a function and not simply storing a value. Per the tidyverse style guide,[4] functions should be written in snake_case and named using **verbs**—after all, they define something that the code will *do*. A function's name should clearly suggest what it does (without becoming too long).

> **Remember**: Although tidyverse functions are written in snake_case, many built-in R functions use a dot . to separate words—for example, install.packages() and is.numeric() (which determines whether a value is a number and not, for example, a character string).

A function includes several different parts:

- **Arguments**: The value assigned to the function name uses the syntax function(...) to indicate that you are creating a function (as opposed to a number or character string). The words put between the parentheses are names for variables that will contain the values passed in as arguments. For example, when you call make_full_name("Alice", "Kim"), the value of the first argument ("Alice") will be assigned to the first variable (first_name), and the value of the second argument ("Kim") will be assigned to the second variable (last_name).

  Importantly, you can make the argument names anything you want (name_first, given_name, and so on), just as long as you then use that variable name to refer to the argument inside the function body. Moreover, these argument variables are available only while inside the function. You can think of them as being "nicknames" for the values. The variables first_name, last_name, and full_name exist only within this particular function; that is, they are accessible within the **scope** of the function.

- **Body**: The body of the function is a *block of code* that falls between curly braces {} (a "block" is represented by curly braces surrounding code statements). The cleanest style is to put the opening { immediately after the arguments list, and the closing } on its own line.

---

[4] **tidyverse style Guide**: http://style.tidyverse.org/functions.html

The function body specifies all the instructions (lines of code) that your function will perform. A function can contain as many lines of code as you want. You will usually want more than 1 line to make the effort of creating the function worthwhile, but if you have more than 20 lines, you might want to break it up into separate functions. You can use the argument variables in here, create new variables, call other functions, and so on. Basically, any code that you would write outside of a function can be written inside of one as well!

- **Return value**: A function will return (output) whatever value is evaluated in the last statement (line) of that function. In the preceding example, the final `full_name` statement will be returned.

It is also possible to explicitly state what value to return by using the **return()** function, passing it the value that you wish your function to return:

```
# A function to calculate the area of a rectangle
calculate_rect_area <- function(width, height){
    return(width * height) # return a specific result
}
```

However, it is considered good style to use the `return()` statement only when you wish to return a value before the final statement is executed (see Section 6.5). As such, you can place the value you wish to return as the last line of the function, and it will be returned:

```
# A function to calculate the area of a rectangle
calculate_rect_area <- function(width, height){
    # Store a value in a variable, then return that value
    area <- width * height # calculate area
    area # return this value from the function
}
```

```
# A function to calculate the area of a rectangle
calculate_rect_area <- function(width, height){
    # Equivalently, return a value anonymously (without first storing it)
    width * height # return this value from the function
}
```

You can call (execute) a function you defined the same way you call built-in functions. When you do so, R will take the arguments you pass in (e.g., "Alice" and "Kim") and assign them to the argument variables. It then executes each line of code in the function body one at a time. When it gets to the last line (or the `return()` call), it will end the function and return the last expression, which could be assigned to a different variable outside of the function.

Overall, writing functions is an effective way to group lines of code together, creating an abstraction for those statements. Instead of needing to think about doing four or five steps at once, you can just think about a single step: calling the function! This makes it easier to understand your code and the analysis you need to perform.

6.4.1 Debugging Functions

A central part of writing functions is fixing the (inevitable) errors that you introduce in the process. Identifying errors within the functions you write is more complex than resolving an issue with a single line of code because you will need to search across the entire function to find the source of the error! The best technique for honing in on and identifying the line of code with the error is to run each line of code *one at a time*. While it is possible to execute each line individually in RStudio (using cmd+enter), this process requires further work when functions require *arguments*.

For example, consider a function that calculates a person's body mass index (BMI):

```
# Calculate body mass index (kg/m^2) given the input in pounds (lbs) and
# inches (inches)
calculate_bmi <- function(lbs, inches) {
  height_in_meters <- inches * 0.0254
  weight_in_kg <- lbs * 0.453592
  bmi <- weight_in_kg / height_in_meters ^ 2
  bmi
}

# Calculate the BMI of a person who is 180 pounds and 70 inches tall
calculate_bmi(180, 70)
```

Recall that when you execute a function, R evaluates each line of code, replacing the arguments of that function with the values you supply. When you execute the function (e.g., by calling calculate_bmi(180, 70)), you are essentially *replacing* the variable lbs with the value 180, and replacing the variable inches with the value 70 throughout the function.

But if you try to run each statement in the function one at a time, then the variables lbs and inches won't have values (because you never actually called the function)! Thus a strategy for debugging functions is to assign sample values to your arguments, and then run through the function line by line. For example, you could do the following (either within the function, in another part of the script, or just in the console):

```
# Set sample values for the `lbs` and `inches` variables
lbs <- 180
inches <- 70
```

With those variables assigned, you can run each statement inside the function one at a time, checking the intermediate results to see where your code makes a mistake—and then you can fix that line and retest the function! Be sure to delete the temporary variables when you're done.

Note that while this will identify *syntax errors*, it will not help you identify *logical* errors. For example, this strategy will not help if you use the incorrect conversion between inches and meters, or pass the arguments to your function in the incorrect order. For example, calculate_bmi(70, 180) won't return an error, but it will return a *very* different BMI than calculate_bmi(180, 70).

> **Remember:** When you pass arguments to functions, *order matters!* Be sure that you are passing in values in the order expected by the function.

6.5   Using Conditional Statements

Functions are a way to organize and control the flow of execution of your code (e.g., which lines of code get run in which order). In R, as in other languages, you can also control program flow by specifying different instructions that can be run based on a different set of conditions. **Conditional statements** allow you to specify different blocks of code to run when given different contexts, which is often valuable within functions.

In an abstract sense, a conditional statement is saying:

```
IF something is true
  do some lines of code
OTHERWISE
  do some other lines of code
```

In R, you write these conditional statements using the keywords **if** and **else** and the following syntax:

```
# A generic conditional statement
if (condition) {
  # lines of code to run if `condition` is TRUE
} else {
  # lines of code to run if `condition` is FALSE
}
```

Note that the else needs to be on the same line as the closing curly brace (}) of the if block. It is also possible to omit the else and its block, in case you don't want to do anything when the condition isn't met.

The condition can be any variable or expression that resolves to a logical value (TRUE or FALSE). Thus both of the following conditional statements are valid:

```
# Evaluate conditional statements based on the temperature of porridge

# Set an initial temperature value for the porridge
porridge_temp <- 125 # in degrees F

# If the porridge temperature exceeds a given threshold, enter the code block
if (porridge_temp > 120) { # expression is true
  print("This porridge is too hot!") # will be executed
}

# Alternatively, you can store a condition (as a TRUE/FALSE value)
# in a variable
too_cold <- porridge_temp < 70 # a logical value

# If the condition `too_cold` is TRUE, enter the code block
if (too_cold) { # expression is false
  print("This porridge is too cold!") # will not be executed
}
```

You can further extend the set of conditions evaluated using an `else if` statement (e.g., an `if` immediately after an `else`). For example:

```r
# Function to determine if you should eat porridge
test_food_temp <- function(temp) {
  if (temp > 120) {
    status <- "This porridge is too hot!"
  } else if (temp < 70) {
    status <- "This porridge is too cold!"
  } else {
    status <- "This porridge is just right!"
  }
  status # return the status
}


# Use the function on different temperatures
test_food_temp(150) # "This porridge is too hot!"
test_food_temp(60)  # "This porridge is too cold!"
test_food_temp(119) # "This porridge is just right!"
```

Note that a set of conditional statements causes the code to *branch*—that is, only one block of the code will be executed. As such, you may want to have one block return a specific value from a function, while the other block might keep going (or return something else). This is when you would want to use the `return()` function:

```r
# Function to add a title to someone's name
add_title <- function(full_name, title) {
  # If the name begins with the title, just return the name
  if (startsWith(full_name, title)) {
    return(full_name) # no need to prepend the title
  }

  name_with_title <- paste(title, full_name) # prepend the title
  name_with_title # last argument gets returned
}
```

Note that this example didn't use an explicit `else` clause, but rather just let the function "keep going" when the `if` condition wasn't met. While both approaches would be valid (achieve the same desired result), it's better code design to avoid `else` statements when possible and to instead view the `if` conditional as just handling a "special case."

Overall, conditionals and functions are ways to *organize* the flow of code in your program: to explicitly tell the R interpreter in which order lines of code should be executed. These structures become particularly useful as programs get large, or when you need to combine code from multiple script files. For practice using and writing functions, see the set of accompanying book exercises.[5]

---

[5]Function exercises: https://github.com/programming-for-data-science/chapter-06-exercises

Vectors

As you move from practicing R basics to interacting with data, you will need to understand how that data is stored, and to carefully consider the appropriate structure for the organization, analysis, and visualization of your data. This chapter covers the foundational concepts for working with vectors in R. Vectors are *the* fundamental data type in R, so understanding these concepts is key to effectively programming in the language. This chapter discusses how R stores information in vectors, the way in which operations are executed in *vectorized* form, and how to extract data from vectors.

7.1 What Is a Vector?

Vectors are *one-dimensional collections of values* that are all stored in a single variable. For example, you can make a vector people that contains the character strings "Sarah", "Amit", and "Zhang". Alternatively, you could make a vector one_to_seventy that stores the numbers from 1 to 70. Each value in a vector is referred to as an **element** of that vector; thus the people vector would have three elements: "Sarah", "Amit", and "Zhang".

> **Remember**: All the elements in a vector need to have the same *type* (e.g., numeric, character, logical). You can't have a vector whose elements include both numbers and character strings.

7.1.1 Creating Vectors

The easiest and most common syntax for creating vectors is to use the built-in **c()** function, which is used to *combine* values into a vector. The c() function takes in any number of **arguments** of the same type (separated by commas as usual), and **returns** a vector that contains those elements:

```
# Use the `c()` function to create a vector of character values
people <- c("Sarah", "Amit", "Zhang")
print(people)
# [1] "Sarah" "Amit" "Zhang"
```

```
# Use the `c()` function to create a vector of numeric values
numbers <- c(1, 2, 3, 4, 5)
print(numbers)
# [1] 1 2 3 4 5
```

When you print out a variable in R, the interpreter prints out a [1] *before* the value you have stored in your variable. This is R telling you that it is printing from the *first* element in your vector (more on element indexing later in this chapter). When R prints a vector, it prints the elements separated with spaces (technically *tabs*), not commas.

You can use the **length()** function to determine how many **elements** are in a vector:

```
# Create and measure the length of a vector of character elements
people <- c("Sarah", "Amit", "Zhang")
people_length <- length(people)
print(people_length)
# [1] 3

# Create and measure the length of a vector of numeric elements
numbers <- c(1, 2, 3, 4, 5)
print(length(numbers))
# [1] 5
```

Other functions can also help with creating vectors. For example, the **seq()** function mentioned in Chapter 6 takes two arguments and produces a vector of the integers between them. An optional third argument specifies how many numbers to skip in each step:

```
# Use the `seq()` function to create a vector of numbers 1 through 70
# (inclusive)
one_to_seventy <- seq(1, 70)
print(one_to_seventy)
# [1] 1 2 3 4 5 .....

# Make vector of numbers 1 through 10, counting by 2
odds <- seq(1, 10, 2)
print(odds)
# [1] 1 3 5 7 9
```

As a shorthand, you can produce a sequence with the **colon operator (a:b)**, which returns a vector from a to b with the element values being incremented by 1:

```
# Use the colon operator (:) as a shortcut for the `seq()` function
one_to_seventy <- 1:70
```

When you print out one_to_seventy (as in Figure 7.1), in addition to the leading [1] that you've seen in all printed results, there are bracketed numbers at the start of each line. These bracketed numbers tell you the starting position (**index**) of elements printed on that line. Thus the [1] means that the printed line shows elements starting at element number 1, a [28] means that the printed line shows elements starting at element number 28, and so on. This information is

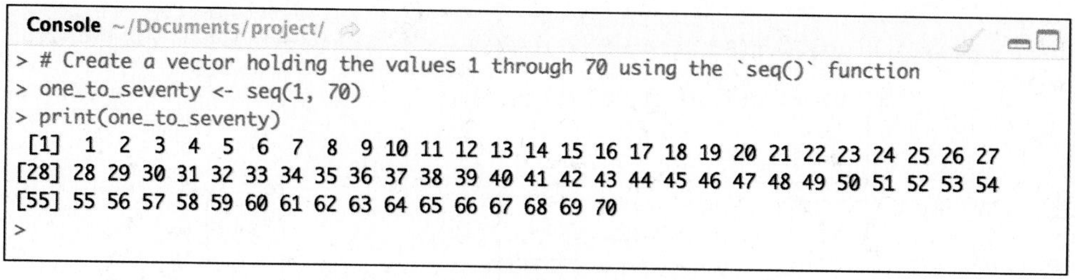

```
Console ~/Documents/project/
> # Create a vector holding the values 1 through 70 using the `seq()` function
> one_to_seventy <- seq(1, 70)
> print(one_to_seventy)
 [1]  1  2  3  4  5  6  7  8  9 10 11 12 13 14 15 16 17 18 19 20 21 22 23 24 25 26 27
[28] 28 29 30 31 32 33 34 35 36 37 38 39 40 41 42 43 44 45 46 47 48 49 50 51 52 53 54
[55] 55 56 57 58 59 60 61 62 63 64 65 66 67 68 69 70
>
```

Figure 7.1   Creating a vector using the `seq()` function and printing the results in the RStudio terminal.

intended to help make the output more readable, so you know where in the vector you are when looking at a printed line of elements.

7.2   Vectorized Operations

When performing operations (such as mathematical operations +, -, and so on) on vectors, the operation is applied to vector elements **element-wise**. This means that each element from the first vector operand is modified by the element in the same corresponding position in the second vector operand. This will produce the value at the corresponding position of the resulting vector. In other words, if you want to add two vectors, then the value of the first element in the result will be the sum of the first elements in each vector, the second element in the result will be the sum of the second elements in each vector, and so on.

Figure 7.2 demonstrates the *element-wise* nature of the vectorized operations shown in the following code:

```
# Create two vectors to combine
v1 <- c(3, 1, 4, 1, 5)
v2 <- c(1, 6, 1, 8, 0)

# Create arithmetic combinations of the vectors
v1 + v2 # returns 4 7 5 9 5
v1 - v2 # returns 2 -5 3 -7 5
v1 * v2 # returns 3 6 4 8 0
v1 / v2 # returns 3 0.167 4 0.125 Inf

# Add a vector to itself (why not?)
v3 <- v2 + v2 # returns 2 12 2 16 0

# Perform more advanced arithmetic!
v4 <- (v1 + v2) / (v1 + v1) # returns 0.67 3.5 0.625 4.5 0.5
```

Vectors support any operators that apply to their "type" (i.e., numeric or character). While you can't apply mathematical operators (namely, +) to combine vectors of character strings, you can use functions like `paste()` to concatenate the elements of two vectors, as described in Section 7.2.3.

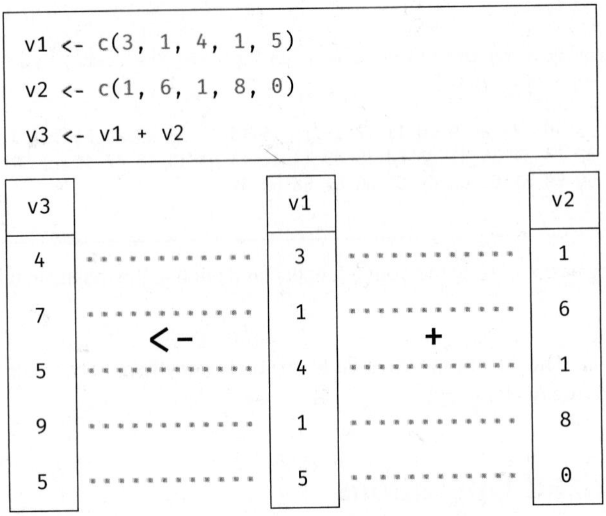

```
v1 <- c(3, 1, 4, 1, 5)

v2 <- c(1, 6, 1, 8, 0)

v3 <- v1 + v2
```

v3		v1		v2
4	. . . .	3	. . . .	1
7	. . . .	1	. . . .	6
5	. . . .	4	. . . .	1
9	. . . .	1	. . . .	8
5	. . . .	5	. . . .	0

<-   +

Figure 7.2   Vector operations are applied *element-wise*: the first element in the resulting vector (v3) is the sum of the first element in the first vector (v1) and the first element in the second vector (v2).

7.2.1 Recycling

Recycling refers to what R does in cases when there are an unequal number of elements in two operand vectors. If R is tasked with performing a vectorized operation with two vectors of unequal length, it will reuse (*recycle*) elements from the shorter vector. For example:

```
# Create vectors to combine
v1 <- c(1, 3, 5, 1, 5)
v2 <- c(1, 2)

# Add vectors
v3 <- v1 + v2 # returns 2 5 6 3 6
```

In this example, R first combined the elements in the first position of each vector (1 + 1 = 2). Then, it combined elements from the second position (3 + 2 = 5). When it got to the third element (which was present only in v1), it went back to the *beginning* of v2 to select a value, yielding 5 + 1 = 6. This recycling is illustrated in Figure 7.3.

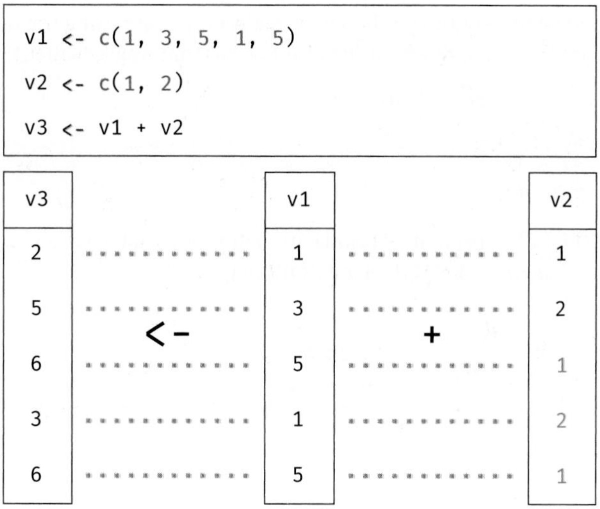

```
v1 <- c(1, 3, 5, 1, 5)
v2 <- c(1, 2)
v3 <- v1 + v2
```

v3		v1		v2
2	· · · · · · · · · · · ·	1	· · · · · · · · · · · ·	1
5	· · · · · · · · · · · ·	3	· · · · · · · · · · · ·	2
6	**< −**	5	**+**	1
3	· · · · · · · · · · · ·	1	· · · · · · · · · · · ·	2
6	· · · · · · · · · · · ·	5	· · · · · · · · · · · ·	1

Figure 7.3   Recycling values in vector addition. If one vector is shorter than another (e.g., v2), the values will be repeated (*recycled*) to match the length of the longer vector. Recycled values are in red.

Remember: Recycling will occur no matter whether the longer vector is the first or the second operand. In either case, R will provide a warning message if the length of the longer vector is not a multiple of the shorter (so that there would be elements "left over" from recycling). This warning doesn't necessarily mean you did something wrong, but you should pay attention to it because it may be indicative of an error (i.e., you thought the vectors were of the same length, but made a mistake somewhere).

7.2.2   Most Everything Is a Vector!

What happens if you try to add a vector and a "regular" single value (a **scalar**)?

```
# Add a single value to a vector of values
v1 <- 1:5 # create vector of numbers 1 to 5
result <- v1 + 4 # add scalar to vector
print(result)
# [1] 5 6 7 8 9
```

As you can see (and probably expected), the operation added 4 to every element in the vector.

This sensible behavior occurs because R stores all character, numeric, and boolean values as vectors. Even when you thought you were creating a single value (a scalar), you were actually creating a vector with a single element (length 1). When you create a variable storing the number 7 (e.g., with x <- 7), R creates a vector of length 1 with the number 7 as that single element.

```
# Confirm that basic types are stored in vectors
is.vector(18) # TRUE
is.vector("hello") # TRUE
is.vector(TRUE) # TRUE
```

This is why R prints the [1] in front of all results: it's telling you that it's showing a vector (which happens to have one element) starting at element number 1.

```
# Create a vector of length 1 in a variable `x`
x <- 7 # equivalent to `x <- c(7)`

# Print out `x`: R displays the vector index (1) in the console
print(x)
# [1] 7
```

This behavior explains why you can't use the length() function to get the length of a character string; it just returns the length of the vector containing that string (which is 1). Instead, you would use the nchar() function to get the number of characters in a character string.

Thus when you add a "scalar" such as 4 to a vector, what you're really doing is adding a vector with a single element 4. As such, the same recycling principle applies, so that the single element is recycled and applied to each element of the first operand.

7.2.3   Vectorized Functions

Because all basic data types are stored as vectors, almost every function you've encountered so far in this book can be applied to vectors, not just to single values. These **vectorized functions** are both more idiomatic and efficient than non-vector approaches. You will find that functions work the same way for vectors as they do for single values, because single values are just instances of vectors!

This means that you can use nearly any function on a vector, and it will act in the same vectorized, element-wise manner: the function will result in a new vector where the function's transformation has been applied to each individual element in order.

For example, consider the round() function described in Chapter 6. This function rounds the given argument to the nearest whole number (or number of decimal places if specified).

```
# Round the number 1.67 to 1 decimal place
round(1.67, 1) # returns 1.7
```

But recall that the 1.67 in the preceding example is actually a vector of length 1. If you instead pass a vector containing multiple values as an argument, the function will perform the same rounding on each element in the vector.

```
# Create a vector of numbers
nums <- c(3.98, 8, 10.8, 3.27, 5.21)

# Perform the vectorized operation
rounded_nums <- round(nums, 1)

# Print the results (each element is rounded)
print(rounded_nums)
# [1]  4.0  8.0  10.8  3.3  5.2
```

Vectorized operations such as these are also possible with character data. For example, the nchar() function, which returns the number of characters in a string, can be used equivalently for a vector of length 1 or a vector with many elements inside of it:

```
# Create a character variable `introduction`, then count the number
# of characters
introduction <- "Hello"
nchar(introduction) # returns 5

# Create a vector of `introductions`, then count the characters in
# each element
introductions <- c("Hi", "Hello", "Howdy")
nchar(introductions) # returns 2  5  5
```

> **Remember:** When you use a function on a vector, you're using that function *on each item* in the vector!

You can even use vectorized functions in which *each argument* is a vector. For example, the following code uses the paste() function to paste together elements in two different vectors. Just as the plus operator (+) performed element-wise addition, other vectorized functions such as paste() are also implemented element-wise:

```
# Create a vector of two colors
colors <- c("Green", "Blue")

# Create a vector of two locations
locations <- c("sky", "grass")

# Use the vectorized paste() operation to paste together the vectors above
band <- paste(colors, locations, sep = "") # returns "Greensky" "Bluegrass"
```

Notice the same element-wise combination is occurring: the `paste()` function is applied to the first elements, then to the second elements, and so on.

This vectorization process is *extremely powerful*, and is a significant factor in what makes R an efficient language for working with large data sets (particularly in comparison to languages that require explicit iteration through elements in a collection).[1] To write really effective R code, you will need to be comfortable applying functions to vectors of data, and getting vectors of data back as results.

> **Going Further**: As with other programming languages, R does support explicit iteration in the form of **loops**. For example, if you wanted to take an action *for* each element in a vector, you could do that using a `for` loop. However, because operations are vectorized in R, there is no need to explicitly iterate through vectors. While you are able to write loops in R, they are almost entirely unnecessary for writing the language and therefore are not discussed in this text.

7.3 Vector Indices

Vectors are the fundamental structure for storing collections of data. Yet, you often want to work with just *some* of the data in a vector. This section discusses a few ways that you can get a **subset** of elements in a vector.

The simplest way that you can refer to individual elements in a vector by their **index**, which is the number of their position in the vector. For example, in the vector

```
vowels <- c("a", "e", "i", "o", "u")
```

the `"a"` (the first element) is at index 1, `"e"` (the second element) is at index 2, and so on.

> **Remember**: In R, vector elements are indexed starting with 1. This is distinct from most other programming languages, which are *zero-indexed* and so reference the first element in a set at index 0.

You can retrieve a value from a vector using **bracket notation**. With this approach, you refer to the element at a particular index of a vector by writing the name of the vector, followed by square brackets (`[]`) that contain the index of interest:

```
# Create the people vector
people <- c("Sarah", "Amit", "Zhang")

# Access the element at index 1
first_person <- people[1]
print(first_person)
# [1] "Sarah"
```

---

[1] **Vectorization in R: Why?** is a blog post by Noam Ross with detailed discussion about the underlying mechanics of vectorization: http://www.noamross.net/blog/2014/4/16/vectorization-in-r--why.html

```
# Access the element at index 2
second_person <- people[2]
print(second_person)
# [1] "Amit"

# You can also use variables inside the brackets
last_index <- length(people) # last index is the length of the vector!
last_person <- people[last_index] # returns "Zhang"
```

> **Caution:** Don't get confused by the [1] in the printed output. It doesn't refer to which index you got from people, but rather to the index in the *extracted* result (e.g., stored in second_person) that is being printed!

If you specify an index that is **out-of-bounds** (e.g., greater than the number of elements in the vector) in the square brackets, you will get back the special value NA, which stands for **n**ot **a**vailable. Note that this is *not* the character string "NA", but rather a specific logical value.

```
# Create a vector of vowels
vowels <- c("a", "e", "i", "o", "u")

# Attempt to access the 10th element
vowels[10] # returns NA
```

If you specify a **negative index** in the square brackets, R will return all elements *except* the (negative) index specified:

```
vowels <- c("a", "e", "i", "o", "u")

# Return all elements EXCEPT that at index 2
all_but_e <- vowels[-2]
print(all_but_e)
# [1] "a" "i" "o" "u"
```

7.3.1   Multiple Indices

Recall that in R, all numbers are stored in vectors. This means that when you specify an index by putting a single number inside the square brackets, you're actually putting a *vector containing a single element* into the brackets. In fact, what you're really doing is specifying a **vector of indices** that you want R to extract from the vector. As such, you can put a vector of any length inside the brackets, and R will extract *all* the elements with those indices from the vector (producing a **subset** of the vector elements):

```
# Create a `colors` vector
colors <- c("red", "green", "blue", "yellow", "purple")

# Vector of indices (to extract from the `colors` vector)
indices <- c(1, 3, 4)
```

```
# Retrieve the colors at those indices
extracted <- colors[indices]
print(extracted)
# [1] "red" "blue" "yellow"

# Specify the index vector anonymously
others <- colors[c(2, 5)]
print(others)
# [1] "green" "purple"
```

It's common practice to use the **colon operator** to quickly specify a range of indices to extract:

```
# Create a `colors` vector
colors <- c("red", "green", "blue", "yellow", "purple")

# Retrieve values in positions 2 through 5
print(colors[2:5])
# [1] "green" "blue" "yellow" "purple"
```

This reads as "a vector of the elements in positions 2 through 5."

7.4   Vector Filtering

The previous examples used a vector of indices (*numeric* values) to retrieve a subset of elements from a vector. Alternatively, you can put a **vector of logical (boolean) values** (e.g., TRUE or FALSE) inside the square brackets to specify which elements you want to return—TRUE in the *corresponding position* means return that element and FALSE means don't return that element:

```
# Create a vector of shoe sizes
shoe_sizes <- c(5.5, 11, 7, 8, 4)

# Vector of booleans (to filter the `shoe_sizes` vector)
filter <- c(TRUE, FALSE, FALSE, FALSE, TRUE)

# Extract every element in an index that is TRUE
print(shoe_sizes[filter])
# [1]  5.5 4
```

R will go through the boolean vector and extract every item at the same position as a TRUE. In the preceding example, since filter is TRUE at indices 1 and 5, then shoe_sizes[filter] returns a vector with the elements from indices 1 and 5.

This may seem a bit strange, but it is actually incredibly powerful because it lets you select elements from a vector that meet a certain criteria—a process called **filtering**. You perform this **filtering operation** by first creating a vector of boolean values that correspond with the indices meeting that criteria, and then put that filter vector inside the square brackets to return the values of interest:

```
# Create a vector of shoe sizes
shoe_sizes <- c(5.5, 11, 7, 8, 4)

# Create a boolean vector that indicates if a shoe size is less than 6.5
shoe_is_small <- shoe_sizes < 6.5 # returns T F F F T

# Use the `shoe_is_small` vector to select small shoes
small_shoes <- shoe_sizes[shoe_is_small] # returns 5.5 4
```

The magic here is that you are once again using recycling: the relational operator < is vectorized, meaning that the shorter vector (6.5) is recycled and applied to each element in the shoe_sizes vector, thus producing the boolean vector that you want!

You can even combine the second and third lines of code into a single statement. You can think of the following as saying shoe_sizes *where* shoe_sizes *is less than 6.5*:

```
# Create a vector of shoe sizes
shoe_sizes <- c(5.5, 11, 7, 8, 4)

# Select shoe sizes that are smaller than 6.5
shoe_sizes[shoe_sizes < 6.5] # returns 5.5 4
```

This is a valid statement because the expression inside of the square brackets (shoe_sizes < 6.5) is evaluated first, producing a boolean vector (a vector of TRUEs and FALSEs) that is then used to filter the shoe_sizes vector. Figure 7.4 diagrams this evaluation. This kind of filtering is crucial for being able to ask real-world questions of data sets.

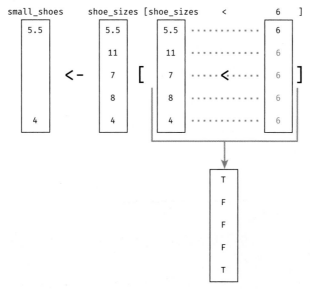

Figure 7.4 A demonstration of vector filtering using relational operators. The value 6 is recycled to match the length of the shoe_sizes vector. The resulting boolean values are used to filter the vector.

7.5   Modifying Vectors

Most operations applied to vectors will create a *new* vector with the modified values. This is the most common process you will use in R. However, it is also possible to manipulate the contents of an existing vector in various ways.

You can assign an element at a particular vector index a *new value* by specifying the index on the *left-hand side* of the operation:

```
# Create a vector `prices`
prices <- c(25, 28, 30)

# Change the first price to 20
prices[1] <- 20
print(prices)
# [1] 20 28 30
```

To create a new element in your vector, you need to specify the index in which you want to store the new value:

```
# Create a vector `prices`
prices <- c(25, 28, 30)

# Add a fourth price
prices[4] <- 32

# Add a new price (35) to the end of the vector
new_index <- length(prices) + 1 # the "end" is 1 after the last element
prices[new_index] <- 35
```

Of course, there's no reason that you can't select multiple elements on the left-hand side and assign them multiple values. The assignment operator is also vectorized!

```
# Create a vector of school supplies
school_supplies <- c("Backpack", "Laptop", "Pen")

# Replace  "Laptop" with "Tablet", and "Pen" with "Pencil"
school_supplies[c(2, 3)] <- c("Tablet", "Pencil")
```

If you try to modify an element at an index that is greater than the length of the vector, R will fill the vector with NA values:

```
# Create a vector `prices`
prices <- c(25, 28, 30)

# Set the sixth element in the vector to have the value 60
prices[6] <- 60
print(prices)
# [1] 25 28 30 NA NA 60
```

Since keeping track of indices can be difficult (and may easily change with your data, making the code *fragile*), a better approach for adding information at the end of a vector is to create a new vector by combining an existing vector with new elements(s):

```
# Use the combine (`c()`) function to create a vector
people <- c("Sarah", "Amit", "Zhang")

# Use the `c()` function to combine the `people` vector and the name "Josh"
more_people <- c(people, "Josh")
print(more_people)
# [1] "Sarah" "Amit" "Zhang" "Josh"
```

Finally, vector modification can be combined with *vector filtering* to allow you to replace a specific subset of values. For example, you could replace all values in a vector that were greater than 10 with the number 10 (to "cap" the values). Because the assignment operator is vectorized, you can leverage recycling to assign a single value to each element that has been filtered from the vector:

```
# Create a vector of values
v1 <- c(1, 5, 55, 1, 3, 11, 4, 27)

# Replace all values greater than 10 with 10
v1[v1 > 10] <- 10 # returns 1 5 10 1 3 10 4 10
```

In this example, the number 10 gets recycled for each element in which v1 is greater than 10 (v1[v1 > 10]).

This technique is particularly powerful when wrangling and cleaning data, as it will allow you to identify and manipulate invalid values or other outliers.

Overall, vectors provide a powerful way of organizing and grouping data for analysis, and will be used throughout your programming with R. For practice working with vectors in R, see the set of accompanying book exercises.[2]

---

[2] **Vector exercises:** https://github.com/programming-for-data-science/chapter-07-exercises

8

Lists

This chapter covers an additional R data type called a **list**. Lists are somewhat similar to vectors, but can store more types of data and usually include more details about that data (with some cost). Lists are R's version of a map, which is a common and extremely useful way of organizing data in a computer program. Moreover, lists are used to create *data frames*, which are the primary data storage type used for working with sets of real data in R. This chapter covers how to create and access elements in a list, as well as how to apply functions to lists.

8.1 What Is a List?

A **list** is a lot like a vector, in that it is a *one-dimensional collection of data*. However, unlike a vector, you can store elements of *different types* in a list; for example, a list can contain numeric data *and* character string data. Lists can also contain more complex data types—including vectors and even other lists!

Elements in a list can also be **tagged** with names that you can use to easily refer to them. For example, rather than talking about the list's "element #1," you can talk about the list's "first_name element." This feature allows you to use lists to create a type of map. In computer programming, a **map** (or "mapping") is a way of associating one value with another. The most common real-world example of a map is a dictionary or encyclopedia. A dictionary associates each word with its definition: you can "look up" a definition by using the word itself, rather than needing to look up the 3891st definition in the book. In fact, this same data structure is called a dictionary in the Python programming language!

> **Caution:** The definition of a list in the R language is distinct from how some other languages use the term "list." When you begin to explore other languages, don't assume that the same terminology implies the same capabilities.

As a result, lists are extremely useful for organizing data. They allow you to group together data like a person's name (characters), job title (characters), salary (number), and whether the person is a member of a union (logical)—and you don't have to remember whether the person's name or title was the first element!

> **Remember:** If you want to label elements in a collection, *use a list*. While vector elements can also be tagged with names, that practice is somewhat uncommon and requires a more verbose syntax for accessing the elements.

8.2   Creating Lists

You create a list by using the **list( )** function and passing it any number of **arguments** (separated by commas) that you want to make up that list—similar to the c( ) function for vectors.

However, you can (and should) specify the *tags* for each element in the list by putting the name of the tag (which is like a variable name), followed by an equals symbol (=), followed by the value you want to go in the list and be associated with that tag. This is similar to how named arguments are specified for functions (see Section 6.2.1). For example:

```
# Create a `person` variable storing information about someone
# Code is shown on multiple lines for readability (which is valid R code!)
person <- list(
  first_name = "Ada",
  job = "Programmer",
  salary = 78000,
  in_union = TRUE
)
```

This creates a list of four elements: "Ada", which is tagged with first_name; "Programmer", which is tagged with job; 78000, which is tagged with salary; and TRUE, which is tagged with in_union.

> **Remember:** You can have vectors as elements of a list. In fact, each scalar value in the preceding example is really a vector (of length 1).

It is possible to create a list without tagging the elements:

```
# Create a list without tagged elements. NOT the suggested usage.
person_alt <- list("Ada", "Programmer", 78000, TRUE)
```

However, tags make it easier and less error-prone to access specific elements. In addition, tags help other programmers read and understand the code—tags let them know what each element in the list represents, similar to an informative variable name. Thus it is recommended to always tag lists you create.

> **Tip:** You can get a vector of the names of your list items using the names( ) function. This is useful for understanding the structure of variables that may have come from other data sources.

Because lists can store elements of different types, they can store values that are lists themselves. For example, consider adding a list of favorite items to the person list in the previous example:

```
# Create a `person` list that has a list of favorite items
person <- list(
  first_name = "Ada",
  job = "Programmer",
  salary = 78000,
  in_union = TRUE,
  favorites = list(
    music = "jazz",
    food = "pizza"
  )
)
```

This data structure (a *list of lists*) is a common way to represent data that is typically stored in *JavaScript Object Notation* (JSON). For more information on working with JSON data, see Chapter 14.

8.3 Accessing List Elements

Once you store information in a list, you will likely want to retrieve or reference that information in the future. Consider the output of printing the person list, as shown in Figure 8.1. Notice that the output includes each tag name prepended with a dollar sign (**$**) symbol, and then on the following line prints the element itself.

Because list elements are (usually) tagged, you can access them by their tag name rather than by the index number you used with vectors. You do this by using **dollar notation**: refer to the element

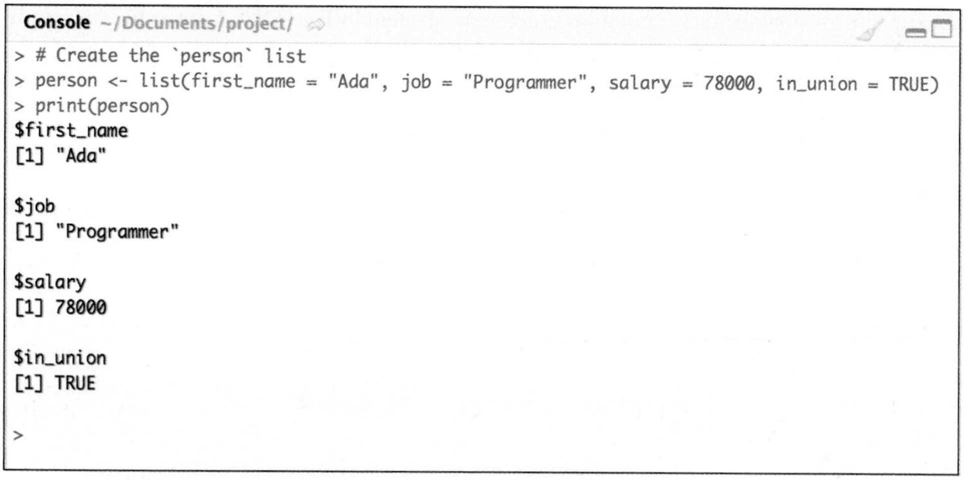

```
Console  ~/Documents/project/
> # Create the `person` list
> person <- list(first_name = "Ada", job = "Programmer", salary = 78000, in_union = TRUE)
> print(person)
$first_name
[1] "Ada"

$job
[1] "Programmer"

$salary
[1] 78000

$in_union
[1] TRUE

>
```

Figure 8.1   Creating and printing a list element in RStudio.

with a particular tag in a list by writing the name of the list, followed by a **$**, followed by the element's tag (a syntax unavailable to named vectors):

```
# Create the `person` list
person <- list(
  first_name = "Ada",
  job = "Programmer",
  salary = 78000,
  in_union = TRUE
)


# Reference specific tags in the `person` list
person$first_name # [1] "Ada"
person$salary     # [1] 78000
```

You can almost read the dollar sign as if it were an "apostrophe s" (possessive) in English. Thus, person$salary would mean "the person list's salary value."

Regardless of whether a list element has a tag, you can also access it by its numeric index (i.e., if it is the first, second, and so on item in the list). You do this by using **double-bracket notation**. With this notation, you refer to the element at a particular index of a list by writing the name of the list, followed by double square brackets (**[[ ]]**) that contain the index of interest:

```
# This is a list (not a vector!), even though elements have the same type
animals <- list("Aardvark", "Baboon", "Camel")

animals[[1]] # [1] "Aardvark"
animals[[3]] # [1] "Camel"
animals[[4]] # Error: subscript out of bounds!
```

You can also use double-bracket notation to access an element by its tag if you put a character string of the tag name inside the brackets. This is particularly useful in cases when the tag name is stored in a variable:

```
# Create the `person` list with an additional `last_name` attribute
person <- list(
  first_name = "Ada",
  last_name = "Gomez",
  job = "Programmer",
  salary = 78000,
  in_union = TRUE
)


# Retrieve values stored in list elements using strings
person[["first_name"]] # [1] "Ada"
person[["salary"]]     # [1] 78000
```

```
# Retrieve values stored in list elements
# using strings that are stored in variables
name_to_use <- "last_name"  # choose name (i.e., based on formality)
person[[name_to_use]]        # [1] "Gomez"
name_to_use <- "first_name" # change name to use
person[[name_to_use]]        # [1] "Ada"

# You can use also indices for tagged elements
# (but they're difficult to keep track of)
person[[1]] # [1] "Ada"
person[[5]] # [1] TRUE
```

Remember that lists can contain complex values (including other lists). Accessing these elements with either dollar or double-bracket notation will return that "nested" list, allowing you to access its elements:

```
# Create a list that stores a vector and a list. `job_post` has
# a *list* of qualifications and a *vector* of responsibilities.
job_post <- list(
  qualifications = list(
    experience = "5 years",
    bachelors_degree = TRUE
  ),
  responsibilities = c("Team Management", "Data Analysis", "Visualization")
)

# Extract the `qualifications` elements (a list) and store it in a variable
job_qualifications <- job_post$qualifications

# Because `job_qualifications` is a list, you can access its elements
job_qualifications$experience # "5 years"
```

In this example, job_qualifications is a variable that refers to a list, so its elements can be accessed via dollar notation. But as with any operator or function, it is also possible to use dollar notation on an *anonymous value* (e.g., a literal value that has not been assigned to a variable). That is, because job_post$qualifications is a list, you can use bracket or dollar notation to refer to an element of that list without assigning it to a variable first:

```
# Access the `qualifications` list's `experience` element
job_post$qualifications$experience # "5 years"

# Access the `responsibilities` vector's first element
# Remember, `job_post$responsibilities` is a vector!
job_post$responsibilities[1] # "Team Management"
```

This example of "chaining" together dollar-sign operators allows you to directly access elements in lists with a complex structure: you can use a single expression to refer to the "job-post's qualification's experience" value.

8.4   Modifying Lists

As with vectors, you can add and modify list elements. List elements can be modified by *assigning a new value* to an existing list element. New elements can be added by assigning a value to a new tag (or index). Moreover, list elements can be removed by reassigning the value NULL to an existing list element. All of these operations are demonstrated in the following example:

```r
# Create the `person` list
person <- list(
  first_name = "Ada",
  job = "Programmer",
  salary = 78000,
  in_union = TRUE
)

# There is currently no `age` element (it's NULL)
person$age # NULL

# Assign a value to the (new) `age` tag
person$age <- 40
person$age # [1] 40

# Reassign a value to list's `job` element
person$job <- "Senior Programmer" # a promotion!
print(person$job)
# [1] "Senior Programmer"

# Reassign a value to the `salary` element (using the current value!)
person$salary <- person$salary * 1.15 # a 15% raise!
print(person$salary)
# [1] 89700

# Remove the `first_name` tag to make the person anonymous
person$first_name <- NULL
```

NULL is a special value that means "undefined" (note that it is a special value NULL, not the character string "NULL"). NULL is somewhat similar to the term NA—the difference is that NA is used to refer to a value that is *missing* (such as an empty element in a vector)—that is, a "hole." Conversely, NULL is used to refer to a value that is not defined but doesn't necessarily leave a "hole" in the data. NA values usually result when you are creating or loading data that may have parts missing; NULL can be used to remove values. For more information on the difference between these values, see this R-Bloggers post.[1]

---

[1] R: NA vs. NULL post on R-Bloggers: https://www.r-bloggers.com/r-na-vs-null/

8.4.1 Single versus Double Brackets

> **Remember:** Vectors use *single*-bracket notation for accessing elements by index, but lists use *double*-bracket notation for accessing elements by index!

The single-bracket syntax used with vectors isn't actually selecting values by index; instead, it is **filtering** by whatever vector is inside the brackets (which may be just a single element—the index number to retrieve). In R, single brackets *always* mean to filter a collection. So if you put single brackets after a list, what you're actually doing is getting a filtered **sublist** of the elements that have those indices, just as single brackets on a vector returns a subset of elements from that vector:

```r
# Create the `person` list
person <- list(
  first_name = "Ada",
  job = "Programmer",
  salary = 78000,
  in_union = TRUE
)

# SINGLE brackets return a list
person["first_name"]
            # $first_name
            # [1] "Ada"

# Test if it returns a list
is.list(person["first_name"]) # TRUE

# DOUBLE brackets return a vector
person[["first_name"]] # [1] "Ada"

# Confirm that it *does not* return a list
is.list(person[["first_name"]]) # FALSE

# Use a vector of column names to create a filtered sub-list
person[c("first_name", "job", "salary")]
            # $first_name
            # [1] "Ada"
            #
            # $job
            # [1] "Programmer"
            #
            # $salary
            # [1] 78000
```

Notice that with lists you can filter by a *vector of tag names* (as well as by a vector of element indices).

In short, remember that single brackets return a list, whereas double brackets return a list element. You almost always want to refer to the value itself rather than a list, so you almost always want to use double brackets (or better yet—dollar notation) when accessing lists.

8.5   Applying Functions to Lists with `lapply()`

Since most functions are *vectorized* (e.g., paste( ), round( )), you can pass them a vector as an argument and the function will be applied to each item in the vector. It "just works." But if you want to apply a function to each item in a list, you need to put in a bit more effort.

In particular, you need to use a function called `lapply()` (for *list apply*). This function takes two arguments: a list you want to operate upon, followed by a function you want to "apply" to each item in that list. For example:

```
# Create an untagged list (not a vector!)
people <- list("Sarah", "Amit", "Zhang")

# Apply the `toupper()` function to each element in `people`
people_upper <- lapply(people, toupper)
            # [[1]]
            # [1] "SARAH"
            #
            # [[2]]
            # [1] "AMIT"
            #
            # [[3]]
            # [1] "ZHANG"

# Apply the `paste()` function to each element in `people`,
# with an addition argument `"dances!"` to each call
dance_party <- lapply(people, paste, "dances!")
            # [[1]]
            # [1] "Sarah dances!"
            #
            # [[2]]
            # [1] "Amit dances!"
            #
            # [[3]]
            # [1] "Zhang dances!"
```

Caution: Make sure you pass your actual function to the lapply( ) function, *not* a character string of your function name (i.e., paste, not "paste"). You're also not actually *calling* that function (i.e., paste, not paste( )). Just put the name of the function! After that, you can include any additional arguments you want the applied function to be called with—for example, how many digits to round to, or what value to paste to the end of a string.

The `lapply()` function returns a *new* list; the original one is unmodified.

You commonly use `lapply()` with your own custom functions that define what you want to do to a single element in that list:

```r
# A function that prepends "Hello" to any item
greet <- function(item) {
  paste("Hello", item) # this last expression will be returned
}


# Create an untagged list (not a vector!)
people <- list("Sarah", "Amit", "Zhang")

# Greet each person by applying the `greet()` function
# to each element in the `people` list
greetings <- lapply(people, greet)
        # [[1]]
        # [1] "Hello Sarah"
        #
        # [[2]]
        # [1] "Hello Amit"

        # [[3]]
        # [1] "Hello Zhang"
```

Additionally, `lapply()` is a member of the "*apply()" family of functions. Each member of this set of functions starts with a different letter and is used with a different data structure, but otherwise all work basically the same way. For example, `lapply()` is used for lists, while `sapply()` (simplified apply) works well for vectors. You can use both `lapply()` and `sapply()` on vectors, the difference is what the function returns. As you might imagine, `lapply()` will return a *list*, while `sapply()` will return a *vector*:

```r
# A vector of people
people <- c("Sarah", "Amit", "Zhang")

# Create a vector of uppercase versions of each name, using `sapply`
sapply(people, toupper) # returns the vector "SARAH" "AMIT" "ZHANG"
```

The `sapply()` function is really useful only with functions that you define yourself. Most built-in R functions are vectorized so they will work correctly on vectors when used directly (e.g., `toupper(people)`).

Lists represent an alternative technique to vectors for organizing data in R. In practice, the two data structures will both be used in your programs, and in fact can be combined to create a *data frame* (described in Chapter 10). For practice working with lists in R, see the set of accompanying book exercises.[2]

---

[2]**List exercises:** https://github.com/programming-for-data-science/chapter-08-exercises

IV

Data Wrangling

The following **data wrangling** chapters provide you with the necessary skills for understanding, loading, manipulating, reshaping, and exploring data structures. Perhaps the most time-consuming part of data science is preparing and exploring your data set, and learning how to perform these tasks programmatically can make the process easier and more transparent. Mastering these skills is thus vital to being an effective data scientist.

Understanding Data

Previous chapters have introduced the basic programming fundamentals for working with data, detailing how you can tell a computer to do data processing for you. To use a computer to analyze data, you need to both *access* a data set and *interpret* that data set so that you can ask meaningful questions about it. This will enable you to transform raw data into actionable information.

This chapter provides a high-level overview of how to interpret data sets as you get started doing data science—it details the sources of data you might encounter, the formats that data may take, and strategies for determining which questions to ask of that data. Developing a clear mental model of what the values in a data set signify is a necessary prerequisite before you can program a computer to effectively analyze that data.

9.1 The Data Generation Process

Before beginning to work with data, it's important to understand *where data comes from*. There are a variety of processes for capturing events as data, each of which has its own limitations and assumptions. The primary modes of data collection fall into the following categories:

- **Sensors**: The volume of data being collected by sensors has increased dramatically in the last decade. Sensors that automatically detect and record information, such as pollution sensors that measure air quality, are now entering the personal data management sphere (think of FitBits or other step counters). Assuming these devices have been properly calibrated, they offer a reliable and consistent mechanism for data collection.

- **Surveys**: Data that is less externally measurable, such as people's opinions or personal histories, can be gathered from *surveys*. Because surveys are dependent on individuals' self-reporting of their behavior, the quality of data may vary (across surveys, or across individuals). Depending on the domain, people may have poor recall (i.e., people don't remember what they ate last week) or have incentives to respond in a particular way (i.e., people may over-report healthy behaviors). The biases inherent in survey responses should be recognized and, when possible, adjusted for in your analysis.

- **Record keeping**: In many domains, organizations use both automatic and manual processes to keep track of their activities. For example, a hospital may track the length and result of every surgery it performs (and a governing body may require that hospital to report those results). The reliability of such data will depend on the quality of the systems used to produce it. Scientific experiments also depend on diligent record keeping of results.

- **Secondary data analysis**: Data can be compiled from *existing knowledge artifacts* or measurements, such as counting word occurrences in a historical text (computers can help with this!).

All of these methods of collecting data can lead to potential concerns and biases. For example, sensors may be inaccurate, people may present themselves in particular ways when responding to surveys, record keeping may only focus on particular tasks, and existing artifacts may already exclude perspectives. When working with any data set, it is vital to consider where the data came from (e.g., *who* recorded it, *how*, and *why*) to effectively and meaningfully analyze it.

9.2   Finding Data

Computers' abilities to record and persist data have led to an explosion of available data values that can be analyzed, ranging from personal biological measures (*how many steps have I taken?*) to social network structures (*who are my friends?*) to private information leaked from insecure websites and government agencies (*what are their Social Security numbers?*). In professional environments, you will likely be working with proprietary data collected or managed by your organization. This might be anything from purchase orders of fair trade coffee to the results of medical research—the range is as wide as the types of organizations (since *everyone* now records data and sees a need for data analytics).

Luckily, there are also plenty of free, nonproprietary data sets that you can work with. Organizations will often make large amounts of data available to the public to support experiment duplication, promote transparency, or just see what other people can do with that data. These data sets are great for building your data science skills and portfolio, and are made available in a variety of formats. For example, data may be accessed as downloadable CSV spreadsheets (see Chapter 10), as relational databases (see Chapter 13), or through a web service API (see Chapter 14).

Popular sources of open data sets include:

- **Government publications**: Government organizations (and other bureaucratic systems) produce *a lot* of data as part of their everyday activities, and often make these data sets available in an effort to appear transparent and accountable to the public. You can currently find publicly available data from many countries, such as the United States,[1] Canada,[2] India,[3] and others. Local governments will also make data available: for example, the City of Seattle [4] makes a vast amount of data available in an easy-to-access format. Government data covers a broad range of topics, though it can be influenced by the political situation surrounding its gathering and retention.

---

[1] **U.S. government's open data:** https://www.data.gov
[2] **Government of Canada open data:** https://open.canada.ca/en/open-data
[3] **Open Government Data Platform India:** https://data.gov.in
[4] **City of Seattle open data portal:** https://data.seattle.gov

- **News and journalism**: Journalism remains one of the most important contexts in which data is gathered and analyzed. Journalists do much of the legwork in producing data—searching existing artifacts, questioning and surveying people, or otherwise revealing and connecting previously hidden or ignored information. News media usually publish the analyzed, summative information for consumption, but they also may make the source data available for others to confirm and expand on their work. For example, the *New York Times*[5] makes much of its historical data available through a web service, while the data politics blog *FiveThirtyEight*[6] makes all of the data behind its articles available on GitHub (invalid models and all).

- **Scientific research**: Another excellent source of data is ongoing scientific research, whether performed in academic or industrial settings. Scientific studies are (in theory) well grounded and structured, providing meaningful data when considered within their proper scope. Since science needs to be disseminated and validated by others to be usable, research is often made publicly available for others to study and critique. Some scientific journals, such as the premier journal *Nature*, require authors to make their data available for others to access and investigate (check out its list[7] of scientific data repositories!).

- **Social networks and media organizations**: Some of the largest quantities of data produced occur online, automatically recorded from people's usage of and interactions with social media applications such as Facebook, Twitter, or Google. To better integrate these services into people's everyday lives, social media companies make much of their data programmatically available for other developers to access and use. For example, it is possible to access live data from Twitter,[8] which has been used for a variety of interesting analyses. Google also provides programmatic access[9] to most of its many services (including search and YouTube).

- **Online communities**: As data science has rapidly increased in popularity, so too has the community of data science practitioners. This community and its online spaces are another great source for interesting and varied data sets and analysis. For example, *Kaggle*[10] hosts a number of data sets as well as "challenges" to analyze them. *Socrata*[11] (which powers the Seattle data repository), also collects a variety of data sets (often from professional or government contributors). Somewhat similarly, the *UCI Machine Learning Repository*[12] maintains a collection of data sets used in machine learning, drawn primarily from academic sources. And there are many other online lists of data sources as well—including a dedicated Subreddit `/r/Datasets`.[13]

In short, there are a huge number of real-world data sets available for you to work with—whether you have a specific question you would like to answer, or just want to explore and be inspired.

---

[5] *New York Times* **Developer Network**: https://developer.nytimes.com
[6] *FiveThirtyEight:* **Our Data**: https://data.fivethirtyeight.com
[7] **Nature: Recommended Data Repositories**: https://www.nature.com/sdata/policies/repositories
[8] **Twitter developer platform**: https://developer.twitter.com/en/docs
[9] **Google APIs Explorer**: https://developers.google.com/apis-explorer/
[10] **Kaggle**: "the home of data science and machine learning": https://www.kaggle.com
[11] **Socrata**: data as a service platform: https://opendata.socrata.com
[12] **UCI Machine Learning Repository**: https://archive.ics.uci.edu/ml/index.php
[13] **/r/DataSets**: https://www.reddit.com/r/datasets/

9.3 Types of Data

Once you acquire a data set, you will have to understand its structure and content before (programmatically) investigating it. Understanding the *types* of data you will encounter depends on your ability to discern the level of measurement for a given piece of data, as well as the different structures that are used to hold that data.

9.3.1 Levels of Measurement

Data can be made up of a variety of types of values (represented by the concept of "data type" in R). More generally, data values can also be discussed in terms of their **level of measurement**[14]—a way of classifying data values in terms of how they can be measured and compared to other values.

The field of statistics commonly classifies values into one of four levels, described in Table 9.1.

Nominal data (often equivalently **categorical data**) is data that has no implicit ordering. For example, you cannot say that "apples are more than oranges," though you can indicate that a particular fruit either is an apple or an orange. Nominal data is commonly used to indicate that an observation belongs in a particular category or group. You do not usually perform mathematical analysis on nominal data (e.g., you can't find the "average" fruit), though you can discuss counts or distributions. Nominal data can be represented by strings (such as the name of the fruit), but also by numbers (e.g., "fruit type #1", "fruit type #2"). Just because a value in a data set is a number, that does not mean you can do math upon it! Note that boolean values (TRUE or FALSE) are a type of nominal value.

Ordinal data establishes an *order* for nominal categories. Ordinal data may be used for classification, but it also establishes that some groups are *greater than* or *less than* others. For example, you may have classifications of hotels or restaurants as *5-star*, *4-star*, and so on. There is an ordering to these categories, but the distances between the values may vary. You are able to find the minimum, maximum, and even median values of ordinal variables, but you can't compute a statistical mean (since ordinal values do not define *how much* greater one value is than another). Note that it is possible to treat nominal variables as ordinal by enforcing an ordering, though in

Table 9.1 **Levels of measurement**

Level	Example	Operations
Nominal unordered; used for classification	*Fruits*: apples, bananas, oranges, etc.	==, != *"same or different"*
Ordinal ordered; can sort	*Hotel rating*: 5-star, 4-star, etc.	==, !=, <, > *"bigger or smaller"*
Ratio ordered, fixed "zero"	*Lengths*: 1 inch, 1.5 inches, 2 inches, etc.	==, !=, <, <, +, -, *, / *"twice as big"*
Interval ordered, no set "zero"	*Dates*: 05/15/2012, 04/17/2015, etc.	==, !=, <, >, +, - *"3 units bigger"*

---

[14]Stevens, S. S. (1946). On the theory of scales of measurement. *Science, 103*(2684), 677–680. https://doi.org/10.1126/science.103.2684.677

effect this changes the measurement level of the data. For example, colors are usually *nominal* data—you cannot say that "red is greater than blue." This is despite the conventional ordering based on the colors of a rainbow; when you say that "red comes before blue (in the rainbow)," you're actually replacing the nominal color value with an ordinal value representing its *position in a rainbow* (which itself is dependent on the *ratio* value of its wavelength)! Ordinal data is also considered categorical.

Ratio data (often equivalently **continuous data**) is the most common level of measurement in real-world data: data based on population counts, monetary values, or amount of activity is usually measured at the ratio level. With ratio data, you can find averages, as well as measure the distance between different values (a feature also available with interval data). As you might expect, you can also compare the ratio of two values when working with ratio data (i.e., value x is twice as great as value y).

Interval data is similar to ratio data, except there is no fixed zero point. For example, dates cannot be discussed in *proportional* terms (i.e., you wouldn't say that *Wednesday is twice as Monday*). Therefore, you can compute the distance (interval) between two values (i.e., *2 days apart*), but you cannot compute the *ratio* between two values. Interval data is also considered continuous.

Identifying and understanding the level of measurement of a particular data feature is important when determining how to analyze a data set. In particular, you need to know what kinds of statistical analysis will be valid for that data, as well as how to interpret what that data is measuring.

9.3.2 Data Structures

In practice, you will need to organize the numbers, strings, vectors, and lists of values described in the previous chapters into more complex formats. Data is organized into more robust **structures**—particularly as the data set gets large—to better signify what those numbers and strings represent. To work with real-world data, you will need to be able to understand these structures and the terminology used to discuss them.

In practice, most data sets are structured as **tables** of information, with individual data values arranged into *rows* and *columns* (see Figure 9.1). These tables are similar to how data may be recorded in a spreadsheet (using a program such as Microsoft Excel). In a table, each row represents a **record** or **observation**: an instance of a single thing being measured (e.g., a person, a sports match). Each column represents a **feature**: a particular property or aspect of the thing being measured (e.g., the person's height or weight, the scores in a sports game). Each data value can be referred to as a **cell** in the table.

Viewed in this way, a table is a collection of "things" being measured, each of which has a particular value for a characteristic of that thing. And, because all the observations share the same characteristics (features), it is possible to analyze them comparatively. Moreover, by organizing data into a table, each data value (cell) can be automatically given two associated meanings: which observation it is from as well as which feature it represents. This structure allows you to discern semantic meaning from the numbers: the number 64 in figure Figure 9.1 is not just some value; it's "Ada's height."

The table in Figure 9.1 represents a *small* (even tiny) data set, in that it contains just five observations (rows). The size of a data set is generally measured in terms of its number of

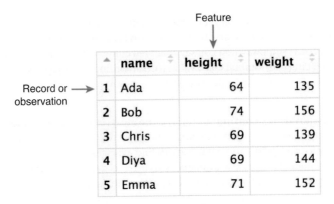

Figure 9.1 A table of data (of people's weights and heights). Rows represent *observations*, while columns represent *features*.

observations: a small data set may contain only a few dozen observations, while a large data set may contain thousands or hundreds of thousands of records. Indeed, "Big Data" is a term that, in part, refers to data sets that are so large that they can't be loaded into the computer's memory without special handling, and may contain billions or even trillions of rows! Yet, even a data set with a relatively small number of observations can contain a large number of cells if they record a lot of features per observations (though these tables can often be "inverted" to have more rows and fewer columns; see Chapter 12). Overall, the number of observations and features (rows and columns) is referred to as the **dimensions** of the data set—not to be confused with referring to a table's "two-dimensional" data structure (because each data value has *two* meanings: observation and feature).

Although it is commonly structured in this way, data need not be represented as a single table. More complex data sets may spread data values across multiple tables (such as in a database; see Chapter 13). In other complex data structures, each individual cell in the table may hold a vector or even its own data table. This can cause the table to no longer be two-dimensional, but three- or more-dimensional. Indeed, many data sets available from web services are structured as "nested tables"; see Chapter 14 for details.

9.4   Interpreting Data

The first thing you will need to do upon encountering a data set (whether one you found online or one that was provided by your organization) is to understand the meaning of the data. This requires understanding the domain you are working in, as well as the specific data schema you are working with.

9.4.1 Acquiring Domain Knowledge

The first step toward being able to understand a data set is to research and understand the data's problem domain. The **problem domain** is the set of topics that are relevant to the problem—that is, the context for that data. Working with data requires **domain knowledge**: you need to have a basic level of understanding of that problem domain to do any sensible analysis of that data. You will need to develop a mental model of what the data values mean. This includes understanding the significance and purpose of any features (so you're not doing math on contextless numbers), the range of expected values for a feature (to detect outliers and other errors), and some of the subtleties that may not be explicit in the data set (such as biases or aggregations that may hide important causalities).

As a specific example, if you wanted to analyze the table shown in Figure 9.1, you would need to first understand what is meant by "height" and "weight" of a person, the implied units of the numbers (inches, centimeters, ... or something else?), an expected range (does Ada's height of 64 mean she is short?), and other external factors that may have influenced the data (e.g., age).

> **Remember:** You do not need to necessarily be an expert in the problem domain (though it wouldn't hurt); you just need to acquire *sufficient* domain knowledge to work within that problem domain!

While people's heights and other data sets discussed in this text should be familiar to most readers, in practice you are quite likely to come across data from problem domains that are outside of your personal domain expertise. Or, more problematically, the data set may be from a problem domain that you *think* you understand but actually have a flawed mental model of (a failure of *meta-cognition*).

For example, consider the data set shown in Figure 9.2, a screenshot taken from the City of Seattle's data repository. This data set presents information on *Land Use Permits*, a somewhat opaque bureaucratic procedure with which you may be unfamiliar. The question becomes: how would you acquire sufficient domain knowledge to understand and analyze this data set?

Gathering domain knowledge almost always requires outside research—you will rarely be able to understand a domain just by looking at a spreadsheet of numbers. To gain general domain knowledge, we recommend you start by consulting a general knowledge reference: *Wikipedia* provides easy access to basic descriptions. Be sure to read any related articles or resources to improve your understanding: sifting through the vast amount of information online requires cross-referencing different resources, and mapping that information to your data set.

That said, the best way to learn about a problem is to find a *domain expert* who can help explain the domain to you. If you want to know about land use permits, try to find someone who has used one in the past. The second best solution is to ask a librarian—librarians are specifically trained to help people discover and acquire basic domain knowledge. Libraries may also support access to more specialized information sources.

9.4.2 Understanding Data Schemas

Once you have a general understanding of the context for a data set, you can begin interpreting the data set itself. You will need to focus on understanding the **data schema** (e.g., what is represented by the rows and columns), as well as the specific context for those values. We suggest you use the following questions to guide your research:

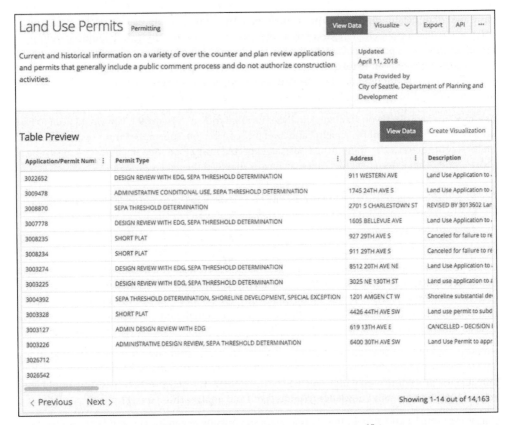

Figure 9.2   A preview of land use permits data from the City of Seattle.[15] Content has been edited for display in this text.

"What meta-data *is available for the data set?"*

Many publicly available data sets come with summative explanations, instructions for access and usage, or even descriptions of individual features. This **meta-data** (data about the data) is the best way to begin to understand what value is represented by each cell in the table, since the information comes directly from the source.

For example, Seattle's land use permits page has a short summary (though you would want to look up what an "over-the-counter review application" is), provides a number of categories and tags, lists the dimensions of the data set (14,200 rows as of this writing), and gives a quick description of each column.

A particularly important piece of meta-data to search for is:

"Who created the data set? Where does it come from?"

---

[15] **City of Seattle: Land Use Permits** (access requires a free account): https://data.seattle.gov/Permitting/Land-Use-Permits/uyyd-8gak

Understanding who generated the data set (and how they did so!) will allow you to know where to find more information about the data—it will let you know who the domain experts are. Moreover, knowing the source and methodology behind the data can help you uncover hidden biases or other subtleties that may not be obvious in the data itself. For example, the Land Use Permits page notes that the data was provided by the "City of Seattle, Department of Planning and Development" (now the Department of Construction & Inspections). If you search for this organization, you can find its website.[16] This website would be a good place to gain further information about the specific data found in the data set.

Once you understand this meta-data, you can begin researching the data set itself:

> *"What features does the data set have?"*

Regardless of the presence of meta-data, you will need to understand the columns of the table to work with it. Go through each column and check if you understand:

1. What "real-world" aspect does each column attempt to capture?
2. For continuous data: what units are the values in?
3. For categorical data: what different categories are represented, and what do those mean?
4. What is the possible range of values?

If the meta-data provides a *key* to the data table, this becomes an easy task. Otherwise, you may need to study the source of the data to determine how to understand the features, sparking additional domain research.

> **Tip:** As you read through a data set—or anything really—you should *write down* the terms and phrases you are not familiar with to look up later. This will discourage you from (inaccurately) guessing a term's meaning, and will help delineate between terms you have and have not yet clarified.

For example, the Land Use Permits data set provides clear descriptions of the columns in the meta-data, but looking at the sample data reveals that some of the values may require additional research. For example, what are the different Permit Types and Decision Types? By going back to the source of the data (the Department of Construction home page), you can navigate to the Permits page and then to the "Permits We Issue (A-Z)" to see a full list of possible permit types. This will let you find out, for example, that "PLAT" refers to "creating or modifying individual parcels of property"—in other words, adjusting lot boundaries.

To understand the features, you will need to look at some sample observations. Open up the spreadsheet or table and look at the first few rows to get a sense for what kind of values they have and what that may say about the data.

Finally, throughout this process, you should continually consider:

> *"What terms do you not know or understand?"*

---

[16] **Seattle Department of Construction & Inspections** (access requires a free account): http://www.seattle.gov/dpd/

Depending on the problem domain, a data set may contain a large amount of *jargon*, both to explain the data and inside the data itself. Making sure you understand all the technical terms used will go a long way toward ensuring you can effectively discuss and analyze the data.

> **Caution:** Watch out for acronyms you are not familiar with, and be sure to look them up!

For example, looking at the "Table Preview," you may notice that many of the values for the "Permit Type" feature use the term "SEPA." Searching for this acronym would lead you to a page describing the *State Policy Environmental Act* (requiring environmental impact to be considered in how land is used), as well as details on the "Threshold Determination" process.

Overall, interpreting a data set will require research and work that is *not* programming. While it may seem like such work is keeping you from making progress in processing the data, having a valid mental model of the data is both useful and necessary to perform data analysis.

9.5   Using Data to Answer Questions

Perhaps the most challenging aspect of data analysis is effectively applying questions of interest to the data set to construct the desired information. Indeed, as a data scientist, it will often be your responsibility to translate from various domain questions to specific observations and features in your data set. Take, for example, a question like:

> *"What is the worst disease in the United States?"*

To answer this question, you will need to understand the problem domain of disease burden measurement and acquire a data set that is well positioned to address the question. For example, one appropriate data set would be the *Global Burden of Disease*[17] study performed by the Institute for Health Metrics and Evaluation, which details the *burden of disease* in the United States and around the world.

Once you have acquired this data set, you will need to **operationalize** the motivating question. Considering each of the key words, you will need to identify a set of *diseases*, and then quantify what is meant by "worst." For example, the question could be more concretely phrased as any of these interpretations:

- Which disease *causes the largest number of deaths* in the United States?
- Which disease *causes the most premature deaths* in the United States?
- Which disease *causes the most disability* in the United States?

Depending on your definition of "worst," you will perform very different computations and analysis, possibly arriving at different answers. You thus need to be able to decide *what precisely is meant by a question*—a task that requires understanding the nuances found in the question's problem domain.

Figure 9.3 shows visualizations that try to answer this very question. The figure contains screenshots of *treemaps* from an online tool called *GBD Compare*.[18] A treemap is like a pie chart that is built with

---

[17] **IHME: Global Burden of Disease:** http://www.healthdata.org/node/835
[18] **GBD Compare:** visualization for global burden of disease: https://vizhub.healthdata.org/gbd-compare/

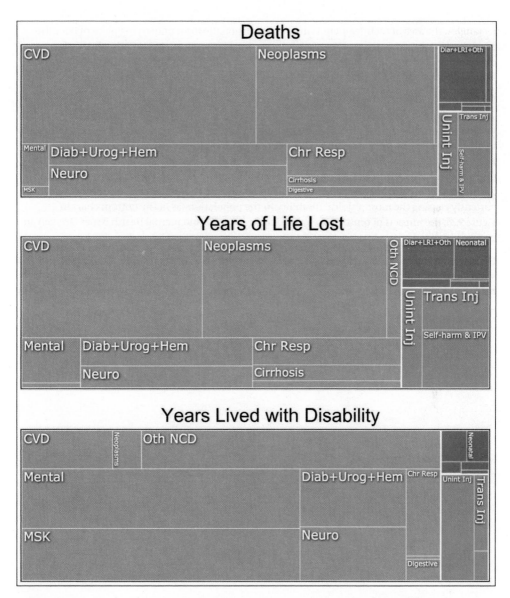

Figure 9.3 Treemaps from the GBD Compare tool showing the proportion of deaths (top), years of life lost (middle), and years lived with disability (bottom) attributable to each disease in the United States.

rectangles: the area of each segment is drawn proportionally to an underlying piece of data. The additional advantage of the treemap is that it can show *hierarchies* of information by *nesting* different levels of rectangles inside of one another. For example, in Figure 9.3, the disease burden from each communicable disease (shown in red) is nested within the same segment of each chart.

Depending on how you choose to operationalize the idea of the "worst disease," different diseases stand out as the most impactful. As you can see in Figure 9.3, almost 90% of all deaths are caused by non-communicable diseases such as cardiovascular diseases (*CVD*) and cancers (*Neoplasms*), shown in blue. When you consider the age of death for each person (computing a metric called *Years of Life Lost*), this value drops to 80%. Moreover, this metric enables you to identify causes of death that disproportionately affect young people, such as traffic accidents (*Trans Inj*) and self-harm, shown in green (see the middle chart in Figure 9.3). Finally, if you consider the "worst" disease to be that currently causing the most physical disability in the population (as in the bottom chart in Figure 9.3), the impacts of musculoskeletal conditions (*MSK*) and mental health issues (*Mental*) are exposed.

Because data analysis is about identifying answers to questions, the first step is to ensure you have a strong understanding of the question of interest and how it is being measured. Only after you have mapped from your questions of interest to specific features (columns) of your data can you perform an effective and meaningful analysis of that data.

10

Data Frames

This chapter introduces data frame values, which are the primary two-dimensional data storage type used in R. In many ways, data frames are similar to the row-and-column table layout that you may be familiar with from spreadsheet programs like Microsoft Excel. Rather than interact with this data structure through a user interface (UI), you will learn how to programmatically and reproducibly perform operations on this data type. This chapter covers ways of creating, describing, and accessing data from data frames in R.

10.1   What Is a Data Frame?

At a practical level, **data frames** act like tables, where data is organized into rows and columns. For example, reconsider the table of names, weights, and heights from Chapter 9, shown in Figure 10.1. In R, you can use data frames to represent these kinds of tables.

Data frames are really just lists (see Chapter 8) in which each element is a vector of the same length. Each vector represents a column, not a row. The elements at corresponding indices in the vectors are considered part of the same row (record). This structure makes sense because each row may have different types of data—such as a person's name (string) and height (number)—and vector elements must all be of the same type.

	name	height	weight
1	Ada	64	135
2	Bob	74	156
3	Chris	69	139
4	Diya	69	144
5	Emma	71	152

Figure 10.1   A table of data (of people's weights and heights) when viewed as a data frame in RStudio.

For example, you can think of the data shown in Figure 10.1 as a *list* of three *vectors*: name, height, and weight. The name, height, and weight of the first person measured are represented by the first elements of the name, height, and weight vectors, respectively.

You can work with data frames as if they were lists, but data frames have additional properties that make them particularly well suited for handling tables of data.

10.2   Working with Data Frames

Many data science questions can be answered by honing in on the desired subset of your data. In this section, you will learn how to create, describe, and access data from data frames.

10.2.1   Creating Data Frames

Typically you will *load* data sets from some external source (see Section 10.3), rather than writing out the data by hand. However, it is also possible to construct a data frame by combining multiple vectors. To accomplish this, you can use the **data.frame()** function, which accepts vectors as arguments, and creates a table with a column for each vector. For example:

```
# Create a data frame by passing vectors to the `data.frame()` function

# A vector of names
name <- c("Ada", "Bob", "Chris", "Diya", "Emma")

# A vector of heights
height <- c(64, 74, 69, 69, 71)

# A vector of weights
weight <- c(135, 156, 139, 144, 152)

# Combine the vectors into a data frame
# Note the names of the variables become the names of the columns!
people <- data.frame(name, height, weight, stringsAsFactors = FALSE)
```

The last argument to the data.frame() function is included because one of the vectors contains strings; it tells R to treat that vector as a typical vector, instead of another data type called a **factor** when constructing the data frame. This is usually what you will want to do—see Section 10.3.2 for more information.

You can also specify data frame column names using the key = value syntax used by *named lists* when you create your data frame:

```
# Create a data frame of names, weights, and heights,
# specifying column names to use
people <- data.frame(
  name = c("Ada", "Bob", "Chris", "Diya", "Emma"),
  height = c(64, 74, 69, 69, 71),
  weight = c(135, 156, 139, 144, 152)
)
```

Because data frame elements are lists, you can access the values from `people` using the same dollar notation and double-bracket notation as you use with lists:

```
# Retrieve information from a data frame using list-like syntax

# Create the same data frame as above
people <- data.frame(name, height, weight, stringsAsFactors = FALSE)

# Retrieve the `weight` column (as a list element); returns a vector
people_weights <- people$weight

# Retrieve the `height` column (as a list element); returns a vector
people_heights <- people[["height"]]
```

For more flexible approaches to accessing data from data frames, see section 10.2.3.

10.2.2   Describing the Structure of Data Frames

While you can interact with data frames as lists, they also offer a number of additional capabilities and functions. For example, Table 10.1 presents a few functions you can use to *inspect* the structure and content of a data frame:

Table 10.1   **Functions for inspecting data frames**

Function	Description
nrow(my_data_frame)	Returns the number of rows in the data frame
ncol(my_data_frame)	Returns the number of columns in the data frame
dim(my_data_frame)	Returns the dimensions (rows, columns) in the data frame
colnames(my_data_frame)	Returns the names of the columns of the data frame
rownames(my_data_frame)	Returns the names of the rows of the data frame
head(my_data_frame)	Returns the first few rows of the data frame (as a new data frame)
tail(my_data_frame)	Returns the last few rows of the data frame (as a new data frame)
View(my_data_frame)	Opens the data frame in a spreadsheet-like viewer (only in RStudio)

```
# Use functions to describe the shape and structure of a data frame

# Create the same data frame as above
people <- data.frame(name, height, weight, stringsAsFactors = F)

# Describe the structure of the data frame
nrow(people) # [1] 5
```

```
ncol(people) # [1] 3
dim(people) # [1] 5 3
colnames(people) # [1] "name" "height" "weight"
rownames(people) # [1] "1" "2" "3" "4" "5"

# Create a vector of new column names
new_col_names <- c("first_name", "how_tall", "how_heavy")

# Assign that vector to be the vector of column names
colnames(people) <- new_col_names
```

Many of these description functions can also be used to modify the structure of a data frame. For example, you can use the `colnames()` functions to assign a new set of column names to a data frame.

10.2.3 Accessing Data Frames

As stated earlier, since data frames are lists, it's possible to use dollar notation (my_df$column_name) or double-bracket notation (my_df[["column_name"]]) to access entire columns. However, R also uses a variation of single-bracket notation that allows you to filter for and access individual data elements (cells) in the table. In this syntax, you put two values separated by a comma (,) inside of single square brackets—the first argument specifies which row(s) you want to extract, while the second argument specifies which column(s) you want to extract.

Table 10.2 summarizes how single-bracket notation can be used to access data frames. Take special note of the fourth option's syntax (for retrieving rows): you still include the comma (,), but because you leave the *which column* value blank, you get all of the columns!

Table 10.2 **Accessing a data frame with single bracket notation**

Syntax	Description	Example
my_df[row_name, col_name]	Element(s) by row and column names	people["Ada", "height"] (element in row *named* Ada and column *named* height)
my_df[row_num, col_num]	Element(s) by row and column indices	people[2, 3] (element in the second row, third column)
my_df[row, col]	Element(s) by row and column; can mix names and indices	people[2, "height"] (second element in the height column)
my_df[row,]	All elements (columns) in row name or index	people[2,] (all columns in the second row)
my_df[, col]	All elements (rows) in a column name or index	people[, "height"] (all rows in the height column; equivalent to list notations)

```
# Assign a set of row names for the vector
# (using the values in the `name` column)
rownames(people) <- people$name

# Extract the row with the name "Ada" (and all columns)
people["Ada", ] # note the comma, indicating all columns

# Extract the second column as a vector
people[, "height"] # note the comma, indicating all rows

# Extract the second column as a data frame (filtering)
people["height"] # without a comma, it returns a data frame
```

Of course, because numbers and strings are stored in vectors, you're actually specifying vectors of names or indices to extract. This allows you to get multiple rows or columns:

```
# Get the `height` and `weight` columns
people[, c("height", "weight")] # note the comma, indicating all rows

# Get the second through fourth rows
people[2:4, ] # note the comma, indicating all columns
```

Additionally, you can use a vector of boolean values to specify your indices of interest (just as you did with vectors):

```
# Get rows where `people$height` is greater than 70 (and all columns)
people[people$height > 70, ] # rows for which `height` is greater than 70
```

> **Remember**: The type of data that is returned when selecting data using single brackets depends on *how many columns* you are selecting. Extracting values from more than one column will produce a data frame; extracting from just one column will produce a vector.

> **Tip**: In general, it's easier, cleaner, and less buggy to filter by column name (character string), rather than by column number, because it's not unusual for column order to change in a data frame. You should almost *never* access data in a data frame by its positional index. Instead, you should use the column name to specify columns, and a filter to specify rows of interest.

> **Going Further**: While data frames are the two-dimensional data structure suggested by this book, they are not the only 2D data structure in R. For example, a *matrix* is a two-dimensional data structure in which all of the values have the same type (usually numeric).
>
> To use all the syntax and functions described in this chapter, first confirm that a data object is a data frame (using is.data.frame( )), and if necessary, *convert* an object to a data frame (such as by using the as.data.frame( ) function).

10.3   Working with CSV Data

Section 10.2 demonstrated constructing your own data frames by "hard-coding" the data values. However, it is much more common to load data from somewhere else, such as a separate file on your computer or a data resource on the internet. R is also able to ingest data from a variety of sources. This section focuses on reading tabular data in **comma-separated value** (CSV) format, usually stored in a file with the extension `.csv`. In this format, each line of the file represents a record (row) of data, while each feature (column) of that record is separated by a comma:

```
name, weight, height
Ada, 64, 135
Bob, 74, 156
Chris, 69, 139
Diya, 69, 144
Emma, 71, 152
```

Most spreadsheet programs, such as Microsoft Excel, Numbers, and Google Sheets, are just interfaces for formatting and interacting with data that is saved in this format. These programs easily import and export `.csv` files. But note that `.csv` files are unable to save the formatting and calculation formulas used in those programs—a `.csv` file stores only the data!

You can load the data from a `.csv` file into R by using the **read.csv()** function:

```
# Read data from the file `my_file.csv` into a data frame `my_df`
my_df <- read.csv("my_file.csv", stringsAsFactors = FALSE)
```

Again, use the `stringsAsFactors` argument to make sure string data is stored as a vector rather than as a *factor* (see Section 10.3.2 for details). This function will return a data frame just as if you had created it yourself.

> **Remember:** If an element is missing from a data frame (which is very common with real-world data), R will fill that cell with the logical value NA, meaning "**n**ot **a**vailable." There are multiple ways[a] to handle this in an analysis; you can filter for those values using bracket notation to replace them, exclude them from your analysis, or impute them using more sophisticated techniques.
>
> ———————————
> [a]See, for example, http://www.statmethods.net/input/missingdata.html

Conversely, you can *write* data to a `.csv` file using the **write.csv()** function, in which you specify the data frame you want to write, the filename of the file you want to write the data to, and other optional arguments:

```
# Write the data in `my_df` to the file `my_new_file.csv`
# The `row.names` argument indicates if the row names should be
# written to the file (usually not)
write.csv(my_df, "my_new_file.csv", row.names = FALSE)
```

Additionally, there are many data sets you can explore that ship with the R software. You can see a list of these data sets using the **data()** function, and begin working with them directly (try

View(mtcars) as an example). Moreover, many packages include data sets that are well suited for demonstrating their functionality. For a robust (though incomplete) list of more than 1,000 data sets that ship with R packages, see this webpage.[1]

10.3.1   Working Directory

The biggest complication when working with .csv files is that the read.csv() function takes as an argument a **path** to a file. Because you want this script to work on any computer (to support collaboration, or so you can code from your personal computer or a computer at a library), you need to be sure to use a **relative path** to the file. The question is: *relative to what?*

Like the command line, the R interpreter (running inside RStudio) has a **current working directory** from which all file paths are relative. The trick is that *the working directory is not necessarily the directory of the current script file!* This makes sense, as you may have many files open in RStudio at the same time, and your R interpreter can have only one working directory.

Just as you can view the current working directory when on the command line (using pwd), you can use an R function to view the current working directory when in R:

```
# Get the absolute path to the current working directory
getwd() # returns a path like /Users/YOUR_NAME/Documents/projects
```

You often will want to change the working directory to be your project's directory (wherever your scripts and data files happen to be; often the root of your project *repository*). It is possible to change the current working directory using the setwd() function. However, this function also takes an absolute path, so doesn't fix the problem of working across machines. You *should not* include this absolute path in your script (though you could use it from the console).

A better solution is to use RStudio itself to change the working directory. This is reasonable because the working directory is a property of the *current running environment*, which is what RStudio makes accessible. The easiest way to do this is to use the Session > Set Working Directory menu option (see Figure 10.2): you can either set the working directory To Source File Location (the folder containing whichever .R script you are currently editing; this is usually what you want), or you can browse for a particular directory with Choose Directory.

As a specific example, consider trying to load the my-data.csv file from the analysis.R script, given the folder structure illustrated in Figure 10.3. In your analysis.R script you want to be able to use a relative path to access your data (my-data.csv). In other words, you don't want to have to specify the **absolute path** (/Users/YOUR_NAME/Documents/projects/analysis-project/data/my-data.csv) to find this. Instead, you want to provide instructions on how your program can find your data file *relative* to where you are working (in your analysis.R file). After setting the session's path to the working directory, you will be able to use the **relative path** to find it:

```
# Load the data using a relative path
# (this works only after setting the working directory,
# most easily with the RStudio UI)
my_data <- read.csv("data/my-data.csv", stringsAsFactors = FALSE)
```

---

[1]**R Package Data Sets:** https://vincentarelbundock.github.io/Rdatasets/datasets.html

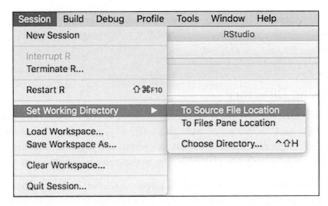

Figure 10.2 Use Session > Set Working Directory to change the working directory through RStudio.

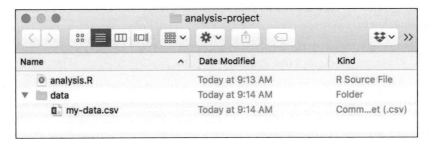

Figure 10.3 The folder structure for a sample project. Once you set the working directory in RStudio, you can access the my-data.csv file from the analysis.R script using the *relative path* data/my-data.csv.

10.3.2 Factor Variables

Remember: You should always include a stringsAsFactors = FALSE argument when either loading or creating data frames. This section explains why you need to do that.

Factors are a data structure for *optimizing* variables that consist of a finite set of categories (i.e., they are categorical variables). For example, imagine that you had a vector of shirt sizes that could take on only the values small, medium, or large. If you were working with a large data set (thousands of shirts!), it would end up taking up a lot of memory to store the character strings (5+ letters per word at 1 or more bytes per letter) for each of those variables.

A factor would instead store a number (called a **level**) for each of these character strings—for example, 1 for small, 2 for medium, or 3 for large (though the order of the numbers may vary). R will remember the relationship between the integers and their labels (the strings). Since each number takes just 2–4 bytes (rather than 1 byte per letter), factors allow R to keep much more information in memory.

To see how factor variables appear similar to (but are actually different from) vectors, you can create a factor variable using **as.factor( )**:

```
# Demonstrate the creation of a factor variable

# Start with a character vector of shirt sizes
shirt_sizes <- c("small", "medium", "small", "large", "medium", "large")

# Create a factor representation of the vector
shirt_sizes_factor <- as.factor(shirt_sizes)

# View the factor and its levels
print(shirt_sizes_factor)
# [1] small  medium small  large  medium large
# Levels: large medium small

# The length of the factor is still the length of the vector,
# not the number of levels
length(shirt_sizes_factor) # 6
```

When you print out the `shirt_sizes_factor` variable, R still (intelligently) prints out the labels that you are presumably interested in. It also indicates the levels, which are the only possible values that elements can take on.

It is worth restating: factors are not vectors. This means that most all the operations and functions you want to use on vectors will not work:

```
# Attempt to apply vector methods to factors variables: it doesn't work!

# Create a factor of numbers (factors need not be strings)
num_factors <- as.factor(c(10, 10, 20, 20, 30, 30, 40, 40))

# Print the factor to see its levels
print(num_factors)
# [1] 10 10 20 20 30 30 40 40
# Levels: 10 20 30 40

# Multiply the numbers by 2
num_factors * 2 # Warning Message: '*' not meaningful for factors
# Returns vector of NA instead

# Changing entry to a level is fine
num_factors[1] <- 40

# Change entry to a value that ISN'T a level fails
num_factors[1] <- 50 # Warning Message: invalid factor level, NA generated
# num_factors[1] is now NA
```

If you create a data frame with a string vector as a column (as happens with `read.csv()`), it will automatically be treated as a factor unless you explicitly tell it not to be:

```
# Attempt to replace a factor with a (new) string: it doesn't work!

# Create a vector of shirt sizes
shirt_size <- c("small", "medium", "small", "large", "medium", "large")

# Create a vector of costs (in dollars)
cost <- c(15.5, 17, 17, 14, 12, 23)

# Data frame of inventory (by default, stringsAsFactors is set to TRUE)
shirts_factor <- data.frame(shirt_size, cost)

# Confirm that the `shirt_size` column is a factor
is.factor(shirts_factor$shirt_size) # TRUE

# Therefore, you are unable to add a new size like "extra-large"
shirts_factor[1, 1] <- "extra-large"
# Warning: invalid factor level, NA generated
```

The NA produced in the preceding example can be avoided if the `stringsAsFactors` option is set to FALSE:

```
# Avoid the creation of factor variables using `stringsAsFactors = FALSE`

# Set `stringsAsFactors` to `FALSE` so that new shirt sizes can be introduced
shirts <- data.frame(shirt_size, cost, stringsAsFactors = FALSE)

# The `shirt_size` column is NOT a factor
is.factor(shirts$shirt_size) # FALSE

# It is possible to add a new size like "extra-large"
shirts[1, 1] <- "extra-large" # no problem!
```

This is not to say that factors can't be useful (beyond just saving memory)! They offer easy ways to group and process data using specialized functions:

```
# Demonstrate the value of factors for "splitting" data into groups
# (while valuable, this is more clearly accomplished through other methods)

# Create vectors of sizes and costs
shirt_size <- c("small", "medium", "small", "large", "medium", "large")
cost <- c(15.5, 17, 17, 14, 12, 23)
```

```
# Data frame of inventory (with factors)
shirts_factor <- data.frame(shirt_size, cost)

# Produce a list of data frames, one for each factor level
#   first argument is the data frame to split
#   second argument the data frame to is the factor to split by
shirt_size_frames <- split(shirts_factor, shirts_factor$shirt_size)

# Apply a function (mean) to each factor level
#   first argument is the vector to apply the function to
#   second argument is the factor to split by
#   third argument is the name of the function
tapply(shirts_factor$cost, shirts_factor$shirt_size, mean)
        # large medium  small
        # 18.50  14.50  16.25
```

While this is a handy use of factors, you can easily do the same type of aggregation without them (as shown in Chapter 11).

In general, the skills associated with this text are more concerned with working with data as vectors. Thus you should always use `stringsAsFactors = FALSE` when creating data frames or loading .csv files that include strings.

This chapter has introduced the data frame as the primary data structure for working with two-dimensional data in R. Moving forward, almost all analysis and visualization work will depend on working with data frames. For practice working with data frames, see the set of accompanying book exercises.[2]

---

[2]**Data frame exercises:** https://github.com/programming-for-data-science/chapter-10-exercises

Manipulating Data with **dplyr**

The **dplyr**[1] ("dee-ply-er") package is the preeminent tool for data wrangling in R (and perhaps in data science more generally). It provides programmers with an intuitive vocabulary for executing data management and analysis tasks. Learning and using this package will make your data preparation and management process faster and easier to understand. This chapter introduces the philosophy behind the package and provides an overview of how to use the package to work with data frames using its expressive and efficient syntax.

11.1 A Grammar of Data Manipulation

Hadley Wickham, the original creator of the dplyr package, fittingly refers to it as a *Grammar of Data Manipulation*. This is because the package provides a set of **verbs** (functions) to describe and perform common data preparation tasks. One of the core challenges in programming is mapping from questions about a data set to specific programming operations. The presence of a data manipulation *grammar* makes this process smoother, as it enables you to use the same vocabulary to both *ask* questions and *write* your program. Specifically, the dplyr grammar lets you easily talk about and perform tasks such as the following:

- **Select** specific features (columns) of interest from a data set

- **Filter** out irrelevant data and keep only observations (rows) of interest

- **Mutate** a data set by adding more features (columns)

- **Arrange** observations (rows) in a particular order

- **Summarize** data in terms of aggregates such as the mean, median, or maximum

- **Join** multiple data sets together into a single data frame

You can use these words when describing the *algorithm* or process for interrogating data, and then use dplyr to write code that will closely follow your "plain language" description because it uses

---

[1] **dplyr**: http://dplyr.tidyverse.org

functions and procedures that share the same language. Indeed, many real-world questions about a data set come down to isolating specific rows/columns of the data set as the "elements of interest" and then performing a basic comparison or computation (e.g., mean, count, max). While it is possible to perform such computation with base R functions (described in the previous chapters), the `dplyr` package makes it much easier to write and read such code.

11.2   Core `dplyr` Functions

The `dplyr` package provides functions that mirror the verbs mentioned previously. Using this package's functions will allow you to quickly and effectively write code to ask questions of your data sets.

Since `dplyr` is an external package, you will need to install it (once per machine) and load it in each script in which you want to use the functions:

```
install.packages("dplyr") # once per machine
library("dplyr")          # in each relevant script
```

> **Fun Fact:** `dplyr` is a key part of the **tidyverse**[a] collection of R packages, which also includes `tidyr` (Chapter 12) and `ggplot2` (Chapter 16). While these packages are discussed individually, you can install and use them all at once by installing and loading the collected `"tidyverse"` package.
>
> ---
> [a]https://www.tidyverse.org

After loading the package, you can call any of the functions just as if they were the built-in functions you've come to know and love.

To demonstrate the usefulness of the `dplyr` package as a tool for asking questions of real data sets, this chapter applies the functions to historical data about U.S. presidential elections. The `presidentialElections` data set is included as part of the `pscl` package, so you will need to install and load that package to access the data:

```
# Install the `pscl` package to use the `presidentialElections` data frame
install.packages("pscl") # once per machine
library("pscl")          # in each relevant script

# You should now be able to interact with the data set
View(presidentialElections)
```

This data set contains the percentage of votes that were cast in each state for the Democratic Party candidate in each presidential election from 1932 to 2016. Each row contains the `state`, `year`, percentage of Democrat votes (`demVote`), and whether each state was a member of the former Confederacy during the Civil War (`south`). For more information, see the `pscl` package reference manual,[2] or use `?presidentialElections` to view the documentation in RStudio.

---
[2]**pscl reference manual:** https://cran.r-project.org/web/packages/pscl/pscl.pdf

11.2.1 Select

The **select()** function allows you to choose and extract columns of interest from your data frame, as illustrated in Figure 11.1.

```
# Select `year` and `demVotes` (percentage of vote won by the Democrat)
# from the `presidentialElections` data frame
votes <- select(presidentialElections, year, demVote)
```

The select() function takes as arguments the data frame to select from, followed by the names of the columns you wish to select (*without quotation marks*)!

This use of select() is equivalent to simply extracting the columns using base R syntax:

```
# Extract columns by name (i.e., "base R" syntax)
votes <- presidentialElections[, c ("year", "demVote")]
```

While this base R syntax achieves the same end, the dplyr approach provides a more **expressive** syntax that is easier to read and write.

> **Remember:** Inside the function argument list (inside the parentheses) of dplyr functions, you specify data frame columns without quotation marks—that is, you just give the column names as *variable names*, rather than as *character strings*. This is referred to as **non-standard evaluation** (NSE).[a] While this capability makes dplyr code easier to write and read, it can occasionally create challenges when trying to work with a column name that is stored in a variable.
>
> If you encounter errors in such situations, you can and should fall back to working with base R syntax (e.g., dollar sign and bracket notation).
>
> ---
> [a]http://dplyr.tidyverse.org/articles/programming.html

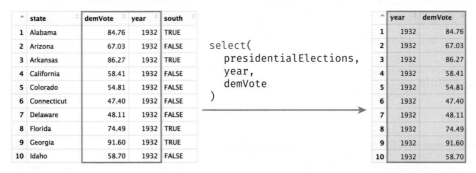

Figure 11.1 Using the select() function to select the columns year and demVote from the presidentialElections data frame.

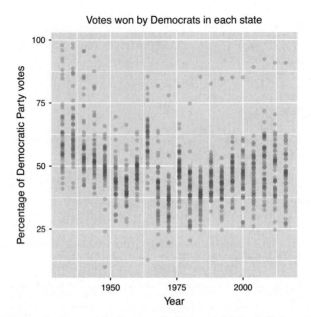

Figure 11.2 Percentage of votes cast for Democratic Party candidates in U.S. presidential elections, built with the `ggplot2` package.

This selection of data could be used to explore trends in voting patterns across states, as shown in Figure 11.2. For an interactive exploration of how state voting patterns have shifted over time, see this piece by the *New York Times*.[3]

Note that the arguments to the `select()` function can also be vectors of column names—you can write exactly what you would specify inside bracket notation, just without calling `c()`. Thus you can both select a range of columns using the `:` operator, and exclude columns using the `-` operator:

```
# Select columns `state` through `year` (i.e., `state`, `demVote`, and `year`)
select(presidentialElections, state:year)

# Select all columns except for `south`
select(presidentialElections, -south)
```

---

[3] **Over the Decades, How States Have Shifted:** https://archive.nytimes.com/www.nytimes.com/interactive/2012/10/15/us/politics/swing-history.html

> **Caution:** Unlike with the use of bracket notation, using `select()` to select a single column will return a data frame, not a vector. If you want to extract a specific column or value from a data frame, you can use the `pull()` function from the `dplyr` package, or use base R syntax. In general, use `dplyr` for manipulating a data frame, and then use base R for referring to specific values in that data.

11.2.2 Filter

The **filter()** function allows you to choose and extract *rows* of interest from your data frame (contrasted with `select()`, which extracts *columns*), as illustrated in Figure 11.3.

```
# Select all rows from the 2008 election
votes_2008 <- filter(presidentialElections, year == 2008)
```

The `filter()` function takes in the data frame to filter, followed by a comma-separated list of conditions that each returned *row* must satisfy. Again, column names must be specified without quotation marks. The preceding `filter()` statement is equivalent to extracting the rows using the following base R syntax:

```
# Select all rows from the 2008 election
votes_2008 <- presidentialElections[presidentialElections$year == 2008, ]
```

The `filter()` function will extract rows that match **all** given conditions. Thus you can specify that you want to filter a data frame for rows that meet the first condition *and* the second condition (and so on). For example, you may be curious about how the state of Colorado voted in 2008:

```
# Extract the row(s) for the state of Colorado in 2008
# Arguments are on separate lines for readability
votes_colorado_2008 <- filter(
  presidentialElections,
  year == 2008,
  state == "Colorado"
)
```

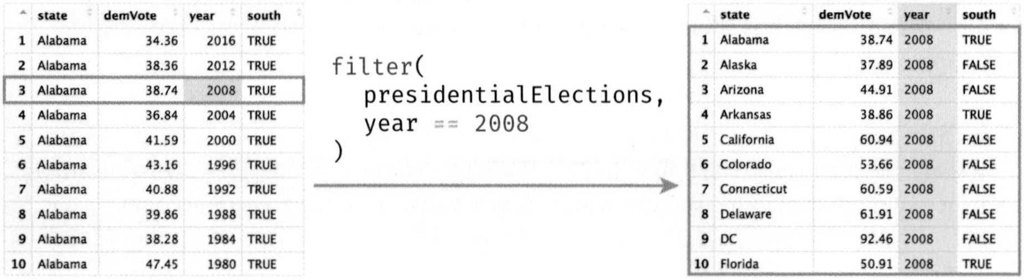

	state	demVote	year	south
1	Alabama	34.36	2016	TRUE
2	Alabama	38.36	2012	TRUE
3	Alabama	38.74	2008	TRUE
4	Alabama	36.84	2004	TRUE
5	Alabama	41.59	2000	TRUE
6	Alabama	43.16	1996	TRUE
7	Alabama	40.88	1992	TRUE
8	Alabama	39.86	1988	TRUE
9	Alabama	38.28	1984	TRUE
10	Alabama	47.45	1980	TRUE

```
filter(
  presidentialElections,
  year == 2008
)
```

	state	demVote	year	south
1	Alabama	38.74	2008	TRUE
2	Alaska	37.89	2008	FALSE
3	Arizona	44.91	2008	FALSE
4	Arkansas	38.86	2008	TRUE
5	California	60.94	2008	FALSE
6	Colorado	53.66	2008	FALSE
7	Connecticut	60.59	2008	FALSE
8	Delaware	61.91	2008	FALSE
9	DC	92.46	2008	FALSE
10	Florida	50.91	2008	TRUE

Figure 11.3 Using the `filter()` function to select observations from the `presidentialElections` data frame in which the `year` column is 2008.

In cases where you are using multiple conditions—and therefore might be writing really long code—you should break the single statement into multiple lines for readability (as in the preceding example). Because you haven't closed the parentheses on the function arguments, R will treat each new line as part of the current statement. See the tidyverse style guide[4] for more details.

> **Caution:** If you are working with a data frame that has row names (`presidentialElections` does not), the `dplyr` functions will remove row names. If you need to retain these names, consider instead making them a column (*feature*) of the data, thereby allowing you to include those names in your wrangling and analysis. You can add row names as a column using the `mutate` function (described in Section 11.2.3):
>
> ```
> # Add row names of a dataframe `df` as a new column called `row_names`
> df <- mutate(df, row_names = rownames(df))
> ```

11.2.3 Mutate

The **mutate( )** function allows you to create additional columns for your data frame, as illustrated in Figure 11.4. For example, it may be useful to add a column to the `presidentialElections` data frame that stores the percentage of votes that went to other candidates:

```
# Add an `other_parties_vote` column that is the percentage of votes
# for other parties
# Also add an `abs_vote_difference` column of the absolute difference
# between percentages
# Note you can use columns as you create them!
presidentialElections <- mutate(
  presidentialElections,
  other_parties_vote = 100 - demVote, # other parties is 100% - Democrat %
  abs_vote_difference = abs(demVote - other_parties_vote)
)
```

The `mutate( )` function takes in the data frame to mutate, followed by a comma-separated list of columns to create using the same `name = vector` syntax you use when creating lists or data frames from scratch. As always, the names of the columns in the data frame are specified without quotation marks. Again, it is common to put each new column declaration on a separate line for spacing and readability.

> **Caution:** Despite the name, the `mutate( )` function doesn't actually change the data frame; instead, it returns a *new* data frame that has the extra columns added. You will often want to replace your old data frame variable with this new value (as in the preceding code).

---

[4]**tidyverse style guide:** http://style.tidyverse.org

	state	demVote	year	south			state	demVote	year	south	other_parties_vote	abs_vote_difference
1	Alabama	84.76	1932	TRUE		1	Alabama	84.76	1932	TRUE	15.24	69.52
2	Arizona	67.03	1932	FALSE		2	Arizona	67.03	1932	FALSE	32.97	34.06
3	Arkansas	86.27	1932	TRUE		3	Arkansas	86.27	1932	TRUE	13.73	72.54
4	California	58.41	1932	FALSE		4	California	58.41	1932	FALSE	41.59	16.82
5	Colorado	54.81	1932	FALSE	→	5	Colorado	54.81	1932	FALSE	45.19	9.62
6	Connecticut	47.40	1932	FALSE		6	Connecticut	47.40	1932	FALSE	52.60	5.20
7	Delaware	48.11	1932	FALSE		7	Delaware	48.11	1932	FALSE	51.89	3.78
8	Florida	74.49	1932	TRUE		8	Florida	74.49	1932	TRUE	25.51	48.98
9	Georgia	91.60	1932	TRUE		9	Georgia	91.60	1932	TRUE	8.40	83.20
10	Idaho	58.70	1932	FALSE		10	Idaho	58.70	1932	FALSE	41.30	17.40

```
mutate(
    presidentialElections,
    other_parties_vote = 100 - demVote,
    abs_vote_difference = abs(demVote - other_parties_vote)
)
```

Figure 11.4 Using the `mutate()` function to create new columns on the `presidentialElections` data frame. Note that the `mutate()` function does not actually change a data frame (you need to assign the result to a variable).

> **Tip:** If you want to rename a particular column rather than adding a new one, you can use the dplyr function `rename()`, which is actually a variation of passing a *named argument* to the `select()` function to select columns aliased to different names.

11.2.4 Arrange

The **arrange()** function allows you to sort the rows of your data frame by some feature (column value), as illustrated in Figure 11.5. For example, you may want to sort the `presidentialElections` data frame by year, and then within each year, sort the rows based on the percentage of votes that went to the Democratic Party candidate:

```
# Arrange rows in decreasing order by `year`, then by `demVote`
# within each `year`
presidentialElections <- arrange(presidentialElections, -year, demVote)
```

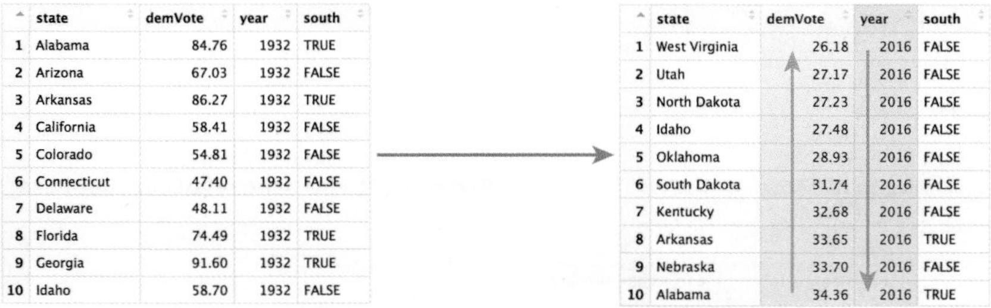

	state	demVote	year	south			state	demVote	year	south
1	Alabama	84.76	1932	TRUE		1	West Virginia	26.18	2016	FALSE
2	Arizona	67.03	1932	FALSE		2	Utah	27.17	2016	FALSE
3	Arkansas	86.27	1932	TRUE		3	North Dakota	27.23	2016	FALSE
4	California	58.41	1932	FALSE		4	Idaho	27.48	2016	FALSE
5	Colorado	54.81	1932	FALSE	→	5	Oklahoma	28.93	2016	FALSE
6	Connecticut	47.40	1932	FALSE		6	South Dakota	31.74	2016	FALSE
7	Delaware	48.11	1932	FALSE		7	Kentucky	32.68	2016	FALSE
8	Florida	74.49	1932	TRUE		8	Arkansas	33.65	2016	TRUE
9	Georgia	91.60	1932	TRUE		9	Nebraska	33.70	2016	FALSE
10	Idaho	58.70	1932	FALSE		10	Alabama	34.36	2016	TRUE

```
arrange(presidentialElections, -year, demVote)
```

Figure 11.5 Using the `arrange()` function to sort the `presidentialElections` data frame. Data is sorted in *decreasing* order by year (`-year`), then sorted by the demVote column *within* each year.

As demonstrated in the preceding code, you can pass multiple arguments into the `arrange()` function (in addition to the data frame to arrange). The data frame will be sorted by the column provided as the second argument, then by the column provided as the third argument (in case of a "tie"), and so on. Like `mutate()`, the `arrange()` function doesn't actually modify the argument data frame; instead, it returns a new data frame that you can store in a variable to use later.

By default, the `arrange()` function will sort rows in *increasing* order. To sort in *reverse* (decreasing) order, place a minus sign (`-`) in front of the column name (e.g., `-year`). You can also use the `desc()` helper function; for example, you can pass `desc(year)` as the argument.

11.2.5 Summarize

The **summarize()** function (equivalently `summarise()` for those using the British spelling) will generate a new data frame that contains a "summary" of a column, computing a single value from the multiple elements in that column. This is an **aggregation** operation (i.e., it will reduce an entire column to a single value—think about taking a sum or average), as illustrated in Figure 11.6. For example, you can calculate the average percentage of votes cast for Democratic Party candidates:

```
# Compute summary statistics for the `presidentialElections` data frame
average_votes <- summarize(
  presidentialElections,
  mean_dem_vote = mean(demVote),
  mean_other_parties = mean(other_parties_vote)
)
```

The `summarize()` function takes in the data frame to aggregate, followed by values that will be computed for the resulting summary table. These values are specified using `name = value` syntax,

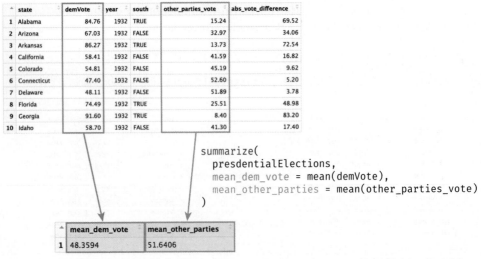

Figure 11.6 Using the `summarize()` function to calculate summary statistics for the `presidentialElections` data frame.

similar to using `mutate()` or defining a list. You can use multiple arguments to include multiple aggregations in the same statement. This will return a data frame with a single row and a different column for each value that is computed by the function, as shown in Figure 11.6.

The `summarize()` function produces a data frame (a table) of summary values. If you want to reference any of those individual aggregates, you will need to extract them from this table using base R syntax or the `dplyr` function `pull()`.

You can use the `summarize()` function to aggregate columns with any function that takes a vector as a parameter and returns a single value. This includes many built-in R functions such as `mean()`, `max()`, and `median()`. Alternatively, you can write your own summary functions. For example, using the `presidentialElections` data frame, you may want to find the *least close election* (i.e., the one in which the `demVote` was furthest from 50% in absolute value). The following code constructs a function to find the *value furthest from 50* in a vector, and then applies the function to the `presidentialElections` data frame using `summarize()`:

```
# A function that returns the value in a vector furthest from 50
furthest_from_50 <- function(vec) {
  # Subtract 50 from each value
  adjusted_values <- vec - 50

  # Return the element with the largest absolute difference from 50
  vec[abs(adjusted_values) == max(abs(adjusted_values))]
}

# Summarize the data frame, generating a column `biggest_landslide`
# that stores the value furthest from 50%
summarize(
  presidentialElections,
  biggest_landslide = furthest_from_50(demVote)
)
```

The true power of the `summarize()` function becomes evident when you are working with data that has been *grouped*. In that case, each different group will be summarized as a different row in the summary table (see Section 11.4).

11.3 Performing Sequential Operations

If you want to do more complex analysis, you will likely want to combine these functions, taking the results from one function call and passing them into another function—this is a very common workflow. One approach to performing this sequence of operations is to create *intermediary variables* for use in your analysis. For example, when working with the `presidentialElections` data set, you may want to ask a question such as the following:

> *"Which state had the highest percentage of votes for the Democratic Party candidate (Barack Obama) in 2008?"*

Answering this seemingly simple question requires a few steps:

1. *Filter* down the data set to only observations from 2008.

2. Of the percentages in 2008, *filter* down to the one with the highest percentage of votes for a Democrat.

3. *Select* the name of the state that meets the above criteria.

You could then implement each step as follows:

```
# Use a sequence of steps to find the state with the highest 2008
# `demVote` percentage

# 1. Filter down to only 2008 votes
votes_2008 <- filter(presidentialElections, year == 2008)

# 2. Filter down to the state with the highest `demVote`
most_dem_votes <- filter(votes_2008, demVote == max(demVote))

# 3. Select name of the state
most_dem_state <- select(most_dem_votes, state)
```

While this approach works, it clutters the work environment with variables you won't need to use again. It *does* help with readability (the result of each step is explicit), but those extra variables make it harder to modify and change the algorithm later (you have to change them in two places).

An alternative to saving each step as a distinct, named variable would be to use **anonymous variables** and **nest** the desired statements within other functions. While this is possible, it quickly becomes difficult to read and write. For example, you could write the preceding algorithm as follows:

```
# Use nested functions to find the state with the highest 2008
# `demVote` percentage
most_dem_state <- select(       # 3. Select name of the state
  filter(                       # 2. Filter down to the highest `demVote`
    filter(                     # 1. Filter down to only 2008 votes
      presidentialElections,    # arguments for the Step 1 `filter`
      year == 2008
    ),
    demVote == max(demVote)     # second argument for the Step 2 `filter`
  ),
  state                         # second argument for the Step 3 `select`
)
```

This version uses anonymous variables—result values that are not assigned to variables (and so are anonymous)—but instead are immediately used as the arguments to other functions. You've used these anonymous variables frequently with the `print()` function and with filters (those vectors of TRUE and FALSE values)—even the `max(demVote)` in the Step 2 filter is an anonymous variable!

This nested approach achieves the same result as the previous example does without creating extra variables. But, even with only three steps, it can get quite complicated to read—in a large part because you have to think about it "inside out," with the code in the middle being evaluated first. This will obviously become undecipherable for more involved operations.

11.3.1 The Pipe Operator

Luckily, dplyr provides a cleaner and more effective way of performing the same task (that is, using the result of one function as an argument to the next). The **pipe operator** (written as **%>%**) takes the result from one function and passes it in as *the first argument* to the next function! You can answer the question asked earlier much more directly using the pipe operator as follows:

```
# Ask the same question of our data using the pipe operator
most_dem_state <- presidentialElections %>% # data frame to start with
  filter(year == 2008) %>% # 1. Filter down to only 2008 votes
  filter(demVote == max(demVote)) %>% # 2. Filter down to the highest `demVote`
  select(state) # 3. Select name of the state
```

Here the presidentialElections data frame is "piped" in as the first argument to the first filter() call; because the argument has been piped in, the filter() call takes in only the remaining arguments (e.g., year == 2008). The result of that function is then piped in as the first argument to the *second* filter() call (which needs to specify only the remaining arguments), and so on. The additional arguments (such as the filter criteria) continue to be passed in as normal, as if no data frame argument is needed.

Because all dplyr functions discussed in this chapter take as a first argument the data frame to manipulate, and then return a manipulated data frame, it is possible to "chain" together any of these functions using a pipe!

Yes, the %>% operator can be awkward to type and takes some getting use to (especially compared to the command line's use of | to pipe). However, you can ease the typing by using the RStudio keyboard shortcut cmd+shift+m.

> **Tip:** You can see all RStudio keyboard shortcuts by navigating to the Tools > Keyboard Shortcuts Help menu, or you can use the keyboard shortcut alt+shift+k (yes, this is the keyboard shortcut to show the keyboard shortcuts menu!).

The pipe operator is loaded when you load the dplyr package (it is available only if you load that package), but it will work with *any* function, not just dplyr ones. This syntax, while slightly odd, can greatly simplify the way you write code to ask questions about your data.

> **Fun Fact**: Many packages load other packages (which are referred to as *dependencies*). For example, the pipe operator is actually part of the `magrittr`[a] package, which is loaded as a dependency of `dplyr`.
>
> ───────────
>
> [a]https://cran.r-project.org/web/packages/magrittr/vignettes/ magrittr.html

Note that as in the preceding example, it is best practice to put each "step" of a pipe sequence on its own line (indented by two spaces). This allows you to easily rearrange the steps (simply by moving lines), as well as to "comment out" particular steps to test and debug your analysis as you go.

11.4   Analyzing Data Frames by Group

`dplyr` functions are powerful, but they are truly awesome when you can apply them to *groups of rows* within a data set. For example, the previously described use of `summarize()` isn't particularly useful since it just gives a single summary for a given column (which you could have done easily using base R functions). However, a **grouped operation** would allow you to compute the same summary measure (e.g., `mean`, `median`, `sum`) automatically for multiple groups of rows, enabling you to ask more nuanced questions about your data set.

The **group_by()** function allows you to create associations among *groups of rows* in a data frame so that you can easily perform such aggregations. It takes as arguments a data frame to do the grouping on, followed by which column(s) you wish to use to group the data—each row in the table will be grouped with other rows that have the same value in that column. For example, you can group all of the data in the `presidentialElections` data set into groups whose rows share the same `state` value:

```
# Group observations by state
grouped <- group_by(presidentialElections, state)
```

The group_by() function returns a **tibble**,[5] which is a version of a data frame used by the "tidyverse"[6] family of packages (which includes `dplyr`). You can think of this as a "special" kind of data frame—one that is able to keep track of "subsets" (groups) within the same variable. While this grouping is not visually apparent (i.e., it does not sort the rows), the tibble keeps track of each row's group for computation, as shown in Figure 11.7.

The group_by() function is useful because it lets you apply operations to groups of data without having to explicitly break your data into different variables (sometimes called *bins* or *chunks*). Once you've used group_by() to group the rows of a data frame, you can apply other verbs (e.g., `summarize()`, `filter()`) to that tibble, and they will be automatically applied to *each* group (as if they were separate data frames). Rather than needing to explicitly extract different sets of data into separate data frames and run the same operations on each, you can use the group_by() function to accomplish all of this with a single command:

───────────

[5]**tibble** package website: http://tibble.tidyverse.org
[6]**tidyverse** website: https://www.tidyverse.org

```
> group_by(presidentialElections, state)
# A tibble: 1,097 x 4
# Groups:    state [51] ◄──── Result is grouped by state
     state          demVote  year south
     <chr>            <dbl> <int> <lgl>
  1 Alabama           84.8  1932 TRUE
  2 Arizona           67.0  1932 FALSE
  3 Arkansas          86.3  1932 TRUE
  4 California        58.4  1932 FALSE
  5 Colorado          54.8  1932 FALSE
  6 Connecticut       47.4  1932 FALSE
  7 Delaware          48.1  1932 FALSE
  8 Florida           74.5  1932 TRUE
  9 Georgia           91.6  1932 TRUE
 10 Idaho             58.7  1932 FALSE
# ... with 1,087 more rows
```

Figure 11.7   A tibble—created by the group_by() function—that stores associations by the grouping variable (state). Red notes are added.

```
# Compute summary statistics by state: average percentages across the years
state_voting_summary <- presidentialElections %>%
  group_by(state) %>%
  summarize(
    mean_dem_vote = mean(demVote),
    mean_other_parties = mean(other_parties_vote)
  )
```

The preceding code will first group the rows together by state, then compute summary information (mean() values) for each one of these groups (i.e., for each state), as illustrated in Figure 11.8. A summary of groups will still return a tibble, where each row is the summary of a

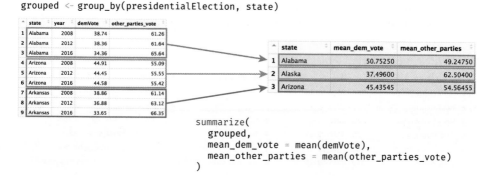

Figure 11.8   Using the group_by() and summarize() functions to calculate summary statistics in the presidentialElections data frame by state.

different group. You can extract values from a tibble using dollar sign or bracket notation, or convert it back into a normal data frame with the `as.data.frame()` function.

This form of grouping can allow you to quickly compare different subsets of your data. In doing so, you're redefining your **unit of analysis**. Grouping lets you frame your analysis question in terms of comparing *groups of observations*, rather than individual observations. This form of abstraction makes it easier to ask and answer complex questions about your data.

11.5   Joining Data Frames Together

When working with real-world data, you will often find that the data is stored across *multiple* files or data frames. This can be done for a number of reasons, such as reducing memory usage. For example, if you had a data frame containing information on a fundraising campaign that tracked donations (e.g., dollar amount, date), you would likely store information about each donor (e.g., email, phone number) in a separate data file (and thus data frame). See Figure 11.9 for an example of what this structure would look like.

This structure has a number of benefits:

1. **Data storage**: Rather than duplicating information about each donor every time that person makes a donation, you can store that information a single time. This will reduce the amount of space your data takes up.

2. **Data updates**: If you need to update information about a donor (e.g., the donor's phone number changes), you can make that change in a *single location*.

This separation and organization of data is a core concern in the design of *relational databases*, which are discussed in Chapter 13.

At some point, you will want to access information from both data sets (e.g., you need to email donors about their contributions), and thus need a way to reference values from both data frames at once—in effect, to *combine* the data frames. This process is called a **join** (because you are "joining" the data frames together). When you perform a join, you identify columns which are present in both tables, and use those columns to "match" corresponding rows to one another. Those column values are used as **identifiers** to determine which rows in each table correspond to one another, and thus will be combined into a single row in the resulting (joined) table.

Donations

	donor_name	amout	date
1	Maria Franca Fissolo	100	2018-02-15
2	Yang Huiyan	50	2018-02-15
3	Maria Franca Fissolo	75	2018-02-15
4	Alice Walton	25	2018-02-16
5	Susanne Klatten	100	2018-02-17
6	Yang Huiyan	150	2018-02-18

Donors

	donor_name	email
1	Alice Walton	alice.walton@gmail.com
2	Jacqueline Mars	jacqueline.mars@gmail.com
3	Maria Franca Fissolo	maria.franca.fissolo@gmail.com
4	Susanne Klatten	susanne.klatten@gmail.com
5	Laurene Powell Jobs	laurene.powell.jobs@gmail.com
6	Francoise Bettencourt Meyers	francoise.bettencourt.meyers@gmail.com

Figure 11.9   An example data frame of donations (left) and donor information (right). Notice that not all donors are present in both data frames.

The **left_join()** function is one example of a join. This function looks for matching columns between two data frames, and then returns a new data frame that is the first ("left") argument with extra columns from the second ("right") argument added on—in effect, "merging" the tables. You specify which columns you want to "match" on by specifying a **by** argument, which takes a vector of columns names (as strings).

For example, because both of the data frames in Figure 11.9 have a donor_name column, you can "match" the rows from the donor table to the donations table by this column and merge them together, producing the joined table illustrated in Figure 11.10.

```
# Combine (join) donations and donors data frames by their shared column
# ("donor_name")
combined_data <- left_join(donations, donors, by = "donor_name")
```

When you perform a **left join** as in the preceding code, the function performs the following steps:

1. It goes through each row in the table on the "left" (the first argument; e.g., donations), considering the values from the shared columns (e.g., donor_name).

2. For each of these values from the left-hand table, the function looks for a row in the right-hand table (e.g., donors) that has the *same* value in the specified column.

3. If it finds such a matching row, it adds any other data values from columns that are in donors but *not* in donations *to that left-hand row* in the resulting table.

4. It repeats steps 1–3 for each row in the left-hand table, until all rows have been given values from their matches on the right (if any).

You can see in Figure 11.10 that there were elements in the left-hand table (donations) that did not match to a row in the right-hand table (donors). This may occur because there are some donations whose donors do not have contact information (there is no matching donor_name entry): those rows will be given NA (*not available*) values, as shown in Figure 11.10.

> **Remember:** A left join returns all of the rows from the *first* table, with all of the columns from *both* tables.

For rows to match, they need to have the same data in *all* specified shared columns. However, if the names of your columns don't match or if you want to match only on specific columns, you can use a *named vector* (one with tags similar to a list) to indicate the different names from each data frame. If you don't specify a by argument, the join will match on *all* shared column names.

```
# An example join in the (hypothetical) case where the tables have
# different identifiers; e.g., if `donations` had a column `donor_name`,
# while `donors` had a column `name`
combined_data <- left_join(donations, donors, by = c("donor_name" = "name"))
```

Donations

	donor_name	amout	date
1	Maria Franca Fissolo	100	2018-02-15
2	Yang Huiyan	50	2018-02-15
3	Maria Franca Fissolo	75	2018-02-15
4	Alice Walton	25	2018-02-16
5	Susanne Klatten	100	2018-02-17
6	Yang Huiyan	150	2018-02-18

Donors

	donor_name	email
1	Alice Walton	alice.walton@gmail.com
2	Jacqueline Mars	jacqueline.mars@gmail.com
3	Maria Franca Fissolo	maria.franca.fissolo@gmail.com
4	Susanne Klatten	susanne.klatten@gmail.com
5	Laurene Powell Jobs	laurene.powell.jobs@gmail.com
6	Francoise Bettencourt Meyers	francoise.bettencourt.meyers@gmail.com

```
left_join(donations, donors, by = "donor_name")
```

	donor_name	amout	date	email
1	Maria Franca Fissolo	100	2018-02-15	maria.franca.fissolo@gmail.com
2	Yang Huiyan	50	2018-02-15	NA
3	Maria Franca Fissolo	75	2018-02-15	maria.franca.fissolo@gmail.com
4	Alice Walton	25	2018-02-16	alice.walton@gmail.com
5	Susanne Klatten	100	2018-02-17	susanne.klatten@gmail.com
6	Yang Huiyan	150	2018-02-18	NA

Figure 11.10 In a left join, columns from the right hand table (Donors) are added to the end of the left-hand table (Donations). Rows are on matched on the shared column (donor_name). Note the observations present in the left-hand table that don't have a corresponding row in the right-hand table (Yang Huiyan).

> **Caution:** Because of how joins are defined, the argument order matters! For example, in a `left_join()`, the resulting table has rows for only the elements in the *left* (first) table; any unmatched elements in the second table are lost.

If you switch the order of the arguments, you will instead keep all of the information from the donors data frame, adding in available information from `donations` (see Figure 11.11).

```
# Combine (join) donations and donors data frames (see Figure 11.11)
combined_data <- left_join(donors, donations, by = "donor_name")
```

Since some `donor_name` values show up multiple times in the right-hand (`donations`) table, the rows from `donors` end up being repeated so that the information can be "merged" with each set of values from `donations`. Again, notice that rows that lack a match in the right-hand table don't get any additional information (representing "donors" who gave their contact information to the organization, but have not yet made a donation).

Because the order of the arguments matters, `dplyr` (and relational database systems in general) provide several different kinds of joins, each influencing *which* rows are included in the final table. Note that in all joins, columns from *both* tables will be present in the resulting table—the join type dictates which rows are included. See Figure 11.12 for a diagram of these joins.

- **`left_join`**: All rows from the first (left) data frame are returned. That is, you get all the data from the left-hand table, with extra column values added from the right-hand table. Left-hand rows without a match will have NA in the right-hand columns.

Donors

	donor_name	email
1	Alice Walton	alice.walton@gmail.com
2	Jacqueline Mars	jacqueline.mars@gmail.com
3	Maria Franca Fissolo	maria.franca.fissolo@gmail.com
4	Susanne Klatten	susanne.klatten@gmail.com
5	Laurene Powell Jobs	laurene.powell.jobs@gmail.com
6	Francoise Bettencourt Meyers	francoise.bettencourt.meyers@gmail.com

Donations

	donor_name	amout	date
1	Maria Franca Fissolo	100	2018-02-15
2	Yang Huiyan	50	2018-02-15
3	Maria Franca Fissolo	75	2018-02-15
4	Alice Walton	25	2018-02-16
5	Susanne Klatten	100	2018-02-17
6	Yang Huiyan	150	2018-02-18

```
left_join(donors, donations, by = "donor_name")
```

	donor_name	email	amout	date
1	Alice Walton	alice.walton@gmail.com	25	2018-02-16
2	Jacqueline Mars	jacqueline.mars@gmail.com	NA	NA
3	Maria Franca Fissolo	maria.franca.fissolo@gmail.com	100	2018-02-15
4	Maria Franca Fissolo	maria.franca.fissolo@gmail.com	75	2018-02-15
5	Susanne Klatten	susanne.klatten@gmail.com	100	2018-02-17
6	Laurene Powell Jobs	laurene.powell.jobs@gmail.com	NA	NA
7	Francoise Bettencourt Meyers	francoise.bettencourt.meyers@gmail.com	NA	NA

Figure 11.11   Switching the order of the tables in a left-hand join (compared to Figure 11.10) returns a different set of rows. All rows from the left-hand table (donors) are returned with additional columns from the right-hand table (donations).

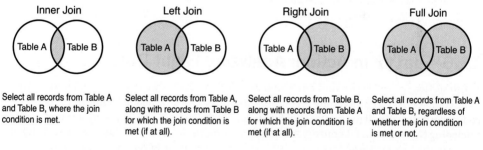

Figure 11.12   A diagram of different join types, downloaded from http://www.sql-join.com/sql-join-types/.

- **right_join**: All rows from the second (right) data frame are returned. That is, you get all the data from the right-hand table, with extra column values added from the left-hand table. Right-hand rows without a match will have NA in the left-hand columns. This is the "opposite" of a left_join, and the equivalent of switching the order of the arguments.

- **inner_join**: Only rows in *both* data frames are returned. That is, you get any rows that had matching observations in both tables, with the column values from both tables. There will be no additional NA values created by the join. Observations from the left that had no match in the right, or observations from the right that had no match in the left, will not be returned at all—the order of arguments *does not matter*.

- **full_join**: All rows from *both* data frames are returned. That is, you get a row for any observation, whether or not it matched. If it happened to match, values from both tables will appear in that row. Observations without a match will have NA in the columns from the other table—the order of arguments *does not matter*.

The key to deciding between these joins is to think about which set of data you want as your set of observations (rows), and which columns you'd be okay with being NA if a record is missing.

> **Tip:** Jenny Bryan has created an excellent "cheatsheet"[a] for `dplyr` join functions that you can reference.
>
> ---
> [a] http://stat545.com/bit001_dplyr-cheatsheet.html

> **Going Further**: All the joins discussed here are *mutating joins*, which add columns from one table to another. `dplyr` also provides *filtering joins*, which exclude rows based on whether they have a matching observation in another table, and *set operations*, which combine observations as if they were set elements. See the package documentation[a] for more detail on these options—but to get started you can focus primarily on the mutating joins.
>
> ---
> [a] https://cran.r-project.org/web/packages/dplyr/vignettes/two- table.html

11.6 `dplyr` in Action: Analyzing Flight Data

In this section, you will learn how `dplyr` functions can be used to ask interesting questions of a more complex data set (the complete code for this analysis is also available online in the book's code repository[7]). You'll use a data set of flights that departed from New York City airports (including Newark, John F. Kennedy, and Laguardia airports) in 2013. This data set is also featured online in the *Introduction to dplyr* vignette,[8] and is drawn from the Bureau of Transportation Statistics database.[9] To load the data set, you will need to install and load the `nycflights13` package. This will load the `flights` data set into your environment.

```
# Load the `nycflights13` package to access the `flights` data frame
install.packages("nycflights13") # once per machine
library("nycflights13")          # in each relevant script
```

Before you can start asking targeted questions of the data set, you will need to understand the structure of the data set a bit better:

---

[7] **dplyr in Action**: https://github.com/programming-for-data-science/in-action/tree/master/dplyr
[8] **Introduction to dplyr**: http://dplyr.tidyverse.org/articles/dplyr.html
[9] **Bureau of Labor Statistics**: air flights data: https://www.transtats.bts.gov/DatabaseInfo.asp?DB_ID=120

```
# Getting to know the `flights` data set
?flights           # read the available documentation
dim(flights)       # check the number of rows/columns
colnames(flights)  # inspect the column names
View(flights)      # look at the data frame in the RStudio Viewer
```

A subset of the flights data frame in RStudio's Viewer is shown in Figure 11.13.

Given this information, you may be interested in asking questions such as the following:

1. Which *airline* has the *highest number of delayed* departures?

2. On *average*, to which *airport* do flights arrive *most early*?

3. In which *month* do flights tend to have the *longest delays*?

Your task here is to map from these questions to specific procedures so that you can write the appropriate dplyr code.

You can begin by asking the first question:

   *"Which **airline** has the **highest number of delayed** departures?"*

This question involves comparing observations (flights) that share a particular feature (airline), so you perform the analysis as follows:

1. Since you want to consider all the flights from a particular airline (based on the carrier feature), you will first want to *group* the data by that feature.

2. You need to figure out the largest number of delayed departures (based on the dep_delay feature)—which means you need to find the flights that were delayed (*filtering* for them).

3. You can take the found flights and aggregate them into a count (*summarize* the different groups).

	year	month	day	dep_time	sched_dep_time	dep_delay	arr_time	sched_arr_time
1	2013	1	1	517	515	2	830	819
2	2013	1	1	533	529	4	850	830
3	2013	1	1	542	540	2	923	850
4	2013	1	1	544	545	-1	1004	1022
5	2013	1	1	554	600	-6	812	837
6	2013	1	1	554	558	-4	740	728
7	2013	1	1	555	600	-5	913	854

Figure 11.13 A subset of the flights data set, which is included as part of the nycflights13 package.

4. You will then need to find which group has the highest count (*filtering*).

5. Finally, you can choose (*select*) the airline of that group.

> **Tip**: When you're trying to find the right operation to answer your question of interest, the phrase *"Find the entry that..."* usually corresponds to a `filter()` operation!

Once you have established this algorithm, you can directly map it to `dplyr` functions:

```
# Identify the airline (`carrier`) that has the highest number of
# delayed flights
has_most_delays <- flights %>%          # start with the flights
  group_by(carrier) %>%                 # group by airline (carrier)
  filter(dep_delay > 0) %>%             # find only the delays
  summarize(num_delay = n()) %>%        # count the observations
  filter(num_delay == max(num_delay)) %>% # find most delayed
  select(carrier)                       # select the airline
```

> **Remember**: Often many approaches can be used to solve the same problem. The preceding code shows one possible approach; as an alternative, you could filter for delayed departures before grouping. The point is to think through how you might solve the problem (by hand) in terms of the *Grammar of Data Manipulation*, and then convert that into `dplyr`!

Unfortunately, the final answer to this question appears to be an abbreviation: UA. To reduce the size of the `flights` data frame, information about each airline is stored in a *separate* data frame called `airlines`. Since you are interested in combining these two data frames (your answer and the airline information), you can use a join:

```
# Get name of the most delayed carrier
most_delayed_name <- has_most_delays %>% # start with the previous answer
  left_join(airlines, by = "carrier") %>% # join on airline ID
  select(name)                           # select the airline name

print(most_delayed_name$name) # access the value from the tibble
# [1] "United Air Lines Inc."
```

After this step, you will have learned that the carrier that had the largest *absolute number* of delays was *United Air Lines Inc.* Before criticizing the airline too strongly, however, keep in mind that you might be interested in the *proportion of flights that are delayed*, which would require a separate analysis.

Next, you can assess the second question:

> *"On average, to which **airport** do flights arrive **most early**?"*

To answer this question, you can follow a similar approach. Because this question pertains to how early flights arrive, the *outcome* (feature) of interest is `arr_delay` (noting that a *negative* amount of delay indicates that the flight arrived *early*). You will want to *group* this information by *destination airport* (dest) where the flight arrived. And then, since you're interested in the *average* arrival delay, you will want to *summarize* those groups to aggregate them:

```
# Calculate the average arrival delay (`arr_delay`) for each destination
# (`dest`)
most_early <- flights %>%
  group_by(dest) %>% # group by destination
  summarize(delay = mean(arr_delay)) # compute mean delay
```

It's always a good idea to check your work as you perform each step of an analysis—don't write a long sequence of manipulations and hope that you got the right answer! By printing out the most_early data frame at this point, you notice that it has *a lot* of NA values, as seen in Figure 11.14.

This kind of unexpected result occurs frequently when doing data programming—and the best way to solve the problem is to work backward. By carefully inspecting the arr_delay column, you may notice that some entries have NA values—the arrival delay is not available for that record. Because you can't take the mean( ) of NA values, you decide to exclude those values from the analysis. You can do this by passing an na.rm = TRUE argument ("NA remove") to the mean( ) function:

```
# Compute the average delay by destination airport, omitting NA results
most_early <- flights %>%
  group_by(dest) %>% # group by destination
  summarize(delay = mean(arr_delay, na.rm = TRUE)) # compute mean delay
```

	dest	delay
1	ABQ	4.381890
2	ACK	NA
3	ALB	NA
4	ANC	-2.500000
5	ATL	NA
6	AUS	NA
7	AVL	NA
8	BDL	NA
9	BGR	NA

Figure 11.14 Average delay by destination in the flights data set. Because NA values are present in the data set, the mean delay for many destinations is calculated as NA. To remove NA values from the mean( ) function, set na.rm = FALSE.

Removing NA values returns numeric results, and you can continue working through your algorithm:

```
# Identify the destination where flights, on average, arrive most early
most_early <- flights %>%
  group_by(dest) %>% # group by destination
  summarize(delay = mean(arr_delay, na.rm = TRUE)) %>% # compute mean delay
  filter(delay == min(delay, na.rm = TRUE)) %>% # filter for least delayed
  select(dest, delay) %>% # select the destination (and delay to store it)
  left_join(airports, by = c("dest" = "faa")) %>% # join on `airports` data
  select(dest, name, delay) # select output variables of interest

print(most_early)
# A tibble: 1 x 3
#   dest  name         delay
#   <chr> <chr>        <dbl>
#1 LEX    Blue Grass   -22
```

Answering this question follows a very similar structure to the first question. The preceding code reduces the steps to a single statement by including the `left_join()` statement in the sequence of piped operations. Note that the column containing the airport code has a different name in the `flights` and `airports` data frames (`dest` and `faa`, respectively), so you use a named vector value for the by argument to specify the match.

As a result, you learn that *LEX—Blue Grass Airport* in Lexington, Kentucky—is the airport with the earliest average arrival time (*22* minutes early!).

A final question is:

> *"In which **month** do flights tend to have the **longest delays?**"*

These kinds of summary questions all follow a similar pattern: *group* the data by a column (feature) of interest, compute a *summary* value for (another) feature of interest for each group, *filter* down to a row of interest, and *select* the columns that answer your question:

```
# Identify the month in which flights tend to have the longest delays
flights %>%
  group_by(month) %>% # group by selected feature
  summarize(delay = mean(arr_delay, na.rm = TRUE)) %>% # summarize delays
  filter(delay == max(delay)) %>% # filter for the record of interest
  select(month) %>% # select the column that answers the question
  print() # print the tibble out directly
# A tibble: 1 x 1
#   month
#   <int>
#1    7
```

If you are okay with the result being in the form of a tibble rather than a vector, you can even pipe the results directly to the `print()` function to view the results in the R console (the answer being

July). Alternatively, you can use a package such as ggplot2 (see Chapter 16) to visually communicate the delays by month, as in Figure 11.15.

```
# Compute delay by month, adding month names for visual display
# Note, `month.name` is a variable built into R
delay_by_month <- flights %>%
  group_by(month) %>%
  summarize(delay = mean(arr_delay, na.rm = TRUE)) %>%
  select(delay) %>%
  mutate(month = month.name)

# Create a plot using the ggplot2 package (described in Chapter 17)
ggplot(data = delay_by_month) +
  geom_point(
    mapping = aes(x = delay, y = month),
    color = "blue",
    alpha = .4,
    size = 3
  ) +
  geom_vline(xintercept = 0, size = .25) +
  xlim(c(-20, 20)) +
  scale_y_discrete(limits = rev(month.name)) +
  labs(title = "Average Delay by Month", y = "", x = "Delay (minutes)")
```

Overall, understanding how to formulate questions, translate them into data manipulation steps (following the *Grammar of Data Manipulation*), and then map those to dplyr functions will enable you to quickly and effectively learn pertinent information about your data set. For practice wrangling data with the dplyr package, see the set of accompanying book exercises.[10]

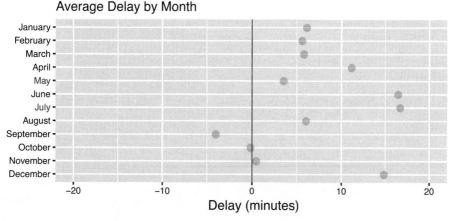

Figure 11.15   Average flight arrival delay in each month, calculated using the flights data set. The plot is built using ggplot2 (discussed in Chapter 16).

---

[10]**dplyr** exercises: https://github.com/programming-for-data-science/chapter-11-exercises

Reshaping Data with **tidyr**

One of the most common data wrangling challenges is adjusting how exactly row and columns are used to represent your data. Structuring (or restructuring) data frames to have the desired shape can be the most difficult part of creating a visualization, running a statistical model, or implementing a machine learning algorithm.

This chapter describes how you can use the **tidyr** ("tidy-er") package to effectively transform your data into an appropriate shape for analysis and visualization.

12.1   What Is "Tidy" Data?

When wrangling data into a data frame for your analysis, you need to decide on the *desired structure* of that data frame. You need to determine what each row and column will represent, so that you can consistently and clearly manipulate that data (e.g., you know what you will be *selecting* and what you will be *filtering*). The tidyr package is used to structure and work with data fames that follow three *principles of tidy data* (as described by the package's documentation[1]):

1. Each variable is in a column.

2. Each observation is a row.

3. Each value is a cell.

Indeed, these principles lead to the data structuring described in Chapter 9: rows represent observations, and columns represent features of that data.

However, asking different questions of a data set may involve different interpretations of what constitutes an "observation." For example, Section 11.6 described working with the flights data set from the nycflights13 package, in which each observation is a *flight*. However, the analysis made comparisons between *airlines*, *airports*, and *months*. Each question worked with a different unit of analysis, implying a different data structure (e.g., what should be represented by each row). While the example somewhat changed the nature of these rows by grouping and joining different data sets, having a more specific data structure where each row represented a *specific* unit of analysis

---

[1]**tidyr**: https://tidyr.tidyverse.org

(e.g., an *airline* or a *month*) may have made much of the wrangling and analysis more straightforward.

To use multiple different definitions of an "observation" when investigating your data, you will need to create multiple representations (i.e., data frames) of the same data set—each with its own configuration of rows and columns.

To demonstrate how you may need to adjust what each observation represents, consider the (fabricated) data set of music concert prices shown in Table 12.1. In this table, each observation (row) represents a *city*, with each city having features (columns) of the ticket price for a specific *band*.

But consider if you wanted to analyze the ticket price across all concerts. You could not do this easily with the data in its current form, since the data is organized by city (not by concert)! You would prefer instead that all of the prices were listed in a single column, as a feature of a row representing a single concert (a city-and-band combination), as in Table 12.2.

Table 12.1   **A "wide" data set of concert ticket price in different cities. Each observation (i.e., unit of analysis) is a city, and each feature is the concert ticket price for a given band.**

city	greensky_bluegrass	trampled_by_turtles	billy_strings	fruition
Seattle	40	30	15	30
Portland	40	20	25	50
Denver	20	40	25	40
Minneapolis	30	100	15	20

Table 12.2   **A "long" data set of concert ticket price by city and band. Each observation (i.e., unit of analysis) is a city–band combination, and each has a single feature that is the ticket price.**

city	band	price
Seattle	greensky_bluegrass	40
Portland	greensky_bluegrass	40
Denver	greensky_bluegrass	20
Minneapolis	greensky_bluegrass	30
Seattle	trampled_by_turtles	30
Portland	trampled_by_turtles	20
Denver	trampled_by_turtles	40
Minneapolis	trampled_by_turtles	100
Seattle	billy_strings	15
Portland	billy_strings	25
Denver	billy_strings	25
Minneapolis	billy_strings	15
Seattle	fruition	30
Portland	fruition	50
Denver	fruition	40
Minneapolis	fruition	20

Both Table 12.1 and Table 12.2 represent the same set of data—they both have prices for 16 different concerts. But by representing that data in terms of different *observations*, they may better support different analyses. These data tables are said to be in a different **orientation**: the price data in Table 12.1 is often referred to being in **wide format** (because it is spread wide across multiple columns), while the price data in Table 12.2 is in **long format** (because it is in one long column). Note that the long format table includes some duplicated data (the names of the cities and bands are repeated), which is part of why the data might instead be stored in wide format in the first place!

12.2   From Columns to Rows: `gather()`

Sometimes you may want to change the structure of your data—how your data is organized in terms of observations and features. To help you do so, the `tidyr` package provides elegant functions for transforming between orientations.

For example, to move from wide format (Table 12.1) to long format (Table 12.2), you need to *gather* all of the prices into a single column. You can do this using the **gather()** function, which collects data values stored across multiple columns into a single new feature (e.g., "price" in Table 12.2), along with an additional new column representing which feature that value was gathered from (e.g., "band" in Table 12.2). In effect, it creates two columns representing *key-value pairs* of the feature and its value from the original data frame.

```
# Reshape by gathering prices into a single feature
band_data_long <- gather(
  band_data_wide, # data frame to gather from
  key = band,     # name for new column listing the gathered features
  value = price,  # name for new column listing the gathered values
  -city           # columns to gather data from, as in dplyr's `select()`
)
```

The `gather()` function takes in a number of arguments, starting with the data frame to gather from. It then takes in a `key` argument giving a name for a column that will contain as values the column names the data was gathered from—for example, a new `band` column that will contains the values `"greensky_bluegrass"`, `"trampled_by_turtles"`, and so on. The third argument is a `value`, which is the name for the column that will contain the gathered values—for example, `price` to contain the price numbers. Finally, the function takes in arguments representing which columns to gather data from, using syntax similar to using `dplyr` to `select()` those columns (in the preceding example, `-city` indicates that it should gather from all columns except `city`). Again, any columns provided as this final set of arguments will have their names listed in the `key` column, and their values listed in the `value` column. This process is illustrated in Figure 12.1. The `gather()` function's syntax can be hard to intuit and remember; try tracing where each value "moves" in the table and diagram.

Note that once data is in long format, you can continue to analyze an individual feature (e.g., a specific band) by filtering for that value. For example, `filter(band_data_long, band == "greensky_bluegrass")` would produce just the prices for a single band.

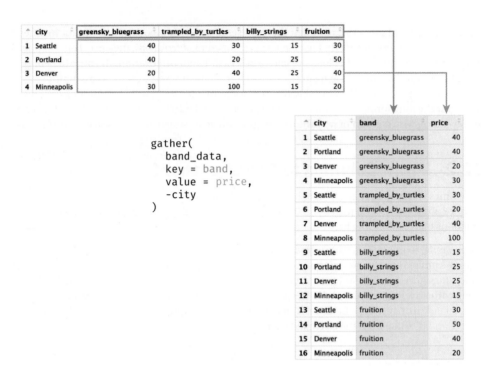

Figure 12.1 The gather() function takes values from multiple columns (greensky_bluegrass, trampled_by_turtles, etc.) and *gathers* them into a (new) single column (price). In doing so, it also creates a new column (band) that stores the names of the columns that were gathered (i.e., the column name in which each value was stored prior to gathering).

12.3   From Rows to Columns: **spread()**

It is also possible to transform a data table from long format into wide format—that is, to *spread* out the prices into multiple columns. Thus, while the gather() function collects multiple features into two columns, the **spread()** function creates multiple features from two existing columns. For example, you can take the long format data shown in Table 12.2 and spread it out so that each observation is a band, as in Table 12.3:

Table 12.3 A "wide" data set of concert ticket prices for a set of bands. Each observation (i.e., unit of analysis) is a band, and each feature is the ticket price in a given city.

band	Denver	Minneapolis	Portland	Seattle
billy_strings	25	15	25	15
fruition	40	20	50	30
greensky_bluegrass	20	30	40	40
trampled_by_turtles	40	100	20	30

```
# Reshape long data (Table 12.2), spreading prices out among multiple features
price_by_band <- spread(
  band_data_long, # data frame to spread from
  key = city,      # column indicating where to get new feature names
  value = price    # column indicating where to get new feature values
)
```

The spread() function takes arguments similar to those passed to the gather() function, but applies them in the opposite direction. In this case, the key and value arguments are where to get the column names and values, respectively. The spread() function will create a new column for each unique value in the provided key column, with values taken from the value feature. In the preceding example, the new column names (e.g., "Denver", "Minneapolis") were taken from the city feature in the long format table, and the values for those columns were taken from the price feature. This process is illustrated in Figure 12.2.

By combining gather() and spread(), you can effectively change the "shape" of your data and what concept is represented by an observation.

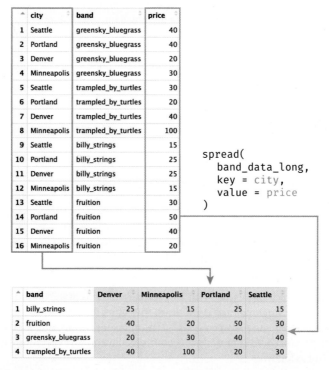

Figure 12.2 The spread() function *spreads out* a single column into multiple columns. It creates a new column for each unique value in the provided key column (city). The values in each new column will be populated with the provided value column (price).

> **Tip:** Before spreading or gathering your data, you will often need to *unite* multiple columns into a single column, or to *separate* a single column into multiple columns. The `tidyr` functions `unite( )`[a] and `separate( )`[b] provide a specific syntax for these common data preparation tasks.
>
> [a]https://tidyr.tidyverse.org/reference/unite.html
> [b]https://tidyr.tidyverse.org/reference/separate.html

12.4 `tidyr` in Action: Exploring Educational Statistics

This section uses a real data set to demonstrate how reshaping your data with `tidyr` is an integral part of the data exploration process. The data in this example was downloaded from the *World Bank Data Explorer*,[2] which is a data collection of hundreds of indicators (measures) of different economic and social development factors. In particular, this example considers *educational indicators*[3] that capture a relevant signal of a country's level of (or investment in) education—for example, government expenditure on education, literacy rates, school enrollment rates, and dozens of other measures of educational attainment. The imperfections of this data set (unnecessary rows at the top of the .csv file, a substantial amount of missing data, long column names with special characters) are representative of the challenges involved in working with real data sets. All graphics in this section were built using the `ggplot2` package, which is described in Chapter 16. The complete code for this analysis is also available online in the book's code repository.[4]

After having downloaded the data, you will need to load it into your R environment:

```
# Load data, skipping the unnecessary first 4 rows
wb_data <- read.csv(
  "data/world_bank_data.csv",
  stringsAsFactors = F,
  skip = 4
)
```

When you first load the data, each observation (row) represents an indicator for a country, with features (columns) that are the values of that indicator in a given year (see Figure 12.3). Notice that many values, particularly for earlier years, are missing (NA). Also, because R does not allow column names to be numbers, the `read.csv( )` function has *prepended* an X to each column name (which is just a number in the raw .csv file).

While in terms of the indicator this data is in long format, in terms of the indicator and year the data is in wide format—a single column contains all the values for a single year. This structure allows you to make comparisons between years for the indicators by filtering for the indicator of interest. For example, you could compare each country's educational expenditure in 1990 to its expenditure in 2014 as follows:

[2]**World Bank Data Explorer:** https://data.worldbank.org
[3]**World Bank education:** http://datatopics.worldbank.org/education
[4]**tidyr in Action**: https://github.com/programming-for-data-science/in-action/tree/master/tidyr

	Country.Name	Country.Code	Indicator.Name	Indicator.Code	X1960	X1961	X1962
1	Aruba	ABW	Population ages 15–64 (% of total)	SP.POP.1564.TO.ZS	53.66992	54.05678	54.38328
2	Aruba	ABW	Population ages 0–14 (% of total)	SP.POP.0014.TO.ZS	43.84719	43.35835	42.92574
3	Aruba	ABW	Unemployment, total (% of total labor force) (modeled...	SL.UEM.TOTL.ZS	NA	NA	NA
4	Aruba	ABW	Unemployment, male (% of male labor force) (modele...	SL.UEM.TOTL.MA.ZS	NA	NA	NA
5	Aruba	ABW	Unemployment, female (% of female labor force) (mod...	SL.UEM.TOTL.FE.ZS	NA	NA	NA
6	Aruba	ABW	Labor force, total	SL.TLF.TOTL.IN	NA	NA	NA
7	Aruba	ABW	Labor force, female (% of total labor force)	SL.TLF.TOTL.FE.ZS	NA	NA	NA

Figure 12.3   Untransformed World Bank educational data used in Section 12.4.

```
# Visually compare expenditures for 1990 and 2014

# Begin by filtering the rows for the indicator of interest
indicator <- "Government expenditure on education, total (% of GDP)"
expenditure_plot_data <- wb_data %>%
  filter(Indicator.Name == indicator)

# Plot the expenditure in 1990 against 2014 using the `ggplot2` package
# See Chapter 16 for details
expenditure_chart <- ggplot(data = expenditure_plot_data) +
  geom_text_repel(
    mapping = aes(x = X1990 / 100, y = X2014 / 100, label = Country.Code)
  ) +
  scale_x_continuous(labels = percent) +
  scale_y_continuous(labels = percent) +
  labs(title = indicator, x = "Expenditure 1990", y = "Expenditure 2014")
```

Figure 12.4 shows that the expenditure (relative to gross domestic product) is fairly correlated between the two time points: countries that spent more in 1990 also spent more in 2014 (specifically, the correlation—calculated in R using the cor( ) function—is .64).

However, if you want to extend your analysis to visually compare how the expenditure across all years varies for a given country, you would need to reshape the data. Instead of having each observation be an indicator for a country, you want each observation to be an indicator for a country for a year—thereby having all of the values for all of the years in a single column and making the data *long(er) format*.

To do this, you can gather( ) the year columns together:

```
# Reshape the data to create a new column for the `year`
long_year_data <- wb_data %>%
  gather(
    key = year, # `year` will be the new key column
    value = value, # `value` will be the new value column
    X1960:X # all columns between `X1960` and `X` will be gathered
  )
```

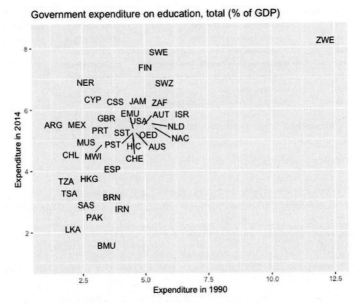

Figure 12.4 A comparison of each country's education expenditures in 1990 and 2014.

	Country.Name	Country.Code	Indicator.Name	year	value
1	Aruba	ABW	Population ages 15-64 (% of total)	X1960	53.66992
2	Aruba	ABW	Population ages 0-14 (% of total)	X1960	43.84719
3	Aruba	ABW	Unemployment, total (% of total labor force) (modeled...	X1960	NA
4	Aruba	ABW	Unemployment, male (% of male labor force) (modele...	X1960	NA
5	Aruba	ABW	Unemployment, female (% of female labor force) (mod...	X1960	NA
6	Aruba	ABW	Labor force, total	X1960	NA
7	Aruba	ABW	Labor force, female (% of total labor force)	X1960	NA
8	Aruba	ABW	Government expenditure on education, total (% of GDP)	X1960	NA
9	Aruba	ABW	Government expenditure on education, total (% of gov...	X1960	NA
10	Aruba	ABW	Expenditure on tertiary education (% of government e...	X1960	NA
11	Aruba	ABW	Government expenditure per student, tertiary (% of G...	X1960	NA

Figure 12.5 Reshaped educational data (long format by `year`). This structure allows you to more easily create visualizations across multiple years.

As shown in Figure 12.5, this `gather()` statement creates a `year` column, so each observation (row) represents the value of an indicator in a particular country in a given year. The expenditure for each year is stored in the `value` column created (coincidentally, this column is given the name `"value"`).

This structure will now allow you to compare fluctuations in an indicator's value over time (across all years):

```
# Filter the rows for the indicator and country of interest
indicator <- "Government expenditure on education, total (% of GDP)"
spain_plot_data <- long_year_data %>%
  filter(
    Indicator.Name == indicator,
    Country.Code == "ESP" # Spain
  ) %>%
  mutate(year = as.numeric(substr(year, 2, 5))) # remove "X" before each year

# Show the educational expenditure over time
chart_title <- paste(indicator, " in Spain")
spain_chart <- ggplot(data = spain_plot_data) +
  geom_line(mapping = aes(x = year, y = value / 100)) +
  scale_y_continuous(labels = percent) +
  labs(title = chart_title, x = "Year", y = "Percent of GDP Expenditure")
```

The resulting chart, shown in Figure 12.6, uses the available data to show a timeline of the fluctuations in government expenditures on education in Spain. This produces a more complete picture of the history of educational investment, and draws attention to major changes as well as the absence of data in particular years.

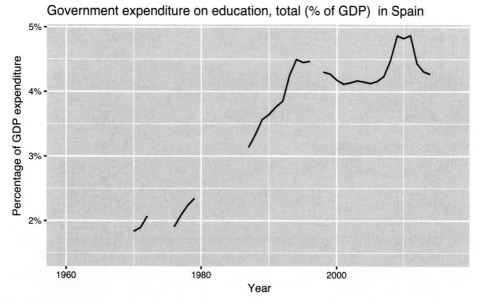

Figure 12.6   Education expenditures over time in Spain.

You may also want to compare two indicators to each other. For example, you may want to assess the relationship between each country's literacy rate (a first indicator) and its unemployment rate (a second indicator). To do this, you would need to reshape the data again so that each observation is a particular country and each column is an indicator. Since indicators are currently in one column, you need to spread them out using the spread( ) function:

```
# Reshape the data to create columns for each indicator
wide_data <- long_year_data %>%
  select(-Indicator.Code) %>% # do not include the `Indicator.Code` column
  spread(
    key = Indicator.Name, # new column names are `Indicator.Name` values
    value = value # populate new columns with values from `value`
  )
```

This wide format data shape allows for comparisons between two different indicators. For example, you can explore the relationship between female unemployment and female literacy rates, as shown in Figure 12.7.

```
# Prepare data and filter for year of interest
x_var <- "Literacy rate, adult female (% of females ages 15 and above)"
y_var <- "Unemployment, female (% of female labor force) (modeled
  ILO estimate)"

lit_plot_data <- wide_data %>%
  mutate(
    lit_percent_2014 = wide_data[, x_var] / 100,
    employ_percent_2014 = wide_data[, y_var] / 100
  ) %>%
  filter(year == "X2014")

# Show the literacy vs. employment rates
lit_chart <- ggplot(data = lit_plot_data) +
  geom_point(mapping = aes(x = lit_percent_2014, y = employ_percent_2014)) +
  scale_x_continuous(labels = percent) +
  scale_y_continuous(labels = percent) +
  labs(
    x = x_var,
    y = "Unemployment, female (% of female labor force)",
    title = "Female Literacy Rate versus Female Unemployment Rate"
  )
```

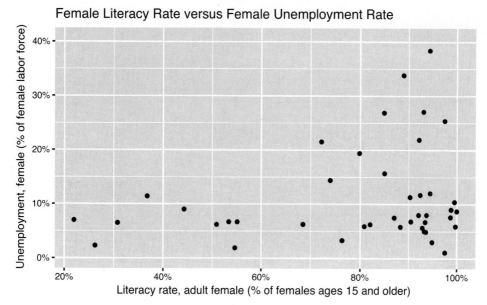

Figure 12.7   Female literacy rate versus unemployment rate in 2014.

Each comparison in this analysis—between two time points, over a full time-series, and between indicators—required a different representation of the data set. Mastering use of the tidyr functions will allow you to quickly transform the shape of your data set, allowing for rapid and effective data analysis. For practice reshaping data with the tidyr package, see the set of accompanying book exercises.[5]

---

[5] **tidyr** exercises: https://github.com/programming-for-data-science/chapter-12-exercises

13

Accessing Databases

This chapter introduces relational databases as a way to structure and organize complex data sets. After introducing the purpose and format of relational databases, it describes the syntax for interacting with them using R. By the end of the chapter you will be able to wrangle data from a database.

13.1   An Overview of Relational Databases

Simple data sets can be stored and loaded from .csv files, and are readily represented in the computer's memory as a data frame. This structure works great for data that is structured just as a set of observations made up of features. However, as data sets become more complex, you run against some limitations.

In particular, your data may not be structured in a way that it can easily and efficiently be represented as a single data frame. For example, imagine you were trying to organize information about music playlists (e.g., on a service such as *Spotify*). If your playlist is the unit of analysis you are interested in, each playlist would be an observation (row) and would have different features (columns) included. One such feature you could be interested in is the songs that appear on the playlist (implying that one of your columns should be songs). However, playlists may have *lots* of different songs, and you may also be tracking further information about each song (e.g., the artist, the genre, the length of the song). Thus you could not easily represent each song as a simple data type such as a number or string. Moreover, because the same song may appear in multiple playlists, such a data set would include a lot of duplicate information (e.g., the title and artist of the song).

To solve this problem, you could use multiple data frames (perhaps loaded from multiple .csv files), joining those data frames together as described in Chapter 11 to ask questions of the data. However, that solution would require you to manage multiple different .csv files, as well as to determine an effective and consistent way of joining them together. Since organizing, tracking, and updating multiple .csv files can be difficult, many large data sets are instead stored in **databases**. Metonymically, a database is a specialized application (called a database management system) used to save, organize, and access information—similar to what git does for versions of code, but in this case for the kind of data that might be found in multiple .csv files. Because many organizations store their data in a database of some kind, you will need to be able to access that data to analyze it.

Moreover, accessing data directly from a database makes it possible to process data sets that are too large to fit into your computer's memory (RAM) at once. The computer will not be required to hold a reference to all the data at once, but instead will be able to apply your data manipulation (e.g., selecting and filtering the data) to the data stored on a computer's hard drive.

13.1.1   What Is a Relational Database?

The most commonly used type of database is a **relational database**. A relational database organizes data into **tables** similar in concept and structure to a data frame. In a table, each **row** (also called a **record**) represents a single "item" or observation, while each **column** (also called a **field**) represents an individual data property of that item. In this way, a database table mirrors an R data frame; you can think of them as somewhat equivalent. However, a relational database may be made up of dozens, if not hundreds or even thousands, of different tables—each one representing a different facet of the data. For example, one table may store information about which music playlists are in the database, another may store information about the individual songs, another may store information about the artists, and so on.

What makes relational databases special is how they specify the *relationships* between these tables. In particular, each record (row) in a table is given a field (column) called the **primary key**. The primary key is a unique value for each row in the table, so it lets you refer to a particular record. Thus even if there were two songs with the same name and artist, you could still distinguish between them by referencing them through their primary key. Primary keys can be any unique identifier, but they are almost always numbers and are frequently automatically generated and assigned by the database. Note that databases can't just use the "row number" as the primary key, because records may be added, removed, or reordered—which means a record won't always be at the same index!

Moreover, each record in one table may be *associated* with a record in another—for example, each record in a songs table might have an associated record in the artists table indicating which artist performed the song. Because each record in the artists table has a unique key, the songs table is able to establish this association by including a field (column) for each record that contains the corresponding key from artists (see Figure 13.1). This is known as a **foreign key** (it is the key from a "foreign" or other table). Foreign keys allow you to join tables together, similar to how you would with dplyr. You can think of foreign keys as a formalized way of defining a consistent column for the join() function's by argument.

Databases can use tables with foreign keys to organize data into complex structures; indeed, a database may have a table that *just* contains foreign keys to link together other tables! For example, if a database needs to represent data such that each playlist can have multiple songs, and songs can be on many playlists (a "many-to-many" relationship), you could introduce a new "bridge table" (e.g., playlists_songs) whose records represent the associations between the two other tables (see Figure 13.2). You can think of this as a "table of lines to draw between the other tables." The database could then join *all three* of the tables to access the information about all of the songs for a particular playlist.

table: `artists`

id	name
10	David Bowie
11	Queen
12	Prince

↑
primary key

foreign key
↓

table: `songs`

id	title	artist_id
80	Bohemian Rhapsody	11
81	Don't Stop Me Now	11
82	Purple Rain	12
83	Starman	10

table: `artists JOIN songs ON artists.id = songs.artist_id`

artists.id	artists.name	songs.id	songs.title	songs.artist_id
10	David Bowie	83	Starman	10
11	Queen	80	Bohemian Rhapsody	11
11	Queen	81	Don't Stop Me Now	11
12	Prince	82	Purple Rain	12

Figure 13.1   An example pair of database tables (top). Each table has a *primary key* column id. The songs table (top right) also has an `artist_id` *foreign key* used to associate it with the `artists` table (top left). The bottom table illustrates how the foreign key can be used when joining the tables.

> **Going Further**: Database design, development, and use is actually its own (very rich) problem domain. The broader question of making databases reliable and efficient is beyond the scope of this book.

13.1.2   Setting Up a Relational Database

To use a relational database on your own computer (e.g., for experimenting or testing your analysis), you will need to install a separate software program to manage that database. This program is called a **relational database management system (RDMS)**. There are a couple of different popular RDMS systems; each of them provides roughly the same syntax (called SQL) for manipulating the tables in the database, though each may support additional specialized features. The most popular RDMSs are described here. You are not required to install any of these RDMSs to work with a database through R; see Section 13.3, below. However, brief installation notes are provided for your reference.

- **SQLite**[1] is the simplest SQL database system, and so is most commonly used for testing and development (though rarely in real-world "production" systems). SQLite databases have the advantage of being highly self-contained: each SQLite database is a single file (with the

---

[1]**SQLite:** https://www.sqlite.org/index.html

table: `playlists`

id	name
100	Awesome Mix
101	Sweet Tunes

table: `songs`

id	title
80	Bohemian Rhapsody
81	Don't Stop Me Now
82	Purple Rain
83	Starman

table: `playlists_songs`

playlist_id	songs_id
100	81
100	82
100	83
101	80
101	82

table: `playlist JOIN playlists_songs ON playlist.id = playlist_id`
`JOIN songs ON songs.id = songs_id`

playlist_id	playlists.name	songs.id	songs.title
100	Awesome Mix	81	Don't Stop Me Now
100	Awesome Mix	82	Purple Rain
100	Awesome Mix	83	Starman
101	Sweet Tunes	80	Bohemian Rhapsody
101	Sweet Tunes	82	Purple Rain

Figure 13.2 An example "bridge table" (top right) used to associate many playlists with many songs. The bottom table illustrates how these three tables might be joined.

`.sqlite` extension) that is formatted to enable the SQLite RDMS to access and manipulate its data. You can almost think of these files as advanced, efficient versions of `.csv` files that can hold multiple tables! Because the database is stored in a single file, this makes it easy to share databases with others or even place one under version control.

To work with an SQLite database you can download and install a command line application[2] for manipulating the data. Alternatively, you can use an application such as *DB Browser for SQLite*,[3] which provides a graphical interface for interacting with the data. This is particularly useful for testing and verifying your SQL and R code.

- **PostgreSQL**[4] (often shortened to "Postgres") is a free open source RDMS, providing a more robust system and set of features (e.g., for speeding up data access and ensuring data integrity) and functions than SQLite. It is often used in real-world production systems, and is the recommended system to use if you need a "full database." Unlike with SQLite, a Postgres database is not isolated to a single file that can easily be shared, though there are ways to export a database.

---

[2]**SQLite download page:** https://www.sqlite.org/download.html; look for "Precompiled Binaries" for your system.
[3]**DB Browser for SQLite:** http://sqlitebrowser.org
[4]**PostgreSQL:** https://www.postgresql.org

You can download and install the Postgres RDMS from its website;[5] follow the instructions in the installation wizard to set up the database system. This application will install the manager on your machine, as well as provide you with a graphical application (*pgAdmin*) to administer your databases. You can also use the provided `psql` command line application if you add it to your PATH; alternatively, the *SQL Shell* application will open the command line interface directly.

- **MySQL**[6] is a free (but closed source) RDMS, providing a similar level of features and structure as Postgres. MySQL is a more popular system than Postgres, so its use is more common, but can be somewhat more difficult to install and set up.

  If you wish to set up and use a MySQL database, we recommend that you install the Community Server Edition from the MySQL website.[7] Note that you do *not* need to sign up for an account (click the smaller "No thanks, just start my download" link instead).

We suggest you use SQLite when you're just experimenting with a database (as it requires the least amount of setup), and recommend Postgres if you need something more full-featured.

13.2 A Taste of SQL

The reason all of the RDMSs described in Section 13.1.2 have "SQL" in their names is because they all use the same syntax—SQL—for manipulating the data stored in the database. **SQL** (Structured Query Language) is a programming language used specifically for managing data in a relational database—a language that is *structured* for *querying* (accessing) that information. SQL provides a relatively small set of commands (referred to as **statements**), each of which is used to interact with a database (similar to the operations described in the *Grammar of Data Manipulation* used by `dplyr`).

This section introduces the most basic of SQL statements: the `SELECT` statement used to access data. Note that it is absolutely possible to access and manipulate a database through R without using SQL; see Section 13.3. However, it is often useful to understand the underlying commands that R is issuing. Moreover, if you eventually need to discuss database manipulations with someone else, this language will provide some common ground.

> **Caution:** Most RDMSs support SQL, though systems often use slightly different "flavors" of SQL. For example, data types may be named differently, or different RDMSs may support additional functions or features.

> **Tip:** For a more thorough introduction to SQL, *w3schools*[a] offers a very newbie-friendly tutorial on SQL syntax and usage. You can also find more information in Forta, *Sams Teach Yourself SQL in 10 Minutes, Fourth Edition* (Sams, 2013), and van der Lans, *Introduction to SQL, Fourth Edition* (Addison-Wesley, 2007).
>
> [a]https://www.w3schools.com/sql/default.asp

[5]**PostgreSQL download page:** https://www.postgresql.org/download
[6]**MySQL:** https://www.mysql.com
[7]**MySQL download Page:** https://dev.mysql.com/downloads/mysql

The most commonly used SQL statement is the **SELECT** statement. The SELECT statement is used to access and extract data from a database (without modifying that data)—this makes it a **query** statement. It performs the same work as the select() function in dplyr. In its simplest form, the SELECT statement has the following format:

```
/* A generic SELECT query statement for accessing data */
SELECT column FROM table
```

(In SQL, comments are written on their own line surrounded by /* */.)

This query will return the data from the specified column in the specified table (keywords like SELECT in SQL are usually written in all-capital letters—though they are not case-sensitive—while column and table names are often lowercase). For example, the following statement would return all of the data from the title column of the songs table (as shown in Figure 13.3):

```
/* Access the `title` column from the `songs` table */
SELECT title FROM songs
```

This would be equivalent to select(songs, title) when using dplyr.

You can select multiple columns by separating the names with commas (,). For example, to select both the id and title columns from the songs table, you would use the following query:

```
/* Access the `id` and `title` columns from the `songs` table */
SELECT id, title FROM songs
```

If you wish to select *all* the columns, you can use the special * symbol to represent "everything"— the same wildcard symbol you use on the command line! The following query will return all columns from the songs table:

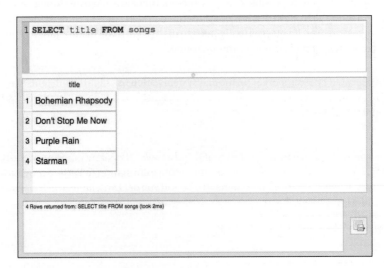

Figure 13.3   A SELECT statement and results shown in the SQLite Browser.

```
/* Access all columns from the `songs` table */
SELECT * FROM songs
```

Using the * wildcard to select data is common practice when you just want to load the entire table from the database.

You can also optionally give the resulting column a new name (similar to a **mutate** manipulation) by using the **AS** keyword. This keyword is placed immediately after the name of the column to be aliased, followed by the new column name. It doesn't actually change the table, just the label of the resulting "subtable" returned by the query.

```
/* Access the `id` column (calling it `song_id`) from the `songs` table */
SELECT id AS song_id FROM songs
```

The SELECT statement performs a select data manipulation. To perform a filter manipulation, you add a **WHERE** clause at the end of the SELECT statement. This clause includes the keyword WHERE followed by a *condition*, similar to the boolean expression you would use with dplyr. For example, to select the title column from the songs table with an artist_id value of 11, you would use the following query (also shown in Figure 13.4):

```
/* Access the `title` column from the `songs` table if `artist_id` is 11 */
SELECT title FROM songs WHERE artist_id = 11
```

This would be the equivalent to the following dplyr statement:

```
# Filter for the rows with a particular `artist_id`, and then select
# the `title` column
filter(songs, artist_id == 11)  %>%
  select(title)
```

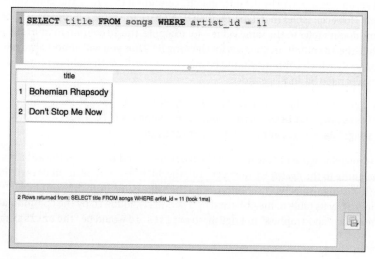

Figure 13.4   A WHERE clause and results shown in the SQLite Browser.

The filter condition is applied to the whole table, not just the selected columns. In SQL, the filtering occurs before the selection.

Note that a WHERE condition uses = (not ==) as the "is equal" operator. Conditions can also use other relational operators (e.g., >, <=), as well as some special keywords such as LIKE, which will check whether the column value is *inside* a string. (String values in SQL must be specified in quotation marks—it's most common to use single quotes.)

You can combine multiple WHERE conditions by using the **AND**, **OR**, and **NOT** keywords as boolean operators:

```
/* Access all columns from `songs` where EITHER condition applies */
SELECT * FROM songs WHERE artist_id = 12 OR title = 'Starman'
```

The statement SELECT columns FROM table WHERE conditions is the most common form of SQL query. But you can also include other keyword clauses to perform further data manipulations. For example, you can include an **ORDER_BY** clause to perform an **arrange** manipulation (by a specified column), or a **GROUP_BY** clause to perform aggregation (typically used with SQL-specific aggregation functions such as MAX( ) or MIN( )). See the official documentation for your database system (e.g., for Postgres[8]) for further details on the many options available when specifying SELECT queries.

The SELECT statements described so far all access data in a single table. However, the entire point of using a database is to be able to store and query data across *multiple* tables. To do this, you use a **join** manipulation similar to that used in dplyr. In SQL, a join is specified by including a **JOIN** clause, which has the following format:

```
/* A generic JOIN between two tables */
SELECT columns FROM table1 JOIN table2
```

As with dplyr, an SQL join will by default "match" columns if they have the same value in the same column. However, tables in databases often don't have the same column names, or the shared column name doesn't refer to the same value—for example, the id column in artists is for the artist ID, while the id column in songs is for the song ID. Thus you will almost always include an **ON** clause to specify which columns should be matched to perform the join (writing the names of the columns separated by an = operator):

```
/* Access artists, song titles, and ID values from two JOINed tables */
SELECT artists.id, artists.name, songs.id, songs.title FROM artists
    JOIN songs ON songs.artist_id = artists.id
```

This query (shown in Figure 13.5) will select the IDs, names, and titles from the artists and songs tables by matching to the *foreign key* (artist_id); the JOIN clause appears on its own line just for readability. To distinguish between columns with the same name from different tables, you specify each column first by its table name, followed by a period (.), followed by the column name. (The dot can be read like "apostrophe s" in English, so artists.id would be "the artists table's id.")

---

[8] PostgreSQL: SELECT: https://www.postgresql.org/docs/current/static/sql-select.html

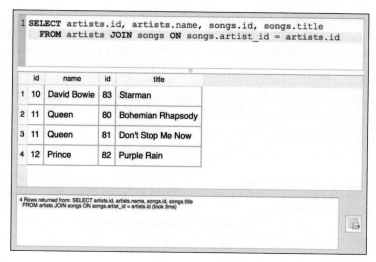

```
1 SELECT artists.id, artists.name, songs.id, songs.title
    FROM artists JOIN songs ON songs.artist_id = artists.id
```

	id	name	id	title
1	10	David Bowie	83	Starman
2	11	Queen	80	Bohemian Rhapsody
3	11	Queen	81	Don't Stop Me Now
4	12	Prince	82	Purple Rain

4 Rows returned from: SELECT artists.id, artists.name, songs.id, songs.title
FROM artists JOIN songs ON songs.artist_id = artists.id (took 3ms)

Figure 13.5   A JOIN statement and results shown in the SQLite Browser.

You can join on multiple conditions by combining them with AND clauses, as with multiple WHERE conditions.

Like dplyr, SQL supports four kinds of joins (see Chapter 11 to review them). By default, the JOIN statement will perform an *inner join*—meaning that only rows that contain matches in both tables will be returned (e.g., the joined table will not have rows that don't match). You can also make this explicit by specifying the join clause with the keywords **INNER JOIN**. Alternatively, you can specify that you want to perform a **LEFT JOIN**, **RIGHT JOIN**, or **OUTER JOIN** (i.e., a *full join*). For example, to perform a *left join* you would use a query such as the following:

```
/* Access artists and song titles, including artists without any songs */
SELECT artists.id, artists.name, songs.id, songs.title FROM artists
    LEFT JOIN songs ON songs.artist_id = artists.id
```

Notice that the statement is written the same way as before, except with an extra word to clarify the type of join.

As with dplyr, deciding on the type of join to use requires that you carefully consider which observations (rows) must be included, and which features (columns) must *not* be missing in the table you produce. Most commonly you are interested in an inner join, which is why that is the default!

13.3 Accessing a Database from R

SQL will allow you to query data from a database; however, you would have to execute such commands through the RDMS itself (which provides an interpreter able to understand the syntax). Luckily, you can instead use R packages to connect to and query a database directly, allowing you to use the same, familiar R syntax and data structures (i.e., data frames) to work with databases. The simplest way to access a database through R is to use the **dbplyr**[9] package, which was developed as part of the tidyverse collection. This package allows you to query a relational database using dplyr functions, avoiding the need to use an external application!

> **Going Further**: RStudio also provides an interface and documentation for connecting to a database through the IDE; see the *Databases Using R* portal.[a]
>
> _____
> [a]https://db.rstudio.com

Because dbplyr is another external package (like dplyr and tidyr), you will need to install it before you can use it. However, because dbplyr is actually a "backend" for dplyr (it provides the behind-the-scenes code that dplyr uses to work with a database), you actually need to use functions from dplyr and so load in the dplyr package instead. However, you will also need to load the DBI package, which is installed along with dbplyr and allows you to connect to the database:

```
install.packages("dbplyr") # once per machine
library("DBI")             # in each relevant script
library("dplyr")           # need dplyr to use its functions on the database!
```

You will also need to install an additional package depending on which kind of database you wish to access. These packages provide a common *interface* (set of functions) across multiple database formats—they will allow you to access an SQLite database and a Postgres database using the same R functions.

```
# To access an SQLite database
install.packages("RSQLite") # once per machine
library("RSQLite")          # in each relevant script

# To access a Postgres database
install.packages("RPostgreSQL") # once per machine
library("RPostgreSQL")          # in each relevant script
```

Remember that databases are managed and accessed through an RDMS, which is a separate program from the R interpreter. Thus, to access databases through R, you will need to "connect" to that external RDMS program and use R to issue statements *through* it. You can connect to an external database using the **dbConnect()** function provided by the DBI package:

```
# Create a "connection" to the RDMS
db_connection <- dbConnect(SQLite(), dbname = "path/to/database.sqlite")

# When finished using the database, be sure to disconnect as well!
dbDisconnect(db_connection)
```

The dbConnect() function takes as a first argument a "connection" interface provided by the relevant database connection package (e.g., RSQLite). The remaining arguments specify the

_____
[9]**dbplyr** repository page: https://github.com/tidyverse/dbplyr

location of the database, and are dependent on where that database is located and what kind of database it is. For example, you use a dbname argument to specify the path to a local SQLite database file, while you use host, user, and password to specify the connection to a database on a remote machine.

> **Caution:** Never include your database password directly in your R script—saving it in plain text will allow others to easily steal it! Instead, dbplyr recommends that you prompt users for the password through RStudio by using the askForPassword()[a] function from the rstudioapi package (which will cause a pop-up window to appear for users to type in their password). See the *dbplyr introduction vignette*[b] for an example.
>
> [a] https://www.rdocumentation.org/packages/rstudioapi/versions/0.7/topics/askForPassword
> [b] https://cran.r-project.org/web/packages/dbplyr/vignettes/dbplyr.html

Once you have a connection to the database, you can use the **dbListTables()** function to get a vector of all the table names. This is useful for checking that you've connected to the database (as well as seeing what data is available to you!).

Since all SQL queries access data FROM a particular table, you will need to start by creating a *reference to* that table in the form of a variable. You can do this by using the **tbl()** function provided by dplyr (*not* dbplyr!). This function takes as arguments the connection to the database and the name of the table you want to reference. For example, to query a songs table as in Figure 13.1, you would use the following:

```
# Create a reference to the "songs" table in the database
songs_table <- tbl(db_connection, "songs")
```

If you print this variable out, you will notice that it looks *mostly* like a normal data frame (specifically a tibble), except that the variable refers to a *remote* source (since the table is in the database, not in R!); see Figure 13.6.

Once you have a reference to the table, you can use the same dplyr functions discussed in Chapter 11 (e.g., select(), filter()). Just use the table in place of the data frame to manipulate!

```
# Construct a query from the `songs_table` for songs by Queen (artist ID 11)
queen_songs_query <- songs_table %>%
  filter(artist_id == 11) %>%
  select(title)
```

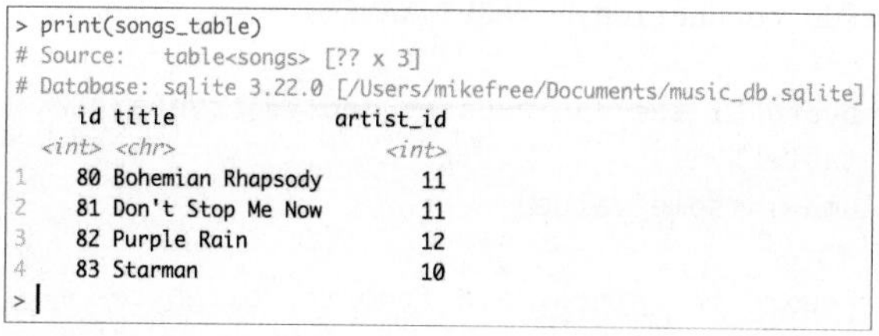

```
> print(songs_table)
# Source:    table<songs> [?? x 3]
# Database: sqlite 3.22.0 [/Users/mikefree/Documents/music_db.sqlite]
      id title            artist_id
   <int> <chr>                <int>
1     80 Bohemian Rhapsody       11
2     81 Don't Stop Me Now       11
3     82 Purple Rain             12
4     83 Starman                 10
> |
```

Figure 13.6   A database tbl, printed in RStudio. This is only a *preview* of the data that will be returned by the data base.

The dbplyr package will automatically convert a sequence of dplyr functions into an equivalent SQL statement, without the need for you to write any SQL! You can see the SQL statement it is generating by using the **show_query( )** function:

```
# Display the SQL syntax stored in the query `queen_songs_query`
show_query(queen_songs_query)
    # <SQL>
    # SELECT `title`
    # FROM `songs`
    # WHERE (`artist_id` = 11.0)
```

Importantly, using dplyr methods on a table does not return a data frame (or even a tibble). In fact, it displays just a small preview of the requested data! Actually querying the data from a database is relatively slow in comparison to accessing data in a data frame, particularly when the database is on a remote computer. Thus dbplyr uses **lazy evaluation**—it actually executes the query on the database only when you explicitly tell it to do so. What is shown when you print the queen_songs_query is just a subset of the data; the results will not include all of the rows returned if there are a large number of them! RStudio very subtly indicates that the data is just a preview of what has been requested—note in Figure 13.6 that the dimensions of the songs_table are unknown (i.e., table<songs> [?? X 3]). Lazy evaluation keeps you from accidentally making a large number of queries and downloading a huge amount of data as you are designing and testing your data manipulation statements (i.e., writing your select( ) and filter( ) calls).

To actually query the database and load the results into memory as a R value you can manipulate, use the **collect( )** function. You can often add this function call as a last step in your pipe of dplyr calls.

```
# Execute the `queen_songs_query` request, returning the *actual data*
# from the database
queen_songs_data <- collect(queen_songs_query) # returns a tibble
```

This tibble is exactly like those described in earlier chapters; you can use as.data.frame( ) to convert it into a data frame. Thus, anytime you want to query data from a database in R, you will need to perform the following steps:

```
# 1. Create a connection to an RDMS, such as a SQLite database
db_connection <- dbConnect(SQLite(), dbname = "path/to/database.sqlite")

# 2. Access a specific table within your database
some_table <- tbl(db_connection, "TABLE_NAME")

# 3. Construct a query of the table using `dplyr` syntax
db_query <- some_table %>%
  filter(some_column == some_value)

# 4. Execute your query to return data from the database
results <- collect(db_query)
```

```
# 5. Disconnect from the database when you're finished
dbDisconnect(db_connection)
```

And with that, you have accessed and queried a database using R! You can now write R code to use the same `dplyr` functions for either a local data frame or a remote database, allowing you to test and then expand your data analysis.

> **Tip:** For more information on using `dbplyr`, check out the introduction vignette.[a]
>
> ----
> [a]https://cran.r-project.org/web/packages/dbplyr/vignettes/dbplyr.html

For practice working with databases, see the set of accompanying book exercises.[10]

----
[10]**Database exercises:** https://github.com/programming-for-data-science/chapter-13-exercises

14

Accessing Web APIs

Previous chapters have described how to access data from local `.csv` files, as well as from local databases. While working with local data is common for many analyses, more complex shared data systems leverage **web services** for data access. Rather than store data on each analyst's computer, data is stored on a *remote server* (i.e., a central computer somewhere on the internet) and accessed similarly to how you access information on the web (via a URL). This allows scripts to always work with the latest data available when performing analysis of data that may be changing rapidly, such as social media data.

In this chapter, you will learn how to use R to programmatically interact with data stored by web services. From an R script, you can read, write, and delete data stored by these services (though this book focuses on the skill of reading data). Web services may make their data accessible to computer programs like R scripts by offering an application programming interface (API). A web service's API specifies *where* and *how* particular data may be accessed, and many web services follow a particular style known as *REpresentational State Transfer (REST)*.[1] This chapter covers how to access and work with data from these RESTful APIs.

14.1 What Is a Web API?

An **interface** is the point at which two different systems meet and *communicate*, exchanging information and instructions. An **application programming interface** (API) thus represents a way of communicating with a computer application by writing a computer program (a set of formal instructions understandable by a machine). APIs commonly take the form of **functions** that can be called to give instructions to programs. For example, the set of functions provided by a package like `dplyr` make up the API for that package.

While some APIs provide an interface for leveraging some *functionality*, other APIs provide an interface for accessing *data*. One of the most common sources of these data APIs are web services—that is, websites that offer an interface for accessing their data.

With web services, the interface (the set of "functions" you can call to access the data) takes the form of **HTTP requests**—that is, requests for data sent following the *HyperText Transfer Protocol*.

---

[1] Fielding, R. T. (2000). *Architectural styles and the design of network-based software architectures.* University of California, Irvine, doctoral dissertation. https://www.ics.uci.edu/~fielding/pubs/dissertation/rest_arch_style.htm. Note that this is the original specification and is very technical.

This is the same protocol (way of communicating) used by your browser to view a webpage! An HTTP request represents a message that your computer sends to a **web server**: another computer on the internet that "serves," or provides, information. That server, upon receiving the request, will determine what data to include in the **response** it sends back to the requesting computer. With a web browser, the response data takes the form of HTML files that the browser can *render* as webpages. With data APIs, the response data will be structured data that you can convert into R data types such as lists or data frames.

In short, loading data from a web API involves sending an HTTP request to a server for a particular piece of data, and then receiving and parsing the response to that request.

Learning how to use web APIs will greatly expand the available data sets you may want to use for analysis. Companies and services with large amounts of data, such as Twitter,[2] iTunes,[3] or Reddit,[4] make (some of) their data publicly accessible through an API. This chapter will use the GitHub API[5] to demonstrate how to work with data stored in a web service.

14.2 RESTful Requests

There are two parts to a request sent to a web API: the name of the resource (data) that you wish to access, and a verb indicating what you want to do with that resource. In a way, the verb is the function you want to call on the API, and the resource is an argument to that function.

14.2.1 URIs

Which resource you want to access is specified with a **Uniform Resource Identifier (URI)**.[6] A URI is a generalization of a URL (Uniform Resource Locator)—what you commonly think of as a "web address." A URI acts a lot like the address on a postal letter sent within a large organization such as a university: you indicate the business address as well as the department and the person to receive the letter, and will get a different response (and different data) from Alice in Accounting than from Sally in Sales.

Like postal letter addresses, URIs have a very specific format used to direct the request to the right resource, illustrated in Figure 14.1.

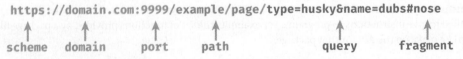

Figure 14.1 The format (schema) of a URI.

---

[2]**Twitter API**: https://developer.twitter.com/en/docs
[3]**iTunes search API**: https://affiliate.itunes.apple.com/resources/documentation/itunes-store-web-service-search-api/
[4]**Reddit API**: https://www.reddit.com/dev/api/
[5]**GitHub API**: https://developer.github.com/v3/
[6]**Uniform Resource Identifier (URI) Generic Syntax** (official technical specification): https://tools.ietf.org/html/rfc3986

Not all parts of the URI are required. For example, you don't necessarily need a `port`, `query`, or `fragment`. Important parts of the URI include:

- `scheme` (`protocol`): The "language" that the computer will use to communicate the request to the API. With web services this is normally `https` (secure HTTP).

- `domain`: The address of the web server to request information from.

- `path`: The identifier of the resource on that web server you wish to access. This may be the name of a file with an extension if you're trying to access a particular file, but with web services it often just looks like a folder path!

- `query`: Extra parameters (arguments) with further details about the resource to access.

The `domain` and `path` usually specify the location of the resource of interest. For example, `www.domain.com/users` might be an *identifier* for a *resource* that serves information about all the users. Web services can also have "subresources" that you can access by adding extra pieces to the path. For example, `www.domain.com/users/layla` might access to the specific resource ("layla") that you are interested in.

With web APIs, the URI is often viewed as being broken up into three parts, as shown in Figure 14.2:

- The **base URI** is the domain that is included on *all* resources. It acts as the "root" for any particular endpoint. For example, the GitHub API has a base URI of `https://api.github.com`. All requests to the GitHub API will have that base.

- An **endpoint** is the location that holds the specific information you want to access. Each API will have many different endpoints at which you can access specific data resources. The GitHub API, for example, has different endpoints for `/users` and `/orgs` so that you can access data about users or organizations, respectively.

  Note that many endpoints support accessing multiple subresources. For example, you can access information about a specific user at the endpoint `/users/:username`. The colon `:` indicates that the subresource name is a *variable*—you can replace that part of the endpoint with whatever string you want. Thus if you were interested in the GitHub user `nbremer`,[7] you would access the `/users/nbremer` endpoint.

  Subresources may have further subresources (which may or may not have variable names). The endpoint `/orgs/:org/repos` refers to the list of repositories belonging to an organization. Variable names in endpoints might alternatively be written inside of curly braces `{}`—for example, `/orgs/{org}/repos`. Neither the colon nor the braces are

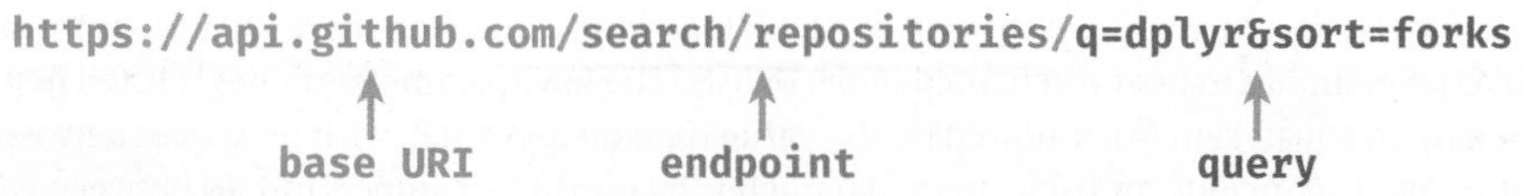

```
https://api.github.com/search/repositories/q=dplyr&sort=forks
                       ↑                    ↑                ↑
                    base URI            endpoint          query
```

Figure 14.2   The anatomy of a web API request URI.

---

[7] **Nadieh Bremer, freelance data visualization designer:** https://www.visualcinnamon.com

programming language syntax; instead, they are common conventions used to communicate how to specify endpoints.

- **Query parameters** allow you to specify additional information about which exact information you want from the endpoint, or how you want it to be organized (see Section 14.2.1.1 for more details).

> **Remember**: One of the biggest challenges in accessing a web API is understanding what resources (data) the web service makes available and which endpoints (URIs) can request those resources. Read the web service's documentation carefully—popular services often include examples of URIs and the data returned from them.

A query is constructed by appending the endpoint and any query parameters to the base URI. For example, so you could access a GitHub user by combining the base URI (`https://api.github.com`) and endpoint (`/users/nbremer`) into a single string: `https://api.github.com/users/nbremer`. Sending a request to that URI will return data about the user—you can send this request from an R program or by visiting that URI in a web browser, as shown in Figure 14.3. In short, you can access a particular data *resource* by sending a request to a particular *endpoint*.

Indeed, one of the easiest ways to make a request to a web API is by navigating to the URI using your web browser. Viewing the information in your browser is a great way to explore the resulting data, and make sure you are requesting information from the proper URI (i.e., that you haven't made a typo in the URI).

> **Tip**: The JSON format (see Section 14.4) of data returned from web APIs can be quite messy when viewed in a web browser. Installing a browser extension such as *JSONView*[a] will format the data in a somewhat more readable way. Figure 14.3 shows data formatted with this extension.
>
> ------
>
> [a] https://chrome.google.com/webstore/detail/jsonview/chklaanhfefbnpoihckbnefhakgolnmc

14.2.1.1   Query Parameters

Web URIs can optionally include **query parameters**, which are used to request a more specific subset of data. You can think of them as additional optional arguments that are given to the request function—for example, a keyword to search for or criteria to order results by.

The query parameters are listed at the end of a URI, following a question mark (**?**) and are formed as *key-value* pairs similar to how you named items in lists. The **key** (parameter name) is listed first, followed by an equals sign (**=**), followed by the **value** (parameter value), with no spaces between anything. You can include multiple query parameters by putting an ampersand (**&**) between each key-value pair. You can see an example of this syntax by looking at the URL bar in a web browser when you use a search engine such as Google or Yahoo, as shown in Figure 14.4. Search engines produce URLs with a lot of query parameters, not all of which are obvious or understandable.

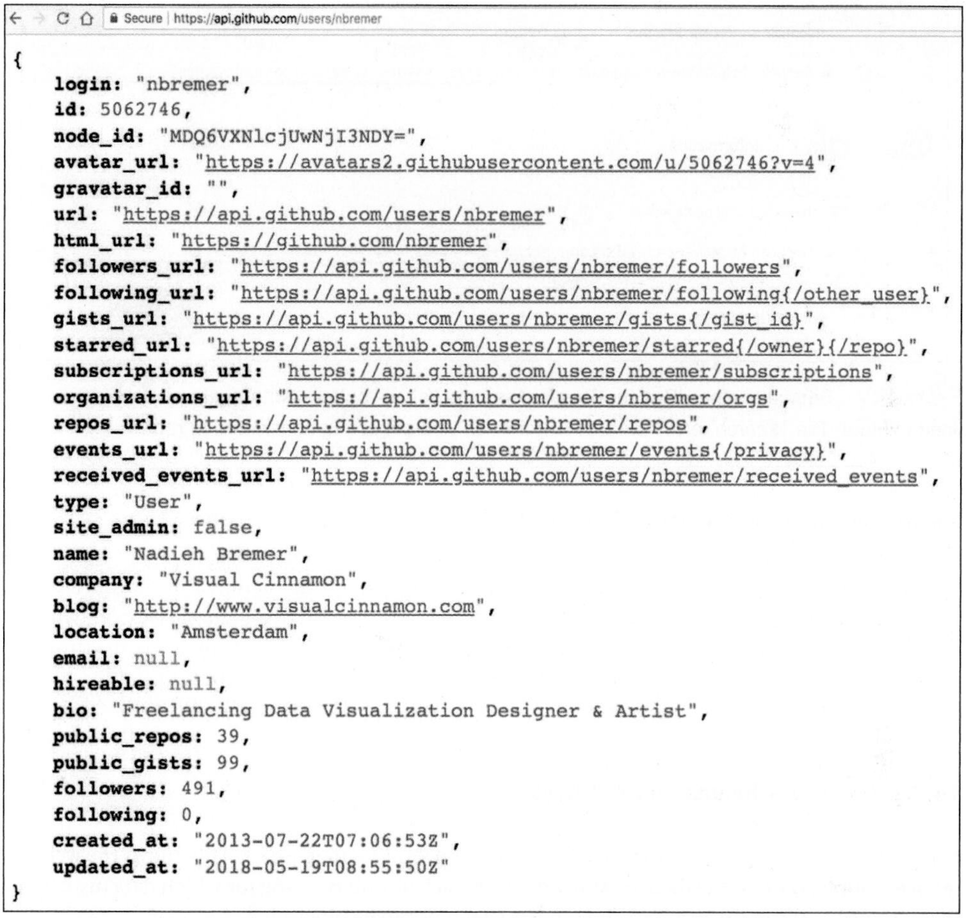


```
{
    login: "nbremer",
    id: 5062746,
    node_id: "MDQ6VXNlcjUwNjI3NDY=",
    avatar_url: "https://avatars2.githubusercontent.com/u/5062746?v=4",
    gravatar_id: "",
    url: "https://api.github.com/users/nbremer",
    html_url: "https://github.com/nbremer",
    followers_url: "https://api.github.com/users/nbremer/followers",
    following_url: "https://api.github.com/users/nbremer/following{/other_user}",
    gists_url: "https://api.github.com/users/nbremer/gists{/gist_id}",
    starred_url: "https://api.github.com/users/nbremer/starred{/owner}{/repo}",
    subscriptions_url: "https://api.github.com/users/nbremer/subscriptions",
    organizations_url: "https://api.github.com/users/nbremer/orgs",
    repos_url: "https://api.github.com/users/nbremer/repos",
    events_url: "https://api.github.com/users/nbremer/events{/privacy}",
    received_events_url: "https://api.github.com/users/nbremer/received_events",
    type: "User",
    site_admin: false,
    name: "Nadieh Bremer",
    company: "Visual Cinnamon",
    blog: "http://www.visualcinnamon.com",
    location: "Amsterdam",
    email: null,
    hireable: null,
    bio: "Freelancing Data Visualization Designer & Artist",
    public_repos: 39,
    public_gists: 99,
    followers: 491,
    following: 0,
    created_at: "2013-07-22T07:06:53Z",
    updated_at: "2018-05-19T08:55:50Z"
}
```


Figure 14.3 GitHub API response returned by the URI `https://api.github.com/users/nbremer`, as displayed in a web browser.

Notice that the exact query parameter name used differs depending on the web service. Google uses a q parameter (likely for "query") to store the search term, while Yahoo uses a p parameter.

Similar to arguments for functions, API endpoints may either require query parameters (e.g., you *must* provide a search term) or optionally allow them (e.g., you *may* provide a sorting order). For example, the GitHub API has a `/search/repositories` endpoint that allows users to *search* for a specific repository: you are required to provide a q parameter for the query, and can optionally provide a `sort` parameter for how to sort the results:

```
# A GitHub API URI with query parameters: search term `q` and sort
# order `sort`
https://api.github.com/search/repositories?q=dplyr&sort=forks
```

Figure 14.4   Search engine URLs for Google (top) and Yahoo (bottom) with query parameters (underlined in blue). The "search term" parameter for each web service is underlined in red.

Results from this request are shown in Figure 14.5.

> **Caution**: Many special characters (e.g., punctuation) cannot be included in a URL. This group includes characters such as spaces! Browsers and many HTTP request packages will automatically *encode* these special characters into a usable format (for example, converting a space into a %20), but sometimes you may need to do this conversion yourself.

14.2.1.2   Access Tokens and API Keys

Many web services require you to register with them to send them requests. This allows the web service to limit access to the data, as well as to keep track of who is asking for which data (usually so that if someone starts "spamming" the service, that user can be blocked).

To facilitate this tracking, many services provide users with **access tokens** (also called **API keys**). These unique strings of letters and numbers identify a particular developer (like a secret password that works just for you). Furthermore, your API key can provide you with additional access to information based on which user you are. For example, when you get an access key for the GitHub API, that key will provide you with additional access and control over your repositories. This enables you to request information about private repos, and even programmatically interact with GitHub through the API (i.e., you can delete a repo[8]—so tread carefully!).

Web services will require you to include your access token in the request, usually as a query parameter; the exact name of the parameter varies, but it often looks like access_token or api_key. When exploring a web service, keep an eye out for whether it requires such tokens.

---

[8]**GitHub API, delete a repository** https://developer.github.com/v3/repos/#delete-a- repository

```
← → C ⌂  🔒 Secure  https://api.github.com/search/repositories?q=dplyr&sort=forks

{
    total_count: 620,
    incomplete_results: false,
  - items: [
      - {
            id: 6427813,
            name: "dplyr",
            full_name: "tidyverse/dplyr",
          - owner: {
                login: "tidyverse",
                id: 22032646,
                avatar_url: "https://avatars2.githubusercontent.com/u/22032646?v=4",
                gravatar_id: "",
                url: "https://api.github.com/users/tidyverse",
                html_url: "https://github.com/tidyverse",
                followers_url: "https://api.github.com/users/tidyverse/followers",
                following_url: "https://api.github.com/users/tidyverse/following{/other_user}",
                gists_url: "https://api.github.com/users/tidyverse/gists{/gist_id}",
                starred_url: "https://api.github.com/users/tidyverse/starred{/owner}{/repo}",
                subscriptions_url: "https://api.github.com/users/tidyverse/subscriptions",
                organizations_url: "https://api.github.com/users/tidyverse/orgs",
                repos_url: "https://api.github.com/users/tidyverse/repos",
                events_url: "https://api.github.com/users/tidyverse/events{/privacy}",
                received_events_url: "https://api.github.com/users/tidyverse/received_events",
                type: "Organization",
                site_admin: false
            },
            private: false,
            html_url: "https://github.com/tidyverse/dplyr",
            description: "Dplyr: A grammar of data manipulation",
            fork: false,
            url: "https://api.github.com/repos/tidyverse/dplyr",
            forks_url: "https://api.github.com/repos/tidyverse/dplyr/forks",
            keys_url: "https://api.github.com/repos/tidyverse/dplyr/keys{/key_id}",
            collaborators_url: "https://api.github.com/repos/tidyverse/dplyr/collaborators{/collaborator}",
            teams_url: "https://api.github.com/repos/tidyverse/dplyr/teams",
            hooks_url: "https://api.github.com/repos/tidyverse/dplyr/hooks",
            issue_events_url: "https://api.github.com/repos/tidyverse/dplyr/issues/events{/number}",
            events_url: "https://api.github.com/repos/tidyverse/dplyr/events",
            assignees_url: "https://api.github.com/repos/tidyverse/dplyr/assignees{/user}",
            branches_url: "https://api.github.com/repos/tidyverse/dplyr/branches{/branch}",
            tags_url: "https://api.github.com/repos/tidyverse/dplyr/tags",
```

Figure 14.5   A subset of the GitHub API response returned by the URI `https://api.github.com/search/repositories?q=dplyr&sort=forks`, as displayed in a web browser.

Caution: Watch out for APIs that mention using an authentication service called **OAuth** when explaining required API keys. OAuth is a system for performing **authentication**—that is, having someone *prove* that they are who they say they are. OAuth is generally used to let someone log into a website from your application (like what a "Log in with Google" button does). OAuth systems require more than one access key, and these keys *must* be kept secret. Moreover, they usually require you to run a web server to use them correctly (which requires significant extra setup; see the full `httr` documentation[a] for details). You can do this in R, but may want to avoid this challenge while learning how to use APIs.

[a] https://cran.r-project.org/web/packages/httr/httr.pdf

Access tokens are a lot like passwords; you will want to keep them secret and not share them with others. This means that you should not include them in any files you commit to git and push to GitHub. The best way to ensure the secrecy of access tokens in R is to create a separate script file in your repo (e.g., api_keys.R) that includes exactly one line, assigning the key to a variable:

```
# Store your API key from a web service in a variable
# It should be in a separate file (e.g., `api_keys.R`)
api_key <- "123456789abcdefg"
```

To access this variable in your "main" script, you can use the **source( )** function to load and run your api_keys.R script (similar to clicking the *Source* button to run a script). This function will execute all lines of code in the specified script file, as if you had "copy-and-pasted" its contents and run them all with ctrl+enter. When you use source( ) to execute the api_keys.R script, it will execute the code statement that defines the api_key variable, making it available in your environment for your use:

```
# In your "main" script, load your API key from another file

# (Make sure working directory is set before running the following code!)

source("api_keys.R") # load the script using a *relative path*
print(api_key) # the key is now available!
```

Anyone else who runs the script will need to provide an api_key variable to access the API using that user's own key. This practice keeps everyone's account separate.

You can keep your api_keys.R file from being committed by including the filename in the **.gitignore** file in your repo; that will keep it from even possibly being committed with your code! See Chapter 3 for details about working with the .gitignore file.

14.2.2 HTTP Verbs

When you send a request to a particular resource, you need to indicate what you want to *do* with that resource. This is achieved by specifying an **HTTP verb** in the request. The HTTP protocol supports the following verbs:

- GET: Return a representation of the current state of the resource.
- POST: Add a new subresource (e.g., insert a record).
- PUT: Update the resource to have a new state.
- PATCH: Update a portion of the resource's state.
- DELETE: Remove the resource.
- OPTIONS: Return the set of methods that can be performed on the resource.

By far the most commonly used verb is **GET**, which is used to "get" (download) data from a web service—this is the type of request that is sent when you enter a URL into a web browser. Thus you would send a GET request for the `/users/nbremer` endpoint to access that data resource.

Taken together, this structure of treating each datum on the web as a resource that you can interact with via HTTP requests is referred to as the **REST architecture** (*REpresentational State Transfer*). Thus, a web service that enables data access through named resources and responds to HTTP requests is known as a **RESTful** service, that has a RESTful API.

14.3 Accessing Web APIs from R

To access a web API, you just need to send an HTTP request to a particular URI. As mentioned earlier, you can easily do this with the browser: navigate to a particular address (base URI + endpoint), and that will cause the browser to send a GET request and display the resulting data. For example, you can send a request to the GitHub API to search for repositories that match the string "dplyr" (see the response in Figure 14.5):

```
# The URI for the `search/repositories` endpoint of the GitHub API: query
# for `dplyr`, sorting by `forks`
https://api.github.com/search/repositories?q=dplyr&sort=forks
```

This query accesses the `/search/repositories` endpoint, and also specifies two query parameters:

- q: The term(s) you are searching for
- sort: The attribute of each repository that you would like to use to sort the results (in this case, the number of forks of the repo)

(Note that the data you will get back is structured in JSON format. See Section 14.4 for details.)

While you can access this information using your browser, you will want to load it into R for analysis. In R, you can send GET requests using the **httr**[9] package. As with dplyr, you will need to install and load this package to use it:

```
install.packages("httr") # once per machine
library("httr")          # in each relevant script
```

This package provides a number of functions that reflect HTTP verbs. For example, the **GET()** function will send an HTTP GET request to the URI:

```
# Make a GET request to the GitHub API's "/search/repositories" endpoint
# Request repositories that match the search "dplyr", and sort the results
# by forks
url <- "https://api.github.com/search/repositories?q=dplyr&sort=forks"
response <- GET(url)
```

This code will make the same request as your web browser, and store the response in a variable called `response`. While it is possible to include query parameters in the URI string (as above), httr

---

[9]**Getting started with httr**: official quickstart guide for httr: https://cran.r-project.org/web/packages/httr/vignettes/quickstart.html

also allows you to include them as a list passed as a `query` argument. Furthermore, if you plan on accessing multiple different endpoints (which is common), you can structure your code a bit more modularly, as described in the following example; this structure makes it easy to set and change variables (instead of needing to do a complex `paste()` operation to produce the correct string):

```
# Restructure the previous request to make it easier to read and update. DO THIS.

# Make a GET request to the GitHub API's "search/repositories" endpoint
# Request repositories that match the search "dplyr", sorted by forks

# Construct your `resource_uri` from a reusable  `base_uri` and an `endpoint`
base_uri <- "https://api.github.com"
endpoint <- "/search/repositories"
resource_uri <- paste0(base_uri, endpoint)

# Store any query parameters you want to use in a list
query_params <- list(q = "dplyr", sort = "forks")

# Make your request, specifying the query parameters via the `query` argument
response <- GET(resource_uri, query = query_params)
```

If you try printing out the `response` variable that is returned by the GET() function, you will first see information about the response:

```
Response [https://api.github.com/search/repositories?q=dplyr&sort=forks]
  Date: 2018-03-14 06:43
  Status: 200
  Content-Type: application/json; charset=utf-8
  Size: 171 kB
```

This is called the **response header**. Each response has two parts: the **header** and the **body**. You can think of the response as an envelope: the header contains meta-data like the address and postage date, while the body contains the actual contents of the letter (the data).

> **Tip:** The URI shown when you print out the `response` variable is a good way to check exactly which URI you sent the request to: copy that into your browser to make sure it goes where you expected!

Since you are almost always interested in working with the response body, you will need to extract that data from the response (e.g., open up the envelope and pull out the letter). You can do this with the `content()` function:

```
# Extract content from `response`, as a text string
response_text <- content(response, type = "text")
```

Note the second argument `type = "text"`; this is needed to keep `httr` from doing its own processing on the response data (you will use other methods to handle that processing).

14.4   Processing JSON Data

Now that you're able to load data into R from an API and extract the content as text, you will need to transform the information into a usable format. Most APIs will return data in **JavaScript Object Notation (JSON)** format. Like CSV, JSON is a format for writing down structured data—but, while .csv files organize data into rows and columns (like a data frame), JSON allows you to organize elements into key–value pairs similar to an R *list*! This allows the data to have much more complex structure, which is useful for web services, but can be challenging for data programming.

In JSON, lists of key–value pairs (called **objects**) are put inside braces (**{ }**), with the key and the value separated by a colon (**:**) and each pair separated by a comma (**,**). Key–value pairs are often written on separate lines for readability, but this isn't required. Note that keys need to be character strings (so, *"in quotes"*), while values can either be character strings, numbers, booleans (written in lowercase as true and false), or even other lists! For example:

```
{
  "first_name": "Ada",
  "job": "Programmer",
  "salary": 78000,
  "in_union": true,
  "favorites": {
    "music": "jazz",
    "food": "pizza",
  }
}
```

The above JSON object is equivalent to the following R list:

```
# Represent the sample JSON data (info about a person) as a list in R
list(
  first_name = "Ada",
  job = "Programmer",
  salary = 78000,
  in_union = TRUE,
  favorites = list(music = "jazz", food = "pizza") # nested list in the list!
)
```

Additionally, JSON supports **arrays** of data. Arrays are like *untagged lists* (or vectors with different types), and are written in square brackets (**[ ]**), with values separated by commas. For example:

```
["Aardvark", "Baboon", "Camel"]
```

which is equivalent to the R list:

```
list("Aardvark", "Baboon", "Camel")
```

Just as R allows you to have nested lists of lists, JSON can have any form of nested objects and arrays. This structure allows you to store arrays (think *vectors*) within objects (think *lists*), such as the following (more complex) set of data about Ada:

```
{
  "first_name": "Ada",
  "job": "Programmer",
  "pets": ["Magnet", "Mocha", "Anni", "Fifi"],
  "favorites": {
    "music": "jazz",
    "food": "pizza",
    "colors": ["green", "blue"]
  }
}
```

The JSON equivalent of a data frame is to store data as an *array of objects*. This is like having a list of lists. For example, the following is an array of objects of FIFA Men's World Cup data[10]:

```
[
  {"country": "Brazil", "titles": 5, "total_wins": 70, "total_losses": 17},
  {"country": "Italy", "titles": 4, "total_wins": 66, "total_losses": 20},
  {"country": "Germany", "titles": 4, "total_wins": 45, "total_losses": 17},
  {"country": "Argentina", "titles": 2, "total_wins": 42, "total_losses": 21},
  {"country": "Uruguay", "titles": 2, "total_wins": 20, "total_losses": 19}
]
```

You could think of this information as a *list of lists* in R:

```
# Represent the sample JSON data (World Cup data) as a list of lists in R
list(
  list(country = "Brazil", titles = 5, total_wins = 70, total_losses = 17),
  list(country = "Italy", titles = 4, total_wins = 66, total_losses = 20),
  list(country = "Germany", titles = 4, total_wins = 45, total_losses = 17),
  list(country = "Argentina", titles = 2, total_wins = 42, total_losses = 21),
  list(country = "Uruguay", titles = 2, total_wins = 20, total_losses = 19)
)
```

This structure is incredibly common in web API data: as long as each object in the array has the same set of keys, then you can easily consider this structure to be a data frame where each object (list) represents an observation (row), and each key represents a feature (column) of that observation. A data frame representation of this data is shown in Figure 14.6.

> **Remember**: In JSON, tables are represented as lists of *rows*, instead of a data frame's list of *columns*.

---

[10] **FIFA World Cup data:** https://www.fifa.com/fifa-tournaments/statistics-and-records/worldcup/teams/index.html

	country	titles	total_wins	total_losses
1	Brazil	5	70	17
2	Italy	4	66	20
3	Germany	4	45	17
4	Argentina	2	42	21
5	Uruguay	2	20	19

```
 1 · [
 2 ·     {
 3             "country": "Brazil",
 4             "titles": 5,
 5             "total_wins": 70,
 6             "total_losses": 17
 7         },
 8 ·     {
 9             "country": "Italy",
10             "titles": 4,
11             "total_wins": 66,
12             "total_losses": 20
13         },
14 ·     {
15             "country": "Germany",
16             "titles": 4,
17             "total_wins": 45,
18             "total_losses": 17
19         },
20 ·     {
21             "country": "Argentina",
22             "titles": 2,
23             "total_wins": 42,
24             "total_losses": 21
25         },
26 ·     {
27             "country": "Uruguay",
28             "titles": 2,
29             "total_wins": 20,
30             "total_losses": 19
31         }
32 ]
```

Figure 14.6   A data frame representation of World Cup statistics (left), which can also be represented as JSON data (right).

14.4.1   Parsing JSON

When working with a web API, the usual goal is to take the JSON data contained in the response and convert it into an R data structure you can use, such as a list or data frame. This will allow you to interact with the data by using the data manipulation skills introduced in earlier chapters. While the httr package is able to parse the JSON body of a response into a list, it doesn't do a very clean job of it (particularly for complex data structures).

A more effective solution for transforming JSON data is to use the jsonlite package.[11] This package provides helpful methods to convert JSON data into R data, and is particularly well suited for converting content into data frames.

As always, you will need to install and load this package:

```
install.packages("jsonlite") # once per machine
library("jsonlite") # in each relevant script
```

The jsonlite package provides a function called **fromJSON()** that allows you to convert from a JSON string into a list—or even a data frame if the intended columns have the same lengths!

---

[11]**Package jsonlite**: full documentation for jsonlite: https://cran.r-project.org/web/packages/jsonlite/jsonlite.pdf

```
# Make a request to a given `uri` with a set of `query_params`
# Then extract and parse the results

# Make the request
response <- GET(uri, query = query_params)

# Extract the content of the response
response_text <- content(response, "text")

# Convert the JSON string to a list
response_data <- fromJSON(response_text)
```

Both the raw JSON data (response_text) and the parsed data structure (response_data) are shown in Figure 14.7. As you can see, the raw string (response_text) is indecipherable. However, once it is transformed using the fromJSON() function, it has a much more operable structure.

The response_data will contain a list built out of the JSON. Depending on the complexity of the JSON, this may already be a data frame you can View()—but more likely you will need to explore the list to locate the "main" data you are interested in. Good strategies for this include the following techniques:

- Use functions such as is.data.frame() to determine whether the data is already structured as a data frame.

- You can print() the data, but that is often hard to read (it requires a lot of scrolling).

- The str() function will return a list's structure, though it can still be hard to read.

- The names() function will return the keys of the list, which is helpful for delving into the data.

```
> response_text
[1] "{\n  \"total_count\": 707,\n  \"incomplete_results\":
false,\n  \"items\": [\n    {\n      \"id\": 6427813,\n
\"node_id\": \"MDEwOlJlcG9zaXRvcnk2NDI3ODEz\",\n         \"na
me\": \"dplyr\",\n        \"full_name\": \"tidyverse/dplyr\"
,\n      \"owner\": {\n           \"login\": \"tidyverse\",\n
\"id\": 22032646,\n        \"node_id\": \"MDEyOk9yZ2FuaXph
dGlvbjIyMDMyNjQ2\",\n      \"avatar_url\": \"https://ava
tars2.githubusercontent.com/u/22032646?v=4\",\n           \"g
ravatar_id\": \"\",\n          \"url\": \"https://api.github
.com/users/tidyverse\",\n        \"html_url\": \"https://g
ithub.com/tidyverse\",\n        \"followers_url\": \"https
://api.github.com/users/tidyverse/followers\",\n        \"
following_url\": \"https://api.github.com/users/tidyverse/
following{/other_user}\",\n        \"gists_url\": \"https:
//api.github.com/users/tidyverse/gists{/gist_id}\",\n
```

```
> fromJSON(response_text)
$total_count
[1] 707

$incomplete_results
[1] FALSE

$items
        id                            node_id
1  6427813        MDEwOlJlcG9zaXRvcnk2NDI3ODEz
2 67845042 MDEwOlJlcG9zaXRvcnk2Nzg0NTA0Mg==
3 59305491 MDEwOlJlcG9zaXRvcnk1OTMwNTQ5MQ==
4 24485567 MDEwOlJlcG9zaXRvcnkyNDQ4NTU2Nw==
5 126367748 MDEwOlJlcG9zaXRvcnkxMjYzNjc3NDg=
6 55175084 MDEwOlJlcG9zaXRvcnk1NTE3NTA4NA==
7 118410287 MDEwOlJlcG9zaXRvcnkxMTg0MTAyODc=
```

Figure 14.7 Parsing the text of an API response using fromJSON(). The untransformed text is shown on the left (response_text), which is transformed into a list (on the right) using the fromJSON() function.

As an example continuing the previous code:

```
# Use various methods to explore and extract information from API results

# Check: is it a data frame already?
is.data.frame(response_data) # FALSE

# Inspect the data!
str(response_data) # view as a formatted string
names(response_data) # "href" "items" "limit" "next" "offset" "previous" "total"

# Looking at the JSON data itself (e.g., in the browser),
# `items` is the key that contains the value you want

# Extract the (useful) data
items <- response_data$items # extract from the list
is.data.frame(items) # TRUE; you can work with that!
```

The set of responses—GitHub repositories that match the search term *"dplry"*—returned from the request and stored in the `response_data$items` key is shown in Figure 14.8.

14.4.2   Flattening Data

Because JSON supports—and in fact encourages—nested lists (lists within lists), parsing a JSON string is likely to produce a data frame whose columns *are themselves data frames*. As an example of what a nested data frame may look like, consider the following code:

```
# A demonstration of the structure of "nested" data frames

# Create a `people` data frame with a `names` column
people <- data.frame(names = c("Ed", "Jessica", "Keagan"))
```

	id	node_id	name	full_name
1	6427813	MDEwOlJlcG9zaXRvcnk2NDI3ODEz	dplyr	tidyverse/dplyr
2	67845042	MDEwOlJlcG9zaXRvcnk2Nzg0NTA0Mg==	m9-dplyr	INFO-201/m9-dplyr
3	59305491	MDEwOlJlcG9zaXRvcnk1OTMwNTQ5MQ==	sparklyr	rstudio/sparklyr
4	24485567	MDEwOlJlcG9zaXRvcnkyNDQ4NTU2Nw==	dplyr-tutorial	justmarkham/dplyr-tutorial
5	126367748	MDEwOlJlcG9zaXRvcnkxMjYzNjc3NDg=	ch10-dplyr	info201/ch10-dplyr
6	55175084	MDEwOlJlcG9zaXRvcnk1NTE3NTA4NA==	tidytext	juliasilge/tidytext
7	118410287	MDEwOlJlcG9zaXRvcnkxMTg0MTAyODc=	ch10-dplyr	info201a-w18/ch10-dplyr
8	50487685	MDEwOlJlcG9zaXRvcnk1MDQ4NzY4NQ==	lecture-8-exercises	INFO-498F/lecture-8-exercises
9	86504302	MDEwOlJlcG9zaXRvcnk4NjUwNDMwMg==	dbplyr	tidyverse/dbplyr
10	84520584	MDEwOlJlcG9zaXRvcnk4NDUyMDU4NA==	Data-Analysis-with-R	susanli2016/Data-Analysis-with-R

Figure 14.8   Data returned by the GitHub API: repositories that match the term *"dplyr"* (stored in the variable `response_data$items` in the code example).

```
# Create a data frame of favorites with two columns
favorites <- data.frame(
  food = c("Pizza", "Pasta", "Salad"),
  music = c("Bluegrass", "Indie", "Electronic")
)

# Store the second data frame as a column of the first -- A BAD IDEA
people$favorites <- favorites # the `favorites` column is a data frame!

# This prints nicely, but is misleading
print(people)
  #      names favorites.food favorites.music
  # 1       Ed          Pizza       Bluegrass
  # 2  Jessica          Pasta           Indie
  # 3   Keagan          Salad      Electronic

# Despite what RStudio prints, there is not actually a column `favorites.food`
people$favorites.food # NULL

# Access the `food` column of the data frame stored in `people$favorites`
people$favorites$food # [1] Pizza Pasta Salad
```

Nested data frames make it hard to work with the data using previously established techniques and syntax. Luckily, the jsonlite package provides a helpful function for addressing this issue, called **flatten()**. This function takes the columns of each nested data frame and converts them into appropriately named columns in the "outer" data frame, as shown in Figure 14.9:

```
# Use `flatten()` to format nested data frames
people <- flatten(people)
people$favorites.food # this just got created! Woo!
```

Note that flatten() works on only values that are already data frames. Thus you may need to find the appropriate element inside of the list—that is, the element that is the data frame you want to flatten.

In practice, you will almost always want to flatten the data returned from a web API. Thus, your algorithm for requesting and parsing data from an API is this:

1. Use GET() to *request the data* from an API, specifying the URI (and any query parameters).

2. Use content() to *extract the data* from your response as a JSON string (as "text").

3. Use fromJSON() to *convert the data* from a JSON string into a list.

4. Explore the returned information to *find your data* of interest.

5. Use flatten() to *flatten your data* into a properly structured data frame.

6. Programmatically analyze your data frame in R (e.g., with dplyr).

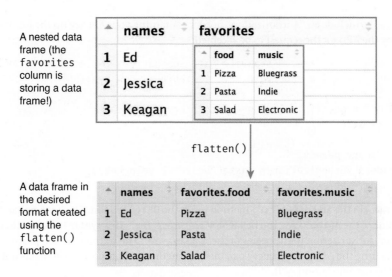

A nested data frame (the `favorites` column is storing a data frame!)

A data frame in the desired format created using the `flatten()` function

Figure 14.9 The `flatten()` function transforming a nested data frame (top) into a usable format (bottom).

14.5 APIs in Action: Finding Cuban Food in Seattle

This section uses the Yelp Fusion API[12] to answer the question:

> *"Where is the best Cuban food in Seattle?"*

Given the geographic nature of this question, this section builds a map of the best-rated Cuban restaurants in Seattle, as shown in Figure 14.12. The complete code for this analysis is also available online in the book's code repository.[13]

To send requests to the Yelp Fusion API, you will need to acquire an API key. You can do this by signing up for an account on the API's website, and registering an application (it is common for APIs to require you to register for access). As described earlier, you should store your API key in a separate file so that it can be kept secret:

```
# Store your API key in a variable: to be done in a separate file
# (i.e., "api_key.R")
yelp_key <- "abcdef123456"
```

This API requires you to use an alternative syntax for specifying your API key in the HTTP request—instead of passing your key as a query parameter, you'll need to add a header to the request that you make to the API. An **HTTP header** provides additional information to the server about *who is sending the request*—it's like extra information on the request's envelope. Specifically,

---

[12] **Yelp Fusion API documentation**: https://www.yelp.com/developers/documentation/v3
[13] **APIs in Action**: https://github.com/programming-for-data-science/in-action/tree/master/apis

you will need to include an "Authorization" header containing your API key (in the format expected by the API) for the request to be accepted:

```
# Load your API key from a separate file so that you can access the API:
source("api_key.R") # the `yelp_key` variable is now available

# Make a GET request, including your API key as a header
response <- GET(
  uri,
  query = query_params,
  add_headers(Authorization = paste("bearer", yelp_key))
)
```

This code invokes the add_headers() method *inside* the GET() request. The header that it adds sets the value of the Authorization header to *"bearer yelp_key"*. This syntax indicates that the API should grant authorization to the bearer of the API key (you). This authentication process is used instead of setting the API key as a query parameter (a method of authentication that is not supported by the Yelp Fusion API).

As with any other API, you can determine the URI to send the request to by reading through the documentation. Given the prompt of *searching* for Cuban restaurants in Seattle, you should focus on the *Business Search* documentation,[14] a section of which is shown in Figure 14.10.

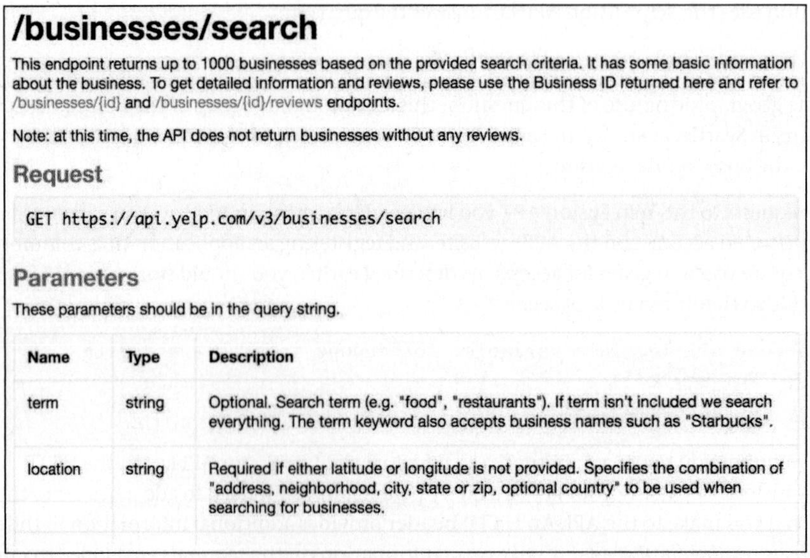

/businesses/search

This endpoint returns up to 1000 businesses based on the provided search criteria. It has some basic information about the business. To get detailed information and reviews, please use the Business ID returned here and refer to /businesses/{id} and /businesses/{id}/reviews endpoints.

Note: at this time, the API does not return businesses without any reviews.

Request

```
GET https://api.yelp.com/v3/businesses/search
```

Parameters
These parameters should be in the query string.

Name	Type	Description
term	string	Optional. Search term (e.g. "food", "restaurants"). If term isn't included we search everything. The term keyword also accepts business names such as "Starbucks".
location	string	Required if either latitude or longitude is not provided. Specifies the combination of "address, neighborhood, city, state or zip, optional country" to be used when searching for businesses.

Figure 14.10   A subset of the Yelp Fusion API *Business Search* documentation.

---

[14] **Yelp Fusion API Business Search** endpoint documentation: https://www.yelp.com/developers/documentation/v3/business_search

As you read through the documentation, it is important to identify the query parameters that you need to specify in your request. In doing so, you are mapping from your question of interest to the specific R code you will need to write. For this question ("Where is the best Cuban food in Seattle?"), you need to figure out how to make the following specifications:

- **Food**: Rather than search all businesses, you need to search for only restaurants. The API makes this available through the `term` parameter.

- **Cuban**: The restaurants you are interested in must be of a certain type. To support this, you can specify the `category` of your search (making sure to specify a supported category, as described elsewhere in the documentation[15]).

- **Seattle**: The restaurant you are looking for must be in Seattle. There are a few ways of specifying a location, the most general of which is to use the `location` parameter. You can further limit your results using the `radius` parameter.

- **Best**: To find the best food, you can control how the results are sorted with the `sort_by` parameter. You'll want to sort the results before you receive them (that is, by using an API parameter and not `dplyr`) to save you some effort and to make sure the API sends only the data you care about.

Often the most time-consuming part of using an API is figuring out how to hone in on your data of interest using the parameters of the API. Once you understand how to control which resource (data) is returned, you can then construct and send an HTTP request to the API:

```r
# Construct a search query for the Yelp Fusion API's Business Search endpoint
base_uri <- "https://api.yelp.com/v3"
endpoint <- "/businesses/search"
search_uri <- paste0(base_uri, endpoint)

# Store a list of query parameters for Cuban restaurants around Seattle
query_params <- list(
  term = "restaurant",
  categories = "cuban",
  location = "Seattle, WA",
  sort_by = "rating",
  radius = 8000 # measured in meters, as detailed in the documentation
)

# Make a GET request, including the API key (as a header) and the list of
# query parameters
response <- GET(
  search_uri,
  query = query_params,
  add_headers(Authorization = paste("bearer", yelp_key))
)
```

---

[15]**Yelp Fusion API Category List:** https://www.yelp.com/developers/documentation/v3/all_category_list

As with any other API response, you will need to use the `content()` method to extract the content from the response, and then format the result using the `fromJSON()` method. You will then need to find the data frame of interest in your response. A great way to start is to use the `names()` function on your result to see what data is available (in this case, you should notice that the `businesses` key stores the desired information). You can `flatten()` this item into a data frame for easy access.

```
# Parse results and isolate data of interest
response_text <- content(response, type = "text")
response_data <- fromJSON(response_text)

# Inspect the response data
names(response_data) # [1] "businesses" "total" "region"

# Flatten the data frame stored in the `businesses` key of the response
restaurants <- flatten(response_data$businesses)
```

The data frame returned by the API is shown in Figure 14.11.

Because the data was requested in sorted format, you can *mutate* the data frame to include a column with the rank number, as well as add a column with a string representation of the name and rank:

```
# Modify the data frame for analysis and presentation
# Generate a rank of each restaurant based on row number
restaurants <- restaurants %>%
  mutate(rank = row_number()) %>%
  mutate(name_and_rank = paste0(rank, ". ", name))
```

The final step is to create a map of the results. The following code uses two different visualization packages (namely, ggmap and `ggplot2`), both of which are explained in more detail in Chapter 16.

	id	alias	name	image_url	is_closed	url
1	Wk9f5Zpnu4T6Vzf6CF5iuA	paseo–caribbean–food–fremont–seattle–2	Paseo Caribbean Food – Fremont	https://s3-media3....	FALSE	https://www.yelp.com/biz...
2	Gn5erxCRML47GgbGYdxzFA	bongos-seattle	Bongos	https://s3-media2....	FALSE	https://www.yelp.com/biz...
3	sjq3–ILJ–QYoHNejt62mYw	geos-cuban-and-creole-cafe-seattle	Geo's Cuban & Creole Cafe	https://s3-media4....	FALSE	https://www.yelp.com/biz...
4	G4j9EqGHRg2TdQVD3wE8EA	el-diablo-coffee-seattle-2	El Diablo Coffee	https://s3-media1....	FALSE	https://www.yelp.com/biz...
5	OJlYzcWkdrHhDyzwji3blQ	mojito-seattle	Mojito	https://s3-media2....	FALSE	https://www.yelp.com/biz...
6	XXO8vKCSqB0cz0rVTg18Jg	un-bien-seattle-seattle	Un Bien – Seattle	https://s3-media2....	FALSE	https://www.yelp.com/biz...
7	o28J-GAetKKTJ8yqz4YI_Q	snout-and-co-seattle-2	Snout & Co.	https://s3-media3....	FALSE	https://www.yelp.com/biz...
8	ZHErhyY2p1xd7vcuTXvbwA	cafe-con-leche-seattle	Cafe Con Leche	https://s3-media2....	FALSE	https://www.yelp.com/biz...
9	rGWsX_7SDtgPYXF6subj3w	paseo-caribbean-food-seattle-8	Paseo Caribbean Food	https://s3-media1....	FALSE	https://www.yelp.com/biz...

Figure 14.11 A subset of the data returned by a request to the Yelp Fusion API for Cuban food in Seattle.

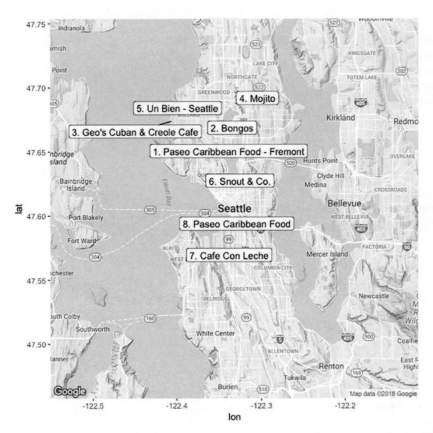

Figure 14.12   A map of the best Cuban restaurants in Seattle, according to the Yelp Fusion API.

```
# Create a base layer for the map (Google Maps image of Seattle)
base_map <- ggmap(get_map(location = "Seattle, WA", zoom = 11))

# Add labels to the map based on the coordinates in the data
base_map +
  geom_label_repel(
    data = response_data,
    aes(x = coordinates.longitude, y = coordinates.latitude, label = name_and_rank)
  )
```

Below is the full script that runs the analysis and creates the map—only 52 lines of clearly commented code to figure out where to go to dinner!

```r
# Yelp API: Where is the best Cuban food in Seattle?
library("httr")
library("jsonlite")
library("dplyr")
library("ggrepel")
library("ggmap")

# Load API key (stored in another file)
source("api_key.R")

# Construct your search query
base_uri <- "https://api.yelp.com/v3/"
endpoint <- "businesses/search"
uri <- paste0(base_uri, endpoint)

# Store a list of query parameters
query_params <- list(
  term = "restaurant",
  categories = "cuban",
  location = "Seattle, WA",
  sort_by = "rating",
  radius = 8000
)

# Make a GET request, including your API key as a header
response <- GET(
  uri,
  query = query_params,
  add_headers(Authorization = paste("bearer", yelp_key))
)

# Parse results and isolate data of interest
response_text <- content(response, type = "text")
response_data <- fromJSON(response_text)

# Save the data frame of interest
restaurants <- flatten(response_data$businesses)

# Modify the data frame for analysis and presentation
restaurants <- restaurants %>%
  mutate(rank = row_number()) %>%
  mutate(name_and_rank = paste0(rank, ". ", name))

# Create a base layer for the map (Google Maps image of Seattle)
base_map <- ggmap(get_map(location = "Seattle, WA", zoom = 11))
```

```
# Add labels to the map based on the coordinates in the data
base_map +
  geom_label_repel(
    data = restaurants,
    aes(x = coordinates.longitude, y = coordinates.latitude, label = name_and_rank)
  )
```

Using this approach, you can use R to load and format data from web APIs, enabling you to analyze and work with a wider variety of data. For practice working with APIs, see the set of accompanying book exercises.[16]

---

[16] **API exercises:** https://github.com/programming-for-data-science/chapter-14-exercises

V

Data Visualization

This section of the book covers the conceptual (design) and technical (programming) skills necessary to construct meaningful visualizations. It provides the necessary visualization theory (Chapter 15) to identify optimal layouts for your data, and includes in-depth descriptions of the most popular visualization packages in R (Chapter 16 and Chapter 17).

15

Designing Data Visualizations

Data visualization, when done well, allows you to reveal patterns in your data and communicate insights to your audience. This chapter describes the conceptual and design skills necessary to craft *effective* and *expressive* visual representations of your data. In doing so, it introduces skills for each of the following steps in the visualization process:

1. Understanding the *purpose* of visualization

2. Selecting a *visual layout* based on your question and data type

3. Choosing optimal *graphical encodings* for your variables

4. Identifying visualizations that are able to *express* your data

5. Improving the *aesthetics* (i.e., making it readable and informative)

15.1   The Purpose of Visualization

"The purpose of visualization is insight, not pictures."[1]

Generating visual displays of your data is a key step in the analytical process. While you should strive to design aesthetically pleasing visuals, it's important to remember that visualization is a *means to an end*. Devising appropriate renderings of your data can help expose underlying patterns in your data that were previously unseen, or that were undetectable by other tests.

---

[1]Card, S. K., Mackinlay, J. D., & Shneiderman, B. (1999). *Readings in information visualization: Using vision to think*. Burlington, MA: Morgan Kaufmann.

To demonstrate how visualization makes a distinct contribution to the data analysis process (beyond statistical tests), consider the canonical data set **Anscombe's Quartet** (which is included with the R software as the data set anscombe). This data set consists of four pairs of x and y data: (x_1, y_1), (x_x, y_2), and so on. The data set is shown in Table 15.1.

The challenge of Anscombe's Quartet is to identify differences between the four pairs of columns. For example, how does the (x_1, y_1) pair differ from the (x_2, y_2) pair? Using a nonvisual approach to answer this question, you could compute a variety of descriptive statistics for each set, as shown in Table 15.2. Given these six statistical assessments, these four data sets *appear* to be identical. However, if you graphically represent the relationship between each x and y pair, as in Figure 15.1, you reveal the distinct nature of their relationships.

While computing summary statistics is an important part of the data exploration process, it is only through visual representations that differences across these sets emerge. The simple graphics in Figure 15.1 expose variations in the **distributions** of x and y values, as well as in the **relationships** between them. Thus the choice of representation becomes paramount when analyzing and presenting data. The following sections introduce basic principles for making that choice.

Table 15.1 **Anscombe's Quartet: four data sets with two features each**

x_1	y_1	x_2	y_2	x_3	y_3	x_4	y_4
10.00	8.04	10.00	9.14	10.00	7.46	8.00	6.58
8.00	6.95	8.00	8.14	8.00	6.77	8.00	5.76
13.00	7.58	13.00	8.74	13.00	12.74	8.00	7.71
9.00	8.81	9.00	8.77	9.00	7.11	8.00	8.84
11.00	8.33	11.00	9.26	11.00	7.81	8.00	8.47
14.00	9.96	14.00	8.10	14.00	8.84	8.00	7.04
6.00	7.24	6.00	6.13	6.00	6.08	8.00	5.25
4.00	4.26	4.00	3.10	4.00	5.39	19.00	12.50
12.00	10.84	12.00	9.13	12.00	8.15	8.00	5.56
7.00	4.82	7.00	7.26	7.00	6.42	8.00	7.91
5.00	5.68	5.00	4.74	5.00	5.73	8.00	6.89

Table 15.2 **Anscombe's Quartet: the (X, Y) pairs share identical summary statistics**

Set	Mean X	Std. Deviation X	Mean Y	Std. Deviation Y	Correlation	Linear Fit
1	9.00	3.32	7.50	2.03	0.82	$y = 3 + 0.5x$
2	9.00	3.32	7.50	2.03	0.82	$y = 3 + 0.5x$
3	9.00	3.32	7.50	2.03	0.82	$y = 3 + 0.5x$
4	9.00	3.32	7.50	2.03	0.82	$y = 3 + 0.5x$

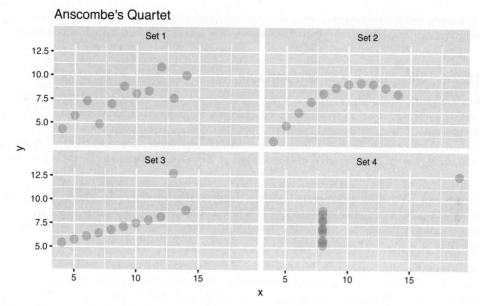

Figure 15.1 Anscombe's Quartet: scatterplots reveal four different (x, y) relationships that are not detectable using descriptive statistics.

15.2   Selecting Visual Layouts

The challenge of visualization, like many design challenges, is to identify an optimal solution (i.e., a visual layout) given a set of constraints. In visualization design, the primary constraints are:

1. The specific *question of interest* you are attempting to answer in your domain

2. The *type of data* you have available for answering that question

3. The limitations of the human *visual processing system*

4. The *spatial limitations* in the medium you are using (pixels on the screen, inches on the page, etc.)

This section focuses on the second of these constraints (data type); the last two constraints are addressed in Section 15.3 and Section 15.4. The first constraint (the question of interest) is closely tied to Chapter 10 on understanding data. Based on your domain, you need to hone in on a question of interest, and identify a data set that is well suited for answering your question. This section will expand upon the same data set and question from Chapter 10:

> *"What is the worst disease in the United States?"*

As with the Anscombe's Quartet example, most basic exploratory data questions can be reduced to investigating *how a variable is distributed* or *how variables are related* to one another. Once you have mapped from your question of interest to a specific data set, your visualization type will largely depend on the data type of your variables. The data type of each column—*nominal, ordinal,* or *continuous*—will dictate how the information can be represented. The following sections describe techniques for visually exploring each variable, as well as making comparisons across variables.

15.2.1   Visualizing a Single Variable

Before assessing relationships *across* variables, it is important to understand how each individual variable (i.e., column or feature) is distributed. The primary question of interest is often *what does this variable look like?* The specific visual layout you choose when answering this question will depend on whether the variable is **categorical** or **continuous**. To use the disease burden data set as an example, you may want to know *what is the range* of the number of deaths attributable to each disease.

For continuous variables, a **histogram** will allow you to see the distribution and range of values, as shown in Figure 15.2. Alternatively, you can use a **box plot** or a **violin plot**, both of which are shown Figure 15.3. Note that **outliers** (extreme values) in the data set have been removed to better express the information in the charts.

While these visualizations display information about the distribution of the number of deaths by cause, they all leave an obvious question unanswered: *what are the names of these diseases?* Figure 15.4 uses a **bar chart** to label the top 10 causes of death, but due to the constraint of the page size, this display is able to express just a small subset of the data. In other words, bar charts don't easily scale to hundreds or thousands of observations because they are inefficient to scan, or won't fit in a given medium.

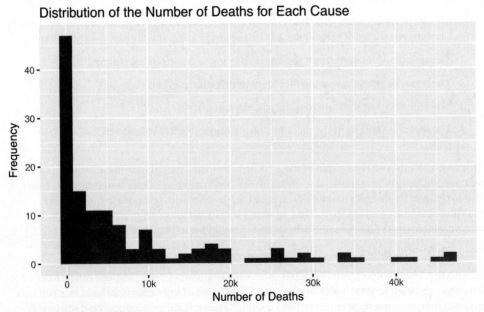

Figure 15.2 The distribution of the number of deaths attributable to each disease in the United States (a continuous variable) using a histogram. Some outliers have been removed for demonstration.

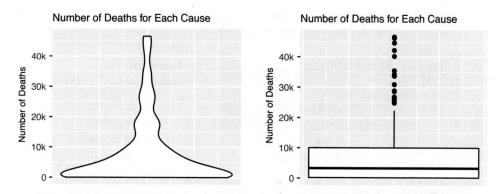

Figure 15.3   Alternative visualizations for showing distributions of the number of deaths in the United States: violin plot (left) and box plot (right). Some outliers have been removed for demonstration.

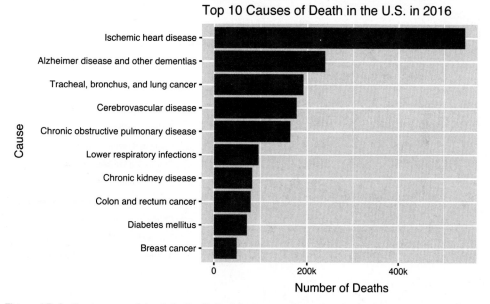

Figure 15.4   Top causes of death in the United States as shown in a bar chart.

15.2.1.1   Proportional Representations

Depending on the data stored in a given column, you may be interested in showing each value relative to the total of the column. For example, using the disease burden data set, you may want to express each value **proportional** to the total number of deaths. This allows you answer the question, *Of all deaths, what percentage is attributable to each disease?* To do this, you can transform the data to percentages, or use a representation that more clearly expresses *parts of a whole*. Figure 15.5 shows the use of a **stacked bar chart** and a **pie chart**, both of which more intuitively express proportionality. You can also use a **treemap**, as shown later in Figure 15.14, though the true

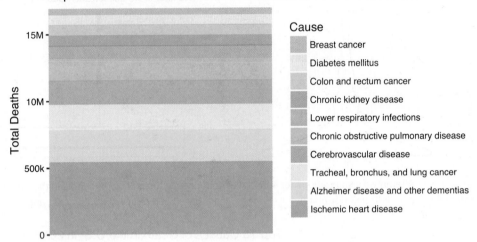

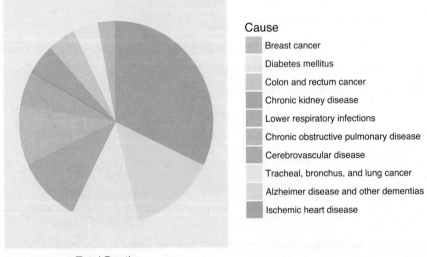

Figure 15.5 Proportional representations of the top causes of death in the United States: stacked bar chart (top) and pie chart (bottom).

benefit of a treemap is expressing hierarchical data (more on this later in the chapter). Later sections explore the trade-offs in *perceptual accuracy* associated with each of these representations.

If your variable of interest is a categorical variable, you will need to *aggregate your data* (e.g., count the number of occurrences of different categories) to ask similar questions about the distribution.

Once doing so, you can use similar techniques to show the data (e.g., bar chart, pie chart, treemap). For example, the diseases in this data set are categorized into three types of diseases: *non-communicable diseases*, such as heart disease or lung cancer; *communicable diseases*, such as tuberculosis or whooping cough; and *injuries*, such as road traffic accidents or self harm. To understand how this categorical variable (disease type) is distributed, you can count the number of rows for each category, then display those quantitative values, as in Figure 15.6.

15.2.2 Visualizing Multiple Variables

Once you have explored each variable independently, you will likely want to assess relationships between or across variables. The type of visual layout necessary for making these comparisons will (again) depend largely on the type of data you have for each variable.

For comparing relationships between two continuous variables, the best choice is a **scatterplot**. The visual processing system is quite good at estimating the linearity in a field of points created by a scatterplot, allowing you to describe how two variables are related. For example, using the disease burden data set, you can compare different metrics for measuring health loss. Figure 15.7 compares the disease burden as measured by the number of *deaths* due to each cause to the number of *years of life lost* (a metric that accounts for the age at death for each individual).

You can extend this approach to multiple continuous variables by creating a **scatterplot matrix** of all continuous features in the data set. Figure 15.8 compares all *pairs of metrics* of disease burden,

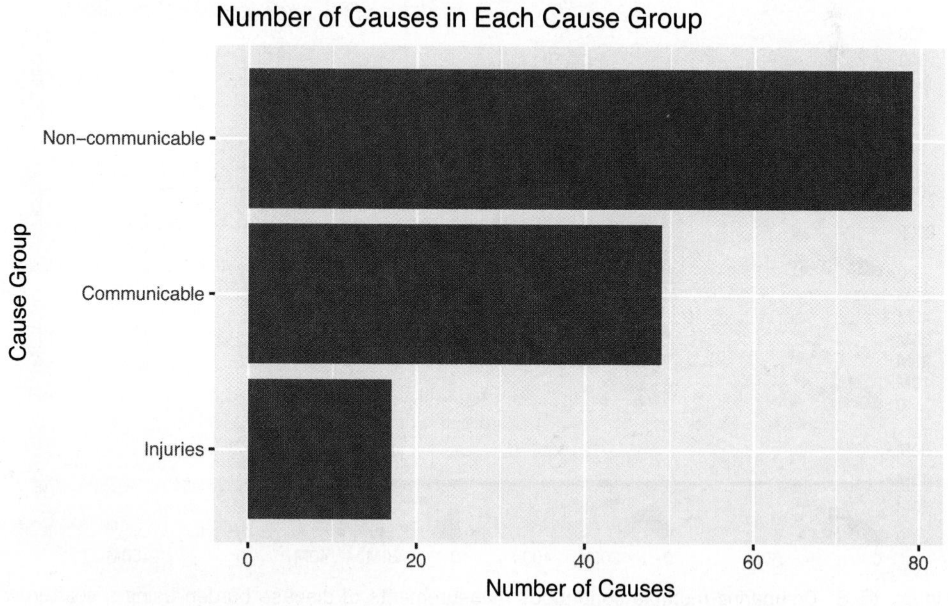

Figure 15.6 A visual representation of the number of causes in each disease category: non-communicable diseases, communicable diseases, and injuries.

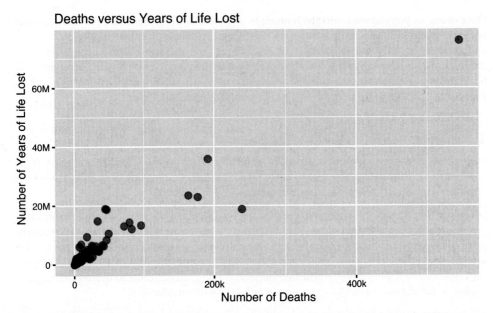

Figure 15.7 Using a scatterplot to compare two continuous variables: the number of deaths versus the years of live lost for each disease in the United States.

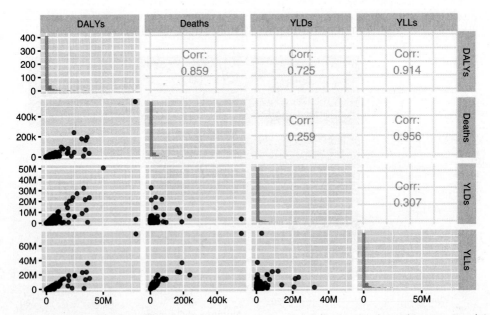

Figure 15.8 Comparing multiple continuous measurements of disease burden using a scatterplot matrix.

including number of deaths, years of life lost (*YLLs*), years lived with disability (*YLDs*, a measure of the disability experienced by the population), and disability-adjusted life years (*DALYs*, a combined measure of life lost and disability).

When comparing relationships between one continuous variable and one categorical variable, you can compute summary statistics for each group (see Figure 15.6), use a violin plot to display distributions for each category (see Figure 15.9), or use **faceting** to show the distribution for each category (see Figure 15.10).

For assessing relationships between two categorical variables, you need a layout that enables you to assess the *co-occurrences* of nominal values (that is, whether an observation contains both values). A great way to do this is to count the co-occurrences and show a **heatmap**. As an example, consider a broader set of population health data that evaluates the leading cause of death in each country (also from the Global Burden of Disease study). Figure 15.11 shows a subset of this data, including the disease type (communicable, non-communicable) for each disease, and the region where each country is found.

One question you may ask about this categorical data is:

> *"In each region, how often is the leading cause of death a communicable disease versus a non-communicable disease?"*

To answer this question, you can aggregate the data by region, and count the number of times each disease category (communicable, non-communicable) appears as the category for the leading cause of death. This aggregated data (shown in Figure 15.12) can then be displayed as a heatmap, as in Figure 15.13.

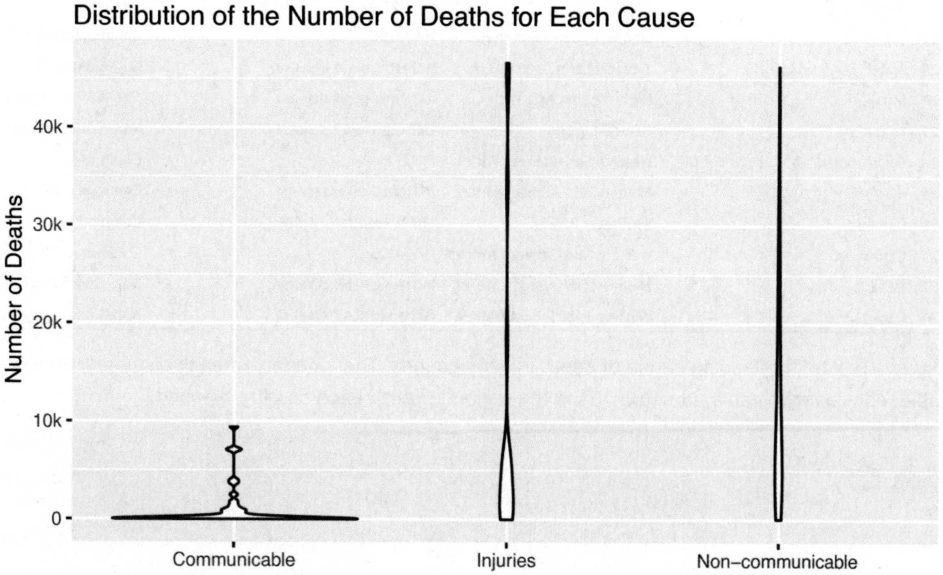

Figure 15.9 A violin plot showing the continuous distributions of the number of deaths for each cause (by category). Some outliers have been removed for demonstration.

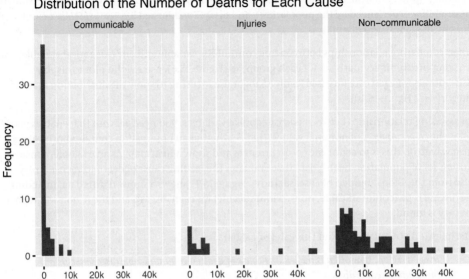

Figure 15.10   A faceted layout of histograms showing the continuous distributions of the number of deaths for each cause (by category). Some outliers have been removed for demonstration.

	country	region	leading_cause_of_death	category
24	Botswana	Southern Sub-Saharan Africa	HIV/AIDS	Communicable
25	Brazil	Tropical Latin America	Ischemic heart disease	Non-Communicable
26	Brunei	High-income Asia Pacific	Ischemic heart disease	Non-Communicable
27	Bulgaria	Central Europe	Ischemic heart disease	Non-Communicable
28	Burkina Faso	Western Sub-Saharan Africa	Malaria	Communicable
29	Burundi	Eastern Sub-Saharan Africa	Diarrheal diseases	Communicable
30	Cambodia	Southeast Asia	Lower respiratory infections	Communicable
31	Cameroon	Western Sub-Saharan Africa	HIV/AIDS	Communicable
32	Canada	High-income North America	Ischemic heart disease	Non-Communicable
33	Cape Verde	Western Sub-Saharan Africa	Ischemic heart disease	Non-Communicable

Figure 15.11   The leading cause of death in each country. The category of each disease (communicable, non-communicable) is shown, as is the region in which each country is found.

	region	category_of_leading_cause	number_of_countries
1	Andean Latin America	Communicable	1
2	Andean Latin America	Non-Communicable	2
3	Australasia	Non-Communicable	2
4	Caribbean	Non-Communicable	18
5	Central Asia	Non-Communicable	9
6	Central Europe	Non-Communicable	13
7	Central Latin America	Communicable	1
8	Central Latin America	Non-Communicable	8

Figure 15.12   Number of countries in each region in which the leading cause of death is communicable/non-communicable.

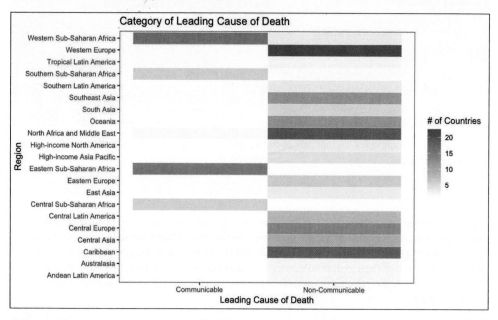

Figure 15.13   A heatmap of the number of countries in each region in which the leading cause of death is communicable/non-communicable.

15.2.3   Visualizing Hierarchical Data

One distinct challenge is showing a hierarchy that exists in your data. If your data naturally has a **nested structure** in which each observation is a member of a group, visually expressing that hierarchy can be critical to your analysis. Note that there may be multiple levels of nesting for each observation (observations may be part of a group, and that group may be part of a larger group). For example, in the disease burden data set, each country is found within a particular region, which can be further categorized into larger groupings called *super-regions*. Similarly, each cause of death

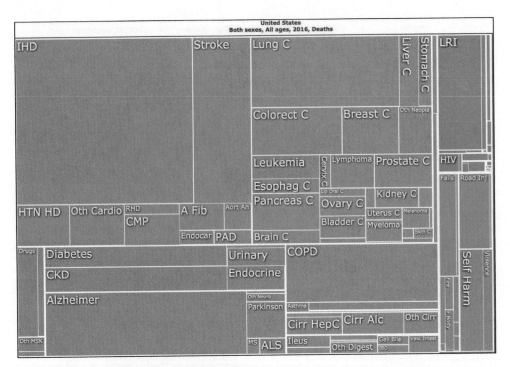

Figure 15.14   A treemap of the number of deaths in the United States from each cause.
Screenshot from GBD Compare, a visualization tool for the global burden of disease
(https://vizhub.healthdata.org/gbd-compare/).

(e.g., lung cancer) is a member of a family of causes (e.g., cancers), which can be further grouped
into overarching categories (e.g., non-communicable diseases). Hierarchical data can be visualized
using treemaps (Figure 15.14), **circle packing** (Figure 15.15), **sunburst diagrams** (Figure 15.16), or
other layouts. Each of these visualizations uses an **area encoding** to represent a numeric value.
These shapes (rectangles, circles, or arcs) are organized in a layout that clearly expresses the
hierarchy of information.

The benefit of visualizing the hierarchy of a data set, however, is not without its costs. As described
in Section 15.3, it is quite difficult to visually decipher and compare values encoded in a treemap
(especially with rectangles of different aspect ratios). However, these displays provide a great
summary overview of hierarchies, which is an important starting point for visually exploring data.

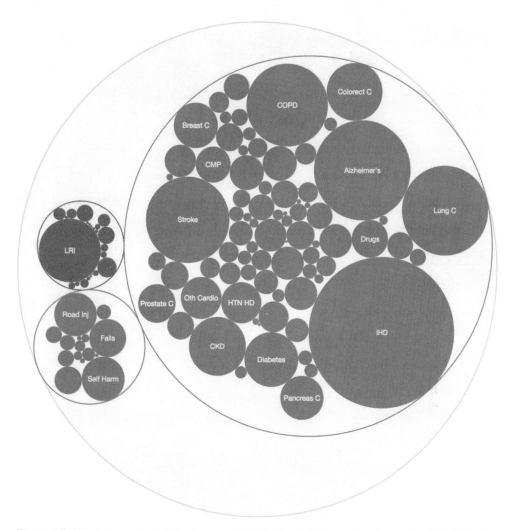

Figure 15.15 A re-creation of the treemap visualization (of disease burden in the United States) using a circle pack layout. Created using the d3.js library https://d3js.org.

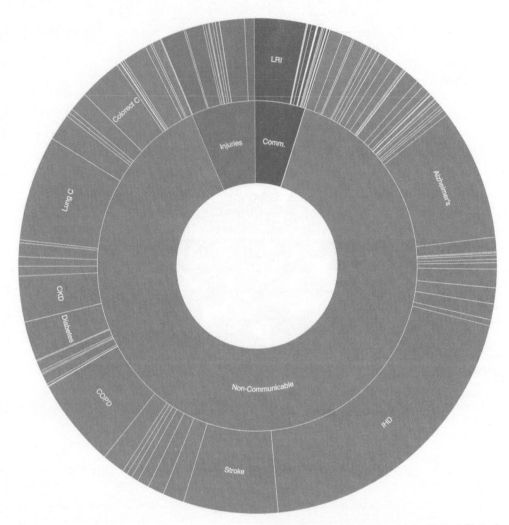

Figure 15.16   A re-creation of the treemap visualization (of disease burden in the United States) using a sunburst diagram. Created using the d3.js library https://d3js.org.

15.3   Choosing Effective Graphical Encodings

While the previously given guidelines for selecting visual layouts based on the data relationship to explore are a good place to start, there are often multiple ways to represent the same data set. Representing data in another format (e.g., visually) is called **encoding** that data. When you encode data, you use a particular "code" such as color or size to represent each value. These visual representations are then visually *de*coded by anyone trying to interpret the underlying values.

Your task is thus to select the encodings that are most accurately decoded by users, answering the question:

> *"What visual form best allows you to exploit the human visual system and available space to accurately display your data values?"*

In designing a visual layout, you should choose the **graphical encodings** that are most accurately visually decoded by your audience. This means that, for every value in your data, your user's interpretation of that value should be as accurate as possible. The accuracy of these perceptions is referred to as the **effectiveness** of a graphical encoding. Academic research[2] measuring the perceptiveness of different visual encodings has established a common set of possible encodings for quantitative information, listed here in order from most effective to least effective:

- **Position**: the horizontal or vertical position of an element along a common scale
- **Length**: the length of a segment, typically used in a stacked bar chart
- **Area**: the area of an element, such as a circle or a rectangle, typically used in a bubble chart (a scatterplot with differently sized markers) or a treemap
- **Angle**: the rotational angle of each marker, typically used in a circular layout like a pie chart
- **Color**: the color of each marker, usually along a continuous color scale
- **Volume**: the volume of a three-dimensional shape, typically used in a 3D bar chart

As an example, consider the very simple data set in Table 15.3. An effective visualization of this data set would enable you to easily distinguish between the values of each group (e.g., between the values 10 and 11). While this identification is simple for a *position* encoding, detecting this 10% difference is very difficult for other encodings. Comparisons between encodings of this data set are shown in Figure 15.17.

Thus when a visualization designer makes a blanket claim like "You should always use a bar chart rather than a pie chart," the designer is really saying, "A bar chart, which uses position encoding along a common scale, is more accurately visually decoded compared to a pie chart (which uses an angle encoding)."

Table 15.3   **A simple data set to demonstrate the perceptiveness of different graphical encodings (shown in Figure 15.17). Users should be able to visually distinguish between these values.**

group	value
a	1
b	10
c	11
d	7
e	8

---

[2]Most notably, Cleveland, W. S., & McGill, R. (1984). Graphical perception: Theory, experimentation, and application to the development of graphical methods. *Journal of the American Statistical Association, 79*(387), 531–554. https://doi.org/10.1080/01621459.1984.10478080

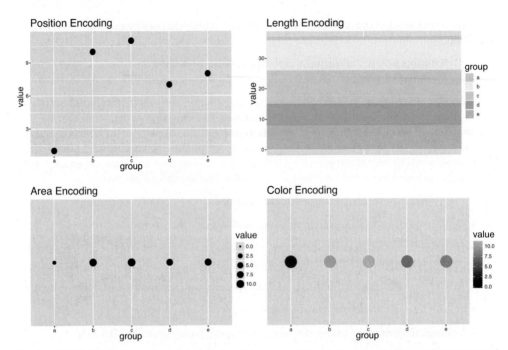

Figure 15.17   Different graphical encodings of the same data. Note the variation in perceptibility of differences between values!

To design your visualization, you should begin by encoding the most important data features with the most accurately decoded visual features (position, then length, then area, and so on). This will provide you with guidance as you compare different chart options and begin to explore more creative layouts.

While these guidelines may feel intuitive, the volume and distribution of your data often make this task more challenging. You may struggle to display all of your data, requiring you to also work to maximize the **expressiveness** of your visualizations (see Section 15.4).

15.3.1   Effective Colors

Color is one of the most prominent visual encodings, so it deserves special consideration. To describe how to use color effectively in visualizations, it is important to understand how color is measured. While there are many different conceptualizations of color spaces, a useful one for visualization is the **hue–saturation–lightness (HSL)** model, which defines a color using three attributes:

- The **hue** of a color, which is likely how you think of describing a color (e.g., "green" or "blue")

- The **saturation** or intensity of a color, which describes how "rich" the color is on a linear scale between gray (0%) and the full display of the hue (100%)

- The **lightness** of the color, which describes how "bright" the color is on a linear scale from black (0%) to white (100%)

This color model can be seen in Figure 15.18, which is an example of an interactive color selector[3] that allows you to manipulate each attribute independently to pick a color. The HSL model provides a good foundation for color selection in data visualization.

When selecting colors for visualization, the data type of your variable should drive your decisions. Depending on the data type (categorical or continuous), the purpose of your encoding will likely be different:

- For categorical variables, a color encoding is used to *distinguish between groups*. Therefore, you should select colors with different hues that are visually distinct and do not imply a rank ordering.

- For continuous variables, a color encoding is used to *estimate values*. Therefore, colors should be picked using a linear *interpolation* between color points (i.e., different lightness values).

Picking colors that most effectively satisfy these goals is trickier than it seems (and beyond the scope of this short section). But as with any other challenge in data science, you can build upon the open source work of other people. One of the most popular tools for picking colors (especially for maps) is Cynthia Brewer's **ColorBrewer**.[4] This tool provides a wonderful set of **color palettes** that differ in hue for categorical data (e.g., "Set3") and in lightness for continuous data (e.g., "Purples"); see Figure 15.19. Moreover, these palettes have been carefully designed to be viewable to people

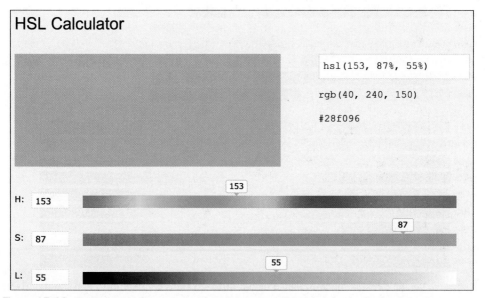

Figure 15.18   An interactive hue–staturation–lightness color picker, from w3schools.

---

[3] **HSL Calculator** by w3schools: https://www.w3schools.com/colors/colors_hsl.asp
[4] **ColorBrewer:** http://colorbrewer2.org

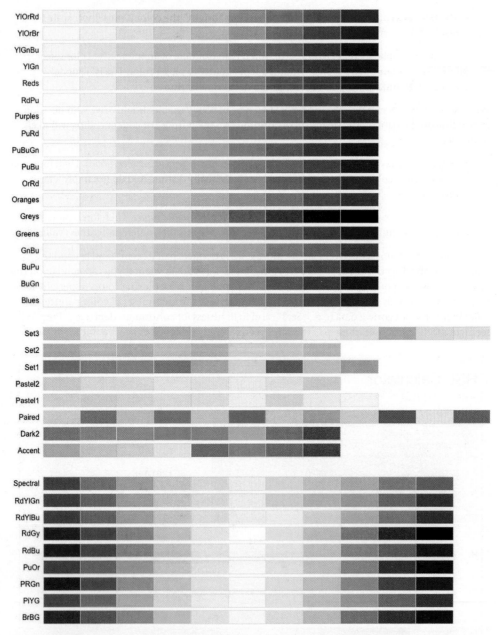

Figure 15.19 All palettes made available by the colorbrewer package in R. Run the display. brewer.all() function to see them in RStudio.

with certain forms of color blindness. These palettes are available in R through the `RColorBrewer` package; see Chapter 16 for details on how to use this package as part of your visualization process.

Selecting between different types of color palettes depends on the *semantic* meaning of the data. This choice is illustrated in Figure 15.20, which shows map visualizations of the population of each county in Washington state. The choice between different types of **continuous color scales** depends on the data:

- **Sequential** color scales are often best for displaying continuous values along a linear scale (e.g., for this population data).

- **Diverging** color scales are most appropriate when the divergence from a center value is meaningful (e.g., the midpoint is zero). For example, if you were showing changes in population over time, you could use a diverging scale to show increases in population using one hue, and decreases in population using another hue.

- **Multi-hue** color scales afford an increase in contrast between colors by providing a broader color range. While this allows for more precise interpretations than a (single hue) sequential color scale, the user may misinterpret or misjudge the differences in hue if the scale is not carefully chosen.

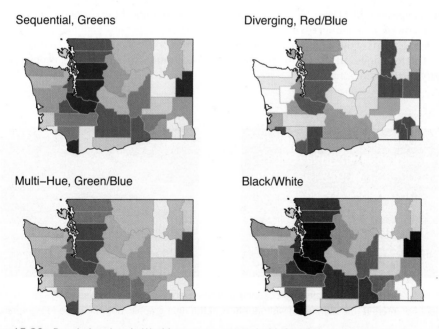

Figure 15.20 Population data in Washington represented with four ColorBrewer scales. The sequential and black/white scales accurately represent continuous data, while the diverging scale (inappropriately) implies divergence from a meaningful center point. Colors in the multi-hue scale may be misinterpreted as having different meanings.

- **Black and white** color scales are equivalent to sequential color scales (just with a hue of gray!) and may be required for your medium (e.g., when printing in a book or newspaper).

Overall, the choice of color will depend on the data. Your goal is to make sure that the color scale chosen enables the viewer to most effectively distinguish between the data's values and meanings.

15.3.2 Leveraging Preattentive Attributes

You often want to draw attention to particular observations in your visualizations. This can help you drive the viewer's focus toward specific instances that best convey the information or intended interpretation (to "tell a story" about the data). The most effective way to do this is to leverage the natural tendencies of the human visual processing system to direct a user's attention. This class of natural tendencies is referred to as **preattentive processing**: the cognitive work that your brain does without you deliberately paying attention to something. More specifically, these are the "[perceptual] tasks that can be performed on large multi-element displays in less than 200 to 250 milliseconds."[5] As detailed by Colin Ware,[6] the visual processing system will automatically process certain stimuli without any conscious effort. As a visualization designer, you want to take advantage of visual attributes that are processed preattentively, making your graphics as rapidly understood as possible.

As an example, consider Figure 15.21, in which you are able to count the occurrences of the number 3 at dramatically different speeds in each graphic. This is possible because your brain naturally identifies elements of the same color (more specifically, opacity) without having to put forth any effort. This technique can be used to drive focus in a visualization, thereby helping people quickly identify pertinent information.

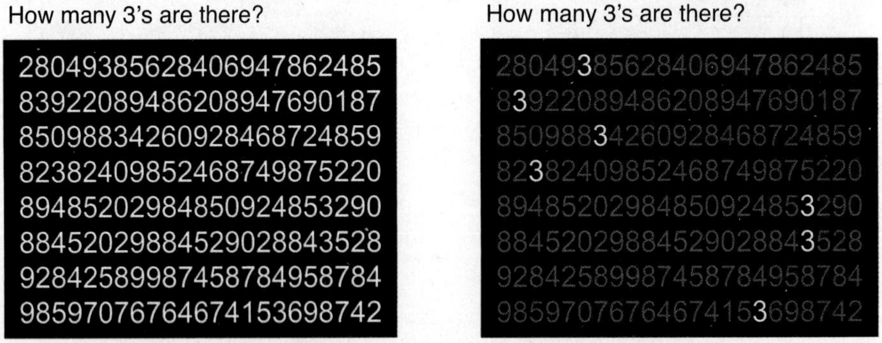

Figure 15.21 Because opacity is processed preattentively, the visual processing system identifies elements of interest (the number 3) without effort in the right graphic, but not in the left graphic.

---

[5]Healey, C. G., & Enns, J. T. (2012). Attention and visual memory in visualization and computer graphics. *IEEE Transactions on Visualization and Computer Graphics, 18*(7), 1170–1188. https://doi.org/10.1109/TVCG.2011.127. Also at: https://www.csc2.ncsu.edu/faculty/healey/PP/

[6]Ware, C. (2012). *Information visualization: Perception for design*. Philadelphia, PA: Elsevier.

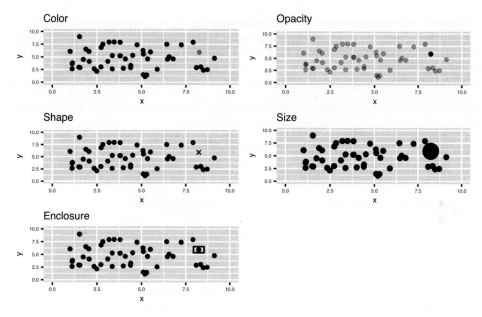

Figure 15.22 Driving focus with preattentive attributes. The selected point is clear in each graph, but especially easy to detect using color.

In addition to color, you can use other visual attributes that help viewers preattentively distinguish observations from those around them, as illustrated in Figure 15.22. Notice how quickly you can identify the "selected" point—though this identification happens more rapidly with some encodings (i.e., color) than with others!

As you can see, color and opacity are two of the most powerful ways to grab attention. However, you may find that you are already using color and opacity to encode a feature of your data, and thus can't also use these encodings to draw attention to particular observations. In that case, you can consider the remaining options (e.g., shape, size, enclosure) to direct attention to a specific set of observations.

15.4   Expressive Data Displays

The other principle you should use to guide your visualization design is to choose layouts that allow you to *express* as much data as possible. This goal was originally articulated as **Mackinlay's Expressiveness Criteria**[7] (clarifications added):

> A set of facts [data] is expressible in a language [visual layout] if that language contains a sentence [form] that
>
> 1. encodes all the facts in the set,
>
> 2. encodes only the facts in the set.

---

[7]Mackinlay, J. (1986). Automating the design of graphical presentations of relational information. *ACM Transactions on Graphics, 5*(2), 110–141. https://doi.org/10.1145/22949.22950. Restatement by Jeffrey Heer.

The prompt of this **expressiveness** aim is to devise visualizations that express all of (and only) the data in your data set. The most common barrier to expressiveness is occlusion (overlapping data points). As an example, consider Figure 15.23, which visualizes the distribution of the number of deaths attributable to different causes in the United States. This chart uses the most visually perceptive visual encoding (position), but fails to express all of the data due to the overlap in values.

There are two common approaches to address the failure of expressiveness caused by overlapping data points:

1. Adjust the *opacity* of each marker to reveal overlapping data.

2. Break the data into different groupings or facets to alleviate the overlap (by showing only a subset of the data at a time).

These approaches are both implemented in combination in Figure 15.24.

Alternatively, you could consider changing the data that you are visualizing by **aggregating** it in an appropriate way. For example, you could group your data by values that have similar number of deaths (putting each into a "bin"), and then use a position encoding to show the number of observations per bin. The result of this is the commonly used layout known as a histogram, as shown in Figure 15.25. While this visualization does communicate summary information to your audience, it is unable to express each individual observation in the data (which would communicate more information through the chart).

At times, the expressiveness and effectiveness principles are at odds with one another. In an attempt to maximize expressiveness (and minimize the overlap of your symbols), you may have to choose a less effective encoding. While there are multiple strategies for this—for example, breaking

Distribution of Number of Deaths

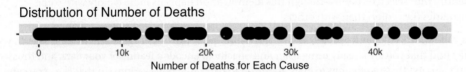

Figure 15.23 Position encoding of the number of deaths from each cause in the United States. Notice how the overlapping points (occlusion) prevent this layout from expressing all of the data. Some outliers have been removed for demonstration.

Distribution of Number of Deaths

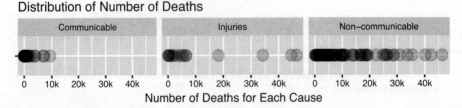

Figure 15.24 Position encoding of the number of deaths from each cause in the United States, faceted by the category of each cause. The use of a lower opacity in conjunction with the faceting enhances the expressiveness of the plots. Some outliers have been removed for demonstration.

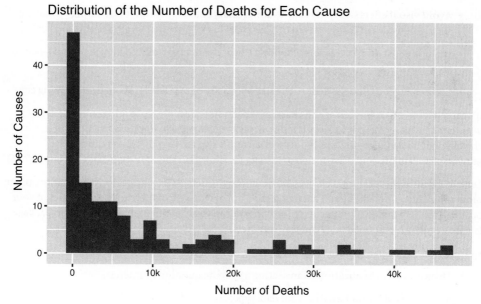

Figure 15.25   Histogram of the number of deaths attributable to each cause.

the data into multiple plots, aggregating the data, and changing the opacity of your symbols—the most appropriate choice will depend on the distribution and volume of your data, as well as the specific question you wish to answer.

15.5   Enhancing Aesthetics

Following the principles described in this chapter will go a long way in helping you devise informative visualizations. But to gain trust and attention from your potential audiences you will also want to spend time investing in the **aesthetics** (i.e., beauty) of your graphics.

> **Tip:** Making beautiful charts is a practice of removing clutter, *not* adding design.

One of the most renowned data visualization theorists, Edward Tufte, frames this idea in terms of the **data–ink ratio**.[8] Tufte argues that in every chart, you should maximize the ink dedicated to displaying the data (and in turn, minimize the non-data ink). This can translate to a number of actions:

- **Remove unnecessary encodings**. For example, if you have a bar chart, the bars should have different colors only if that information isn't otherwise expressed.

---

[8]Tufte, E. R. (1986). *The visual display of quantitative information. Cheshire*, CT: Graphics Press.

- **Avoid visual effects.** Any 3D effects, unnecessary shading, or other distracting formatting should be avoided. Tufte refers to this as "chart junk."

- **Include chart and axis labels.** Provide a title for your chart, as well as meaningful labels for your axes.

- **Lighten legends/labels.** Reduce the size or opacity of axis labels. Avoid using striking colors.

It's easy to look at a chart such as the chart on the left side of Figure 15.26 and claim that it *looks unpleasant*. However, describing *why* it looks distracting and *how* to improve it can be more challenging. If you follow the tips in this section and strive for simplicity, you can remove unnecessary elements and drive focus to the data (as shown on the right-hand side of Figure 15.26).

Luckily, many of these optimal choices are built into the default R packages for visualization, or are otherwise readily implemented. That being said, you may have to adhere to the aesthetics of your organization (or your own preferences!), so choosing an easily configurable visualization package (such as `ggplot2`, described in Chapter 16) is crucial.

As you begin to design and build visualizations, remember the following guidelines:

1. Dedicate each visualization to answering a **specific** question of interest.

2. Select a visual layout based on your data type.

3. Choose optimal graphical encodings based on how well they are visually decoded.

4. Ensure that your layout is able to express your data.

5. Enhance the aesthetics by removing visual effects, and by including clear labels.

These guidelines will be a helpful start, and don't forget that visualizations are about *insights*, not pictures.

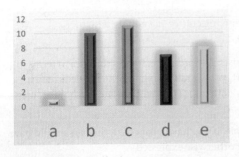 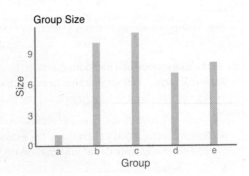

Figure 15.26 Removing distracting and uninformative visual features (left) and adding informative labels to create a cleaner chart (right).

Creating Visualizations with **ggplot2**

The ability to create visualizations (graphical representations) of data is a key step in being able to communicate information and findings to others. In this chapter, you will learn to use the **ggplot2**[1] package to declaratively make beautiful visual representations of your data.

Although R does provide built-in plotting functions, the ggplot2 package is built on the premise of the *Grammar of Graphics* (similar to how dplyr implements a *Grammar of Data Manipulation*; indeed, both packages were originally developed by the same person). This makes the package particularly effective for describing how visualizations should represent data, and has turned it into the preeminent plotting package in R. Learning to use this package will allow you to make nearly any kind of (static) data visualization, customized to your exact specifications.

16.1   A Grammar of Graphics

Just as the grammar of language helps you construct meaningful sentences out of words, the **Grammar of Graphics** helps you construct graphical figures out of different visual elements. This grammar provides a way to talk about parts of a visual plot: all the circles, lines, arrows, and text that are combined into a diagram for visualizing data. Originally developed by Leland Wilkinson, the *Grammar of Graphics* was adapted by Hadley Wickham[2] to describe the *components* of a plot:

- The **data** being plotted

- The **geometric objects** (e.g., circles, lines) that appear on the plot

- The **aesthetics** (appearance) of the geometric objects, and the *mappings* from variables in the data to those aesthetics

- A **position adjustment** for placing elements on the plot so they don't overlap

- A **scale** (e.g., a range of values) for each aesthetic mapping used

---

[1]**ggplot2**: http://ggplot2.tidyverse.org
[2]Wickham, H. (2010). A layered grammar of graphics. *Journal of Computational and Graphical Statistics, 19*(1), 3–28. https://doi.org/10.1198/jcgs.2009.07098. Also at http://vita.had.co.nz/papers/layered-grammar.pdf

- A **coordinate system** used to organize the geometric objects

- The **facets** or groups of data shown in different plots

`ggplot2` further organizes these components into **layers**, where each layer displays a single type of (highly configurable) *geometric object*. Following this grammar, you can think of each plot as a set of layers of images, where each image's appearance is based on some aspect of the data set.

Collectively, this grammar enables you to discuss what plots look like using a standard set of vocabulary. And like with `dplyr` and the *Grammar of Data Manipulation*, `ggplot2` uses this grammar directly to declare plots, allowing you to more easily create specific visual images and tell stories[3] about your data.

16.2 Basic Plotting with `ggplot2`

The `ggplot2` package provides a set of *functions* that mirror the *Grammar of Graphics*, enabling you to efficaciously specify what you want a plot to look like (e.g., what data, geometric objects, aesthetics, scales, and so on you want it to have).

`ggplot2` is yet another external package (like `dplyr`, `httr`, etc.), so you will need to install and load the package to use it:

```
install.packages("ggplot2") # once per machine
library("ggplot2")          # in each relevant script
```

This will make all of the plotting functions you will need available. As a reminder, plots will be rendered in the lower-right quadrant of RStudio, as shown in Figure 16.1.

> **Fun Fact**: Similar to `dplyr`, the `ggplot2` package also comes with a number of built-in data sets. This chapter will use the provided `midwest` data set as an example, described below.

This section uses the `midwest` data set that is included as part of the `ggplot2` package—a subset of the data is shown in Figure 16.2. The data set contains information on each of 437 counties in 5 states in the midwestern United States (specifically, Illinois, Indiana, Michigan, Ohio, and Wisconsin). For each county, there are 28 features that describe the demographics of the county, including racial composition, poverty levels, and education rates. To learn more about the data, you can consult the documentation (`?midwest`).

To create a plot using the `ggplot2` package, you call the **`ggplot()`** function, specifying as an argument the `data` that you wish to plot (i.e., `ggplot(data = SOME_DATA_FRAME)`). This will create a blank canvas upon which you can *layer* different visual markers. Each layer contains a specific *geometry*—think points, lines, and so on—that will be drawn on the canvas. For example, in Figure 16.3 (created using the following code), you can add a layer of points to assess the association between the percentage of people with a college education and the percentage of adults living in poverty in counties in the Midwest.

---

[3]Sander, L. (2016). Telling stories with data using the grammar of graphics. *Code Words*, 6. https://codewords.recurse.com/issues/six/telling-stories-with-data-using-the-grammar-of-graphics

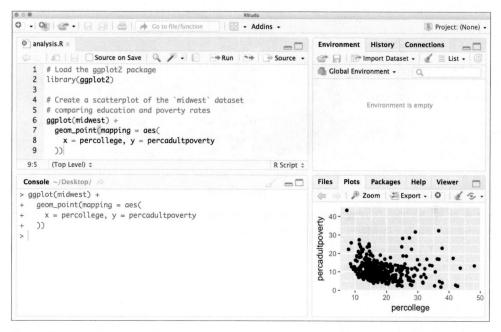

Figure 16.1   ggplot2 graphics will render in the lower-right quadrant of the RStudio window.

	PID	county	state	area	poptotal	popdensity	popwhite	popblack
1	561	ADAMS	IL	0.052	66090	1270.9615	63917	1702
2	562	ALEXANDER	IL	0.014	10626	759.0000	7054	3496
3	563	BOND	IL	0.022	14991	681.4091	14477	429
4	564	BOONE	IL	0.017	30806	1812.1176	29344	127
5	565	BROWN	IL	0.018	5836	324.2222	5264	547
6	566	BUREAU	IL	0.050	35688	713.7600	35157	50
7	567	CALHOUN	IL	0.017	5322	313.0588	5298	1
8	568	CARROLL	IL	0.027	16805	622.4074	16519	111
9	569	CASS	IL	0.024	13437	559.8750	13384	16
10	570	CHAMPAIGN	IL	0.058	173025	2983.1897	146506	16559
11	571	CHRISTIAN	IL	0.042	34418	819.4762	34176	82
12	572	CLARK	IL	0.030	15921	530.7000	15842	10
13	573	CLAY	IL	0.028	14460	516.4286	14403	4

Figure 16.2   A subset of the midwest data set, which captures demographic information on 5 midwestern states. The data set is included as part of the ggplot2 package and used throughout this chapter.

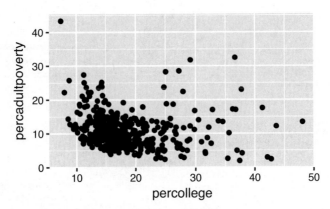

Figure 16.3 A basic use of `ggplot`: comparing the college education rates to adult poverty rates in Midwestern counties by adding a layer of points (thereby creating a scatterplot).

```
# Plot the `midwest` data set, with college education rate on the x-axis and
# percentage of adult poverty on the y-axis
ggplot(data = midwest) +
  geom_point(mapping = aes(x = percollege, y = percadultpoverty))
```

The code for creating a ggplot2 plot involves a few steps:

- The `ggplot()` function is passed the data frame to plot as the named `data` argument (it can also be passed as the first positional argument). Calling this function creates the blank canvas on which the visualization will be created.

- You specify the type of geometric object (sometimes referred to as a "geom") to draw by calling one of the many **geom_ functions**[4]—in this case, `geom_point()`. Functions to render a layer of geometric objects all share a common prefix (`geom_`), followed by the name of the kind of geometry you wish to create. For example, `geom_point()` will create a layer with "point" (dot) elements as the geometry. There are a large number of these functions; more details are provided in Section 16.2.1.

- In each geom_ function, you must specify the **aesthetic mappings**, which specify how data from the data frame will be mapped to the visual aspects of the geometry. These mappings are defined using the **aes()** (*aesthetic*) function. The **aes()** function takes a set of named arguments (like a list), where the argument name is the visual property to map *to*, and the argument value is the data feature (i.e., the column in the data frame) to map *from*. The value returned by the aes() function is passed to the named `mapping` argument (or passed as the first positional argument).

---

[4] **Layer: geoms** function reference: http://ggplot2.tidyverse.org/reference/index.html#section-layer-geoms

> **Caution**: The `aes()` function uses *non-standard evaluation* similar to `dplyr`, so you don't need to put the data frame column names in quotes. This can cause issues if the name of the column you wish to plot is stored as a string in a variable (e.g., `plot_var <- "COLUMN_NAME"`). To handle this situation, you can use the **`aes_string()`** function instead and specify the column names as string values or variables.

- You add layers of geometric objects to the plot by using the addition (**+**) operator.

Thus, you can create a basic plot by specifying a data set, an appropriate `geometry`, and a set of aesthetic mappings.

> **Tip**: The **ggplot2** package includes a **`qplot()`** function[a] for creating "quick plots." This function is a convenient shortcut for making simple, "default"-like plots. While this is a nice starting point, the strength of `ggplot2` lies in its *customizability*, so read on!
>
> _____
> [a]http://www.statmethods.net/advgraphs/ggplot2.html

16.2.1   Specifying Geometries

The most obvious distinction between plots is the geometric objects that they include. `ggplot2` supports the rendering of a variety of geometries, each created using the appropriate `geom_` function. These functions include, but are not limited to, the following:

- **`geom_point()`** for drawing individual points (e.g., for a scatterplot)
- **`geom_line()`** for drawing lines (e.g., for a line chart)
- **`geom_smooth()`** for drawing smoothed lines (e.g., for simple trends or approximations)
- **`geom_col()`** for drawing columns (e.g., for a bar chart)
- **`geom_polygon()`** for drawing arbitrary shapes (e.g., for drawing an area in a coordinate plane)

Each of these `geom_` functions requires as an argument a set of aesthetic mappings (defined using the `aes()` function, described in Section 16.2.2), though the specific *visual properties* that the data will map *to* will vary. For example, you can map a data feature to the `shape` of a `geom_point()` (e.g., if the points should be circles or squares), or you can map a feature to the `linetype` of a `geom_line()` (e.g., if it is solid or dotted), but not vice versa.

Since graphics are two-dimensional representations of data, almost all `geom_` functions *require* an x and y mapping. For example, in Figure 16.4, the bar chart of the number of counties per state (left) is built using the `geom_col()` geometry, while the hexagonal aggregation of the scatterplot from Figure 16.3 (right) is built using the `geom_hex()` function.

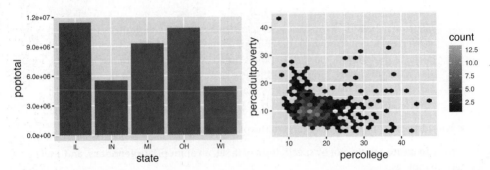

Figure 16.4 Plots with column geometry (left) and binned hexagons (right). The rectangles in the column geometry represent separate observations (counties) that have been automatically stacked on top of each other; see Section 16.3.1 for details.

```
# A bar chart of the total population of each state
# The `state` is mapped to the x-axis, and the `poptotal` is mapped
# to the y-axis
ggplot(data = midwest) +
  geom_col(mapping = aes(x = state, y = poptotal))
```

```
# A hexagonal aggregation that counts the co-occurrence of college
# education rate and percentage of adult poverty
ggplot(data = midwest) +
  geom_hex(mapping = aes(x = percollege, y = percadultpoverty))
```

What makes this really powerful is that you can add *multiple geometries* to a plot. This allows you to create complex graphics showing multiple aspects of your data, as in Figure 16.5.

```
# A plot with both points and a smoothed line
ggplot(data = midwest) +
  geom_point(mapping = aes(x = percollege, y = percadultpoverty)) +
  geom_smooth(mapping = aes(x = percollege, y = percadultpoverty))
```

While the geom_point() and geom_smooth() layers in this code both use the same aesthetic mappings, there's no reason you couldn't assign different aesthetic mappings to each geometry. Note that if the layers *do* share some aesthetic mappings, you can specify those as an argument to the ggplot() function as follows:

```
# A plot with both points and a smoothed line, sharing aesthetic mappings
ggplot(data = midwest, mapping = aes(x = percollege, y = percadultpoverty)) +
  geom_point() + # uses the default x and y mappings
  geom_smooth() + # uses the default x and y mappings
  geom_point(mapping = aes(y = percchildbelowpovert)) # uses own y mapping
```

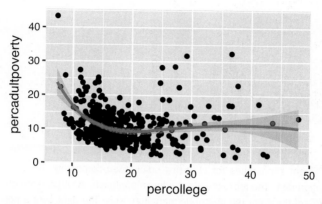

Figure 16.5 A plot comparing the adult poverty rate and the college education rate using multiple geometries. Each layer is added with a different `ggplot2` function: `geom_point()` for points, and `geom_smooth()` for the smoothed line.

Each geometry will use the data and individual aesthetics specified in the `ggplot()` function unless they are overridden by individual specifications.

> **Going Further**: Some geom_ functions also perform a *statistical transformation* on the data, aggregating the data (e.g., counting the number of observations) before mapping that data to an aesthetic. While you can do many of these transformations using the `dplyr` functions `group_by()` and `summarize()`, a statistical transformation allows you to apply some aggregations purely to adjust the data's presentation, without needing to modify the data itself. You can find more information in the documentation.[a]
>
> ———————————
> [a] http://ggplot2.tidyverse.org/reference/index.html#section-layer-stats

16.2.2 Aesthetic Mappings

The aesthetic mappings take properties of the data and use them to influence **visual channels** (graphical encodings), such as position, color, size, or shape. Each visual channel therefore encodes a feature of the data and can be used to express that data. Aesthetic mappings are used for visual features that should be driven by data values, rather than set for all geometric elements. For example, if you want to use a color encoding to express the values in a column, you would use an aesthetic mapping. In contrast, if you want the color of all points to be the same (e.g., blue), you *would not* use an aesthetic mapping (because the color has nothing to do with your data).

The data-driven aesthetics for a plot are specified using the `aes()` function and passed into a particular `geom_` function layer. For example, if you want to know which *state* each county is in, you can add a mapping from the `state` feature of each row to the *color* channel. `ggplot2` will even create a legend for you automatically (as in Figure 16.6)! Note that using the `aes()` function will cause the visual channel to be based on the data specified in the argument.

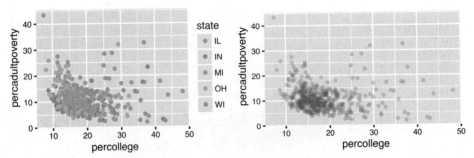

Figure 16.6 Different approaches for choosing color when comparing the adult poverty rate and the college education rate. The left uses a data-driven approach, in which each observation's `state` column is used to set the color (an aesthetic mapping), while the right sets a constant color for all observations. Code is below.

Conversely, if you wish to apply a visual property to an entire geometry, you can set that property on the geometry by passing it as an argument to the `geom_` function, outside of the `aes()` call, as shown in the following code. Figure 16.6 shows both approaches: driving color with the aesthetic (left) and choosing constant styles for each point (right).

```
# Change the color of each point based on the state it is in
ggplot(data = midwest) +
  geom_point(
    mapping = aes(x = percollege, y = percadultpoverty, color = state)
  )

# Set a consistent color ("red") for all points -- not driven by data
ggplot(data = midwest) +
  geom_point(
    mapping = aes(x = percollege, y = percadultpoverty),
    color = "red",
    alpha = .3
  )
```

16.3 Complex Layouts and Customization

Building on these basics, you can use `ggplot2` to create almost any kind of plot you may want. In addition to specifying the geometry and aesthetics, you can further customize plots by using functions that follow from the *Grammar of Graphics*.

16.3.1 Position Adjustments

The plot using `geom_col()` in Figure 16.4 *stacked* all of the observations (rows) per state into a single column. This stacking is the default **position adjustment** for the geometry, which specifies a "rule" as to how different components should be positioned relative to each other to make sure they don't overlap. This positional adjustment can be made more apparent if you map a different variable to the color encoding (using the `fill` aesthetic). In Figure 16.7 you can see the racial breakdown for the population in each state by adding a `fill` to the column geometry:

```
# Load the `dplyr` and `tidyr` libraries for data manipulation
library("dplyr")
library("tidyr")

# Wrangle the data using `tidyr` and `dplyr` -- a common step!
# Select the columns for racial population totals, then
# `gather()` those column values into `race` and `population` columns
state_race_long <- midwest %>%
  select(state, popwhite, popblack, popamerindian, popasian, popother) %>%
  gather(key = race, value = population, -state) # all columns except `state`

# Create a stacked bar chart of the number of people in each state
# Fill the bars using different colors to show racial composition
ggplot(state_race_long) +
  geom_col(mapping = aes(x = state, y = population, fill = race))
```

Remember: You will need to use your `dplyr` and `tidyr` skills to wrangle your data frames into the proper orientation for plotting. Being confident in those skills will make using the ggplot2 library a relatively straightforward process; the hard part is getting your data in the desired shape.

Tip: Use the `fill` aesthetic when coloring in bars or other area shapes (that is, specifying what color to "fill" the area). The `color` aesthetic is instead used for the outline (stroke) of the shapes.

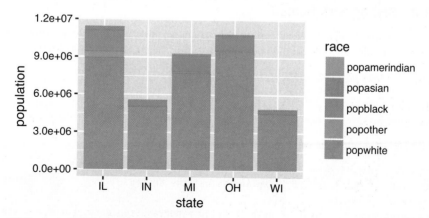

Figure 16.7 A stacked bar chart of the number of people in each state (by race). Colors are added by setting a `fill` aesthetic based on the `race` column.

By default, `ggplot` will adjust the position of each rectangle by stacking the "columns" for each county. The plot thus shows all of the elements instead of causing them to overlap. However, if you wish to specify a different position adjustment, you can use the **position** argument. For example, to see the relative composition (e.g., percentage) of people by race in each state, you can use a `"fill"` position (to *fill* each bar to 100%). To see the relative measures within each state side by side, you can use a `"dodge"` position. To explicitly achieve the default behavior, you can use the `"identity"` position. The first two options are shown in Figure 16.8.

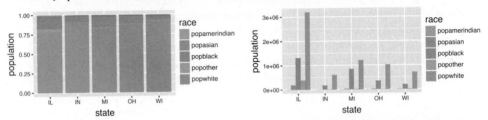

Figure 16.8 Bar charts of state population by race, shown with different position adjustments: filled (left) and dodged (right).

```
# Create a percentage (filled) column of the population (by race) in each state
ggplot(state_race_long) +
  geom_col(
    mapping = aes(x = state, y = population, fill = race), position = "fill"
  )


# Create a grouped (dodged) column of the number of people (by race) in each state
ggplot(state_race_long) +
  geom_col(
    mapping = aes(x = state, y = population, fill = race), position = "dodge"
  )
```

16.3.2 Styling with Scales

Whenever you specify an aesthetic mapping, ggplot2 uses a particular **scale** to determine the *range of values* that the data encoding should be mapped *to*. Thus, when you specify a plot such as:

```
# Plot the `midwest` data set, with college education rate on the x-axis and
# percentage of adult poverty on the y-axis. Color by state.
ggplot(data = midwest) +
  geom_point(mapping = aes(x = percollege, y = percadultpoverty, color = state))
```

ggplot2 automatically adds a scale for each mapping to the plot:

```
# Plot the `midwest` data set, with college education rate and
# percentage of adult poverty. Explicitly set the scales.
ggplot(data = midwest) +
  geom_point(mapping = aes(x = percollege, y = percadultpoverty, color = state)) +
  scale_x_continuous() + # explicitly set a continuous scale for the x-axis
  scale_y_continuous() + # explicitly set a continuous scale for the y-axis
  scale_color_discrete() # explicitly set a discrete scale for the color aesthetic
```

Each scale can be represented by a function named in the following format: scale_, followed by the name of the aesthetic property (e.g., x or color), followed by an _ and the type of the scale (e.g., continuous or discrete). A continuous scale will handle values such as numeric data (where there is a *continuous set* of numbers), whereas a discrete scale will handle values such as colors (since there is a small *discrete* list of distinct colors). Notice also that scales are added to a plot using the + operator, similar to a geom layer.

While the default scales will often suffice for your plots, it is possible to explicitly add different scales to replace the defaults. For example, you can use a scale to change the direction of an axis (scale_x_reverse( )), or plot the data on a *logarithmic scale* (scale_x_log10( )). You can also use scales to specify the range of values on an axis by passing in a limits argument. Explicit limits are useful for making sure that multiple graphs share scales or formats, as well as for customizing the appearance of your visualizations. For example, the following code imposes the same scale across two plots, as shown in Figure 16.9:

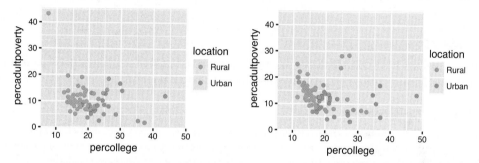

Figure 16.9 Plots of the percent college-educated population versus the percent adult poverty in Wisconsin (left) and Michigan (right). These plots share the same explicit scales (which are not based solely on the plotted data). Notice how it is easy to compare the two data sets to each other because the axes and colors match!

```
# Create a better label for the `inmetro` column
labeled <- midwest %>%
  mutate(location = if_else(inmetro == 0, "Rural", "Urban"))

# Subset data by state
wisconsin_data <- labeled %>% filter(state == "WI")
michigan_data <- labeled %>% filter(state == "MI")

# Define continuous scales based on the entire data set:
# range() produces a (min, max) vector to use as the limits
x_scale <- scale_x_continuous(limits = range(labeled$percollege))
y_scale <- scale_y_continuous(limits = range(labeled$percadultpoverty))

# Define a discrete color scale using the unique set of locations (urban/rural)
color_scale <- scale_color_discrete(limits = unique(labeled$location))
```

```
# Plot the Wisconsin data, explicitly setting the scales
ggplot(data = wisconsin_data) +
  geom_point(
    mapping = aes(x = percollege, y = percadultpoverty, color = location)
  ) +
  x_scale +
  y_scale +
  color_scale

# Plot the Michigan data using the same scales
ggplot(data = michigan_data) +
  geom_point(
    mapping = aes(x = percollege, y = percadultpoverty, color = location)
  ) +
  x_scale +
  y_scale +
  color_scale
```

These scales can also be used to specify the "tick" marks and labels; see the `ggplot2` documentation for details. For further ways of specifying where the data appears on the graph, see Section 16.3.3.

16.3.2.1 Color Scales

One of the most common scales to change is the color scale (i.e., the set of colors used in a plot). While you can use scale functions such as `scale_color_manual()` to specify a specific set of colors for your plot, a more common option is to use one of the predefined ColorBrewer[5] palettes (described in Chapter 15, Figure 15.19). These palettes can be specified as a color scale with the `scale_color_brewer()` function, passing the `palette` as a named argument (see the rendered plot in Figure 16.10).

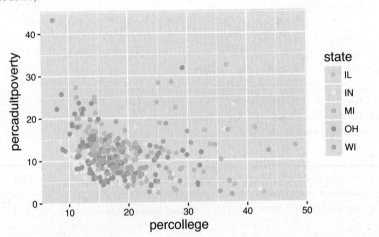

Figure 16.10 A comparison of each county's adult poverty rate and college education rate, using color to show the state each county is in. These colors come from the ColorBrewer `Set3` palette.

---

[5]ColorBrewer: http://colorbrewer2.org

```
# Change the color of each point based on the state it is in
ggplot(data = midwest) +
  geom_point(
    mapping = aes(x = percollege, y = percadultpoverty, color = state)
  ) +
  scale_color_brewer(palette = "Set3") # use the "Set3" color palette
```

If you instead want to define your own color scheme, you can make use of a variety of ggplot2 functions. For discrete color scales[6], you can specify a distinct set of colors to map to using a function such as scale_color_manual( ). For continuous color scales[7], you can specify a range of colors to display using a function such as scale_color_gradient( ).

16.3.3   Coordinate Systems

It is also possible to specify a plot's **coordinate system**, which is used to organize the geometric objects. As with scales, coordinate systems are specified with functions (whose names all start with coord_) and are added to a ggplot. You can use several different coordinate systems,[8] including but not limited to the following:

- **coord_cartesian( )**: The default *Cartesian coordinate* system, where you specify x and y values—x values increase from left to right, and y values increase from bottom to top

- **coord_flip( )**: A Cartesian system with the x and y flipped

- **coord_fixed( )**: A Cartesian system with a "fixed" aspect ratio (e.g., 1.78 for "widescreen")

- **coord_polar( )**: A plot using polar coordinates (i.e., a *pie chart*)

- **coord_quickmap( )**: A coordinate system that approximates a good aspect ratio for maps. See the documentation for more details

The example in Figure 16.11 uses coord_flip( ) to create a horizontal bar chart (a useful layout for making labels more legible). In the geom_col( ) function's aesthetic mapping, you do not change what you assign to the x and y variables to make the bars horizontal; instead, you call the coord_flip( ) function to switch the orientation of the graph. The following code (which generates Figure 16.11) also creates a *factor variable* to sort the bars using the variable of interest:

```
# Create a horizontal bar chart of the most populous counties
# Thoughtful use of `tidyr` and `dplyr` is required for wrangling

# Filter down to top 10 most populous counties
top_10 <- midwest %>%
  top_n(10, wt = poptotal) %>%
  unite(county_state, county, state, sep = ", ") %>% # combine state + county
  arrange(poptotal) %>% # sort the data by population
  mutate(location = factor(county_state, county_state)) # set the row order
```

---

[6]**Gradient color scales** function reference: http://ggplot2.tidyverse.org/reference/scale_gradient.html
[7]**Create your own discrete scale** function reference:http://ggplot2.tidyverse.org/reference/scale_manual.html
[8]**Coordinate systems** function reference: http://ggplot2.tidyverse.org/reference/index.html#section-coordinate-systems

```
# Render a horizontal bar chart of population
ggplot(top_10) +
  geom_col(mapping = aes(x = location, y = poptotal)) +
  coord_flip() # switch the orientation of the x- and y-axes
```

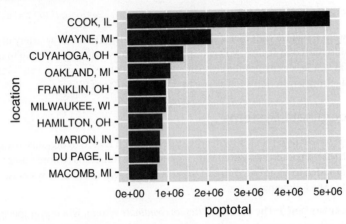

Figure 16.11   A horizontal bar chart of the population in the ten most populous counties. The orientation of the chart is "flipped" by calling the coord_flip() function.

In general, the coordinate system is used to specify where in the plot the x and y axes are placed, while scales are used to determine which values are shown on those axes.

16.3.4 Facets

Facets are ways of grouping a visualization into multiple different pieces (*subplots*). This allows you to view a separate plot for each unique value in a categorical variable. Conceptually, breaking a plot up into facets is similar to using the group_by() verb in dplyr: it creates the same visualization for each group separately (just as summarize() performs the same analysis for each group).

You can construct a plot with multiple facets by using a facet_ function such as facet_wrap(). This function will produce a "row" of subplots, one for each categorical variable (the number of rows can be specified with an additional argument); subplots will "wrap" to the next line if there is not enough space to show them all in a single row. Figure 16.12 demonstrates faceting; as you can see in this plot, using facets is basically an "automatic" way of doing the same kind of grouping performed in Figure 16.9, which shows separate graphs for Wisconsin and Michigan.

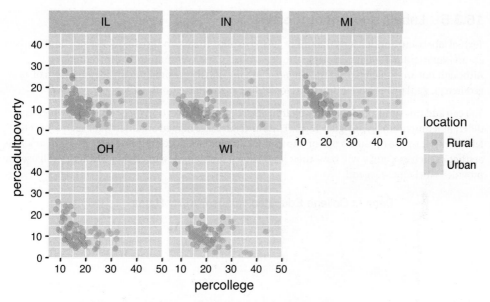

Figure 16.12   A comparison of each county's adult poverty rate and college education rate. A separate plot is created for each state using the `facet_wrap()` function.

```
# Create a better label for the `inmetro` column
labeled <- midwest %>%
  mutate(location = if_else(inmetro == 0, "Rural", "Urban"))

# Create the same chart as Figure 16.9, faceted by state
ggplot(data = labeled) +
  geom_point(
    mapping = aes(x = percollege, y = percadultpoverty, color = location),
    alpha = .6
  ) +
  facet_wrap(~state) # pass the `state` column as a *fomula* to `facet_wrap()`
```

Note that the argument to the `facet_wrap()` function is the column to facet by, with the column name written with a tilde (~) in front of it, turning it into a **formula**.[9] A formula is a bit like an equation in mathematics; that is, it represents a set of operations to perform. The tilde can be read "as a function of." The `facet_` functions take formulas as arguments in order to determine how they should group and divide the subplots. In short, with `facet_wrap()` you need to put a ~ in front of the feature name you want to "group" by. See the official `ggplot2` documentation[10] for `facet_` functions for more details and examples.

---

[9]**Formula** documentation: https://www.rdocumentation.org/packages/stats/versions/3.4.3/topics/formula. See the *Details* in particular.

[10]**ggplot2 facetting**: https://ggplot2.tidyverse.org/reference/#section-facetting

16.3.5 Labels and Annotations

Textual labels and annotations that more clearly express the meaning of axes, legends, and markers are an important part of making a plot understandable and communicating information. Although not an explicit part of the *Grammar of Graphics* (they would be considered a form of geometry), `ggplot2` provides functions for adding such annotations.

You can add titles and axis labels to a chart using the **`labs()`** function (*not* `labels()`, which is a different R function!), as in Figure 16.13. This function takes named arguments for each aspect to label—either `title` (or `subtitle` or `caption`), or the name of the aesthetic (e.g., x, y, color). Axis aesthetics such as x and y will have their label shown on the axis, while other aesthetics will use the provided label for the legend.

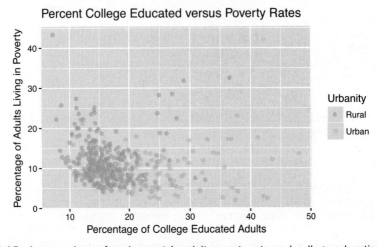

Figure 16.13 A comparison of each county's adult poverty rate and college education rate. The `labs()` function is used to add a title and labels for each aesthetic mapping.

```
# Adding better labels to the plot in Figure 16.10
ggplot(data = labeled) +
  geom_point(
    mapping = aes(x = percollege, y = percadultpoverty, color = location),
    alpha = .6
  ) +

  # Add title and axis labels
  labs(
    title = "Percent College Educated versus Poverty Rates", # plot title
    x = "Percentage of College Educated Adults", # x-axis label
    y = "Percentage of Adults Living in Poverty", # y-axis label
    color = "Urbanity" # legend label for the "color" property
  )
```

You can also add labels into the plot itself (e.g., to label each point or line) by adding a new geom_text( ) (for plain text) or geom_label( ) (for boxed text). In effect, you're plotting an extra set of data values that happen to be the value names. For example, in Figure 16.14, labels are used to identify the county with the highest level of poverty in each state. The background and border for each piece of text is created by using the geom_label_repel( ) function, which provides labels that don't overlap.

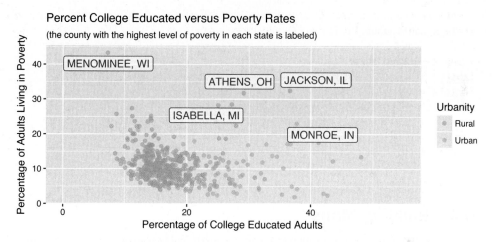

Figure 16.14 Using labels to identify the county in each state with the highest level of poverty. The ggrepel package is used to prevent labels from overlapping.

```
# Load the `ggrepel` package: functions that prevent labels from overlapping
library(ggrepel)

# Find the highest level of poverty in each state
most_poverty <- midwest %>%
  group_by(state) %>% # group by state
  top_n(1, wt = percadultpoverty) %>% # select the highest poverty county
  unite(county_state, county, state, sep = ", ") # for clear labeling

# Store the subtitle in a variable for cleaner graphing code
subtitle <- "(the county with the highest level of poverty
  in each state is labeled)"

# Plot the data with labels
ggplot(data = labeled, mapping = aes(x = percollege, y = percadultpoverty)) +

  # add the point geometry
  geom_point(mapping = aes(color = location), alpha = .6) +
```

```
# add the label geometry
geom_label_repel(
  data = most_poverty, # uses its own specified data set
  mapping = aes(label = county_state),
  alpha = 0.8
) +

# set the scale for the axis
scale_x_continuous(limits = c(0, 55)) +

# add title and axis labels
labs(
  title = "Percent College Educated versus Poverty Rates", # plot title
  subtitle = subtitle, # subtitle
  x = "Percentage of College Educated Adults", # x-axis label
  y = "Percentage of Adults Living in Poverty", # y-axis label
  color = "Urbanity" # legend label for the "color" property
)
```

16.4   Building Maps

In addition to building charts using `ggplot2`, you can use the package to draw geographic maps. Because two-dimensional maps already depend on a coordinate system (*latitude* and *longitude*), you can exploit the `ggplot2` Cartesian layout to create geographic visualizations. Generally speaking, there are two types of maps you will want to create:

- **Choropleth maps**: Maps in which different geographic areas are shaded based on data about each region (as in Figure 16.16). These maps can be used to visualize data that is aggregated to specified geographic areas. For example, you could show the eviction rate in each state using a choropleth map. Choropleth maps are also called *heatmaps*.

- **Dot distribution maps**: Maps in which markers are placed at specific coordinates, as in Figure 16.19. These plots can be used to visualize observations that occur at discrete (latitude/longitude) points. For example, you could show the specific address of each eviction notice filed in a given city.

This section details how to build such maps using `ggplot2` and complementary packages.

16.4.1   Choropleth Maps

To draw a choropleth map, you need to first draw the outline of each geographic unit (e.g., state, country). Because each geography will be an irregular closed shape, `ggplot2` can use the `geom_polygon()` function to draw the outlines. To do this, you will need to load a data file that describes the geometries (outlines) of your areas, appropriately called a **shapefile**. Many shapefiles such as those made available by the U.S. Census Bureau[11] and OpenStreetMap[12] can be freely downloaded and used in R.

To help you get started with mapping, `ggplot2` includes a handful of shapefiles (meaning you don't need to download one). You can load a given shapefile by providing the name of the shapefile you wish to load (e.g., `"usa"`, `"state"`, `"world"`) to the **map_data( )** function. Once you have the desired shapefile in a usable format, you can render a map using the `geom_polygon( )` function. This function plots a shape by drawing lines between each individual pair of x- and y- coordinates (in order), similar to a "connect-the-dots" puzzle. To maintain an appropriate aspect ratio for your map, use the `coord_map( )` coordinate system. The map created by the following code is shown in Figure 16.15.

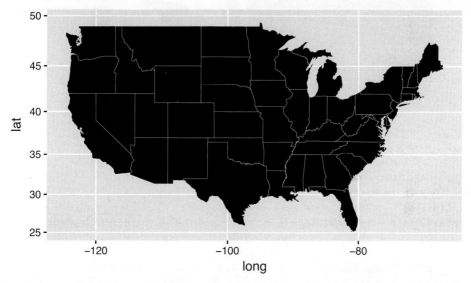

Figure 16.15   A U.S. state map, made with `ggplot2`.

```
# Load a shapefile of U.S. states using ggplot's `map_data()` function
state_shape <- map_data("state")

# Create a blank map of U.S. states
ggplot(state_shape) +
  geom_polygon(
    mapping = aes(x = long, y = lat, group = group),
    color = "white", # show state outlines
    size = .1        # thinly stroked
  ) +
  coord_map() # use a map-based coordinate system
```

---

[11] **U.S. Census: Cartographic Boundary Shapefiles**: https://www.census.gov/geo/maps-data/data/tiger-cart-boundary.html
[12] **OpenStreetMap: Shapefiles**: https://wiki.openstreetmap.org/wiki/Shapefiles

The data in the `state_shape` variable is just a data frame of longitude/latitude points that describe how to draw the outline of each state—the `group` variable indicates which state each point belongs to. If you want each geographic area (in this case, each U.S. state) to express different data through a visual channel such as color, you need to *load* the data, *join* it to the shapefile, and *map* the `fill` of each polygon. As is often the case, the biggest challenge is getting the data in the proper format for visualizing it (not using the visualization package). The map in Figure 16.16, which is built using the following code, shows the eviction rate in each U.S. state in 2016. The data was downloaded from the Eviction Lab at Princeton University.[13]

Figure 16.16 A choropleth map of eviction rates by state, made with `ggplot2`.

```
# Load evictions data
evictions <- read.csv("data/states.csv", stringsAsFactors = FALSE) %>%
  filter(year == 2016) %>% # keep only 2016 data
  mutate(state = tolower(state)) # replace with lowercase for joining

# Join eviction data to the U.S. shapefile
state_shape <- map_data("state") %>% # load state shapefile
  rename(state = region) %>% # rename for joining
  left_join(evictions, by="state") # join eviction data

# Draw the map setting the `fill` of each state using its eviction rate
ggplot(state_shape) +
  geom_polygon(
    mapping = aes(x = long, y = lat, group = group, fill = eviction.rate),
    color = "white", # show state outlines
    size = .1        # thinly stroked
  ) +
  coord_map() + # use a map-based coordinate system
  scale_fill_continuous(low = "#132B43", high = "Red") +
  labs(fill = "Eviction Rate") +
  blank_theme # variable containing map styles (defined in next code snippet)
```

---

[13] **Eviction Lab:** https://evictionlab.org. The Eviction Lab at Princeton University is a project directed by Matthew Desmond and designed by Ashley Gromis, Lavar Edmonds, James Hendrickson, Katie Krywokulski, Lillian Leung, and Adam Porton. The Eviction Lab is funded by the JPB, Gates, and Ford Foundations, as well as the Chan Zuckerberg Initiative.

The beauty and challenge of working with ggplot2 are that nearly every visual feature is configurable. These features can be adjusted using the **theme()** function for any plot (including maps!). Nearly every granular detail—minor grid lines, axis tick color, and more—is available for your manipulation. See the documentation[14] for details. The following is an example set of styles targeted to remove default visual features from maps:

```
# Define a minimalist theme for maps
blank_theme <- theme_bw() +
  theme(
    axis.line = element_blank(),        # remove axis lines
    axis.text = element_blank(),        # remove axis labels
    axis.ticks = element_blank(),       # remove axis ticks
    axis.title = element_blank(),       # remove axis titles
    plot.background = element_blank(),  # remove gray background
    panel.grid.major = element_blank(), # remove major grid lines
    panel.grid.minor = element_blank(), # remove minor grid lines
    panel.border = element_blank()      # remove border around plot
  )
```

16.4.2 Dot Distribution Maps

ggplot also allows you to plot data at discrete locations on a map. Because you are already using a geographic coordinate system, it is somewhat trivial to add discrete points to a map. The following code generates Figure 16.17:

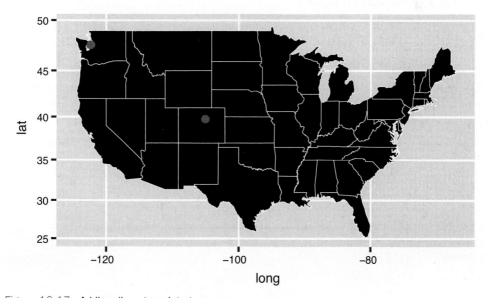

Figure 16.17   Adding discrete points to a map.

---

[14] **ggplot2** themes reference: http://ggplot2.tidyverse.org/reference/index.html#section-themes

```r
# Create a data frame of city coordinates to display
cities <- data.frame(
  city = c("Seattle", "Denver"),
  lat = c(47.6062, 39.7392),
  long = c(-122.3321, -104.9903)
)

# Draw the state outlines, then plot the city points on the map
ggplot(state_shape) +
  geom_polygon(mapping = aes(x = long, y = lat, group = group)) +
  geom_point(
    data = cities, # plots own data set
    mapping = aes(x = long, y = lat), # points are drawn at given coordinates
    color = "red"
  ) +
  coord_map() # use a map-based coordinate system
```

As you seek to increase the granularity of your map visualizations, it may be infeasible to describe every feature with a set of coordinates. This is why many visualizations use images (rather than polygons) to show geographic information such as streets, topography, buildings, and other geographic features. These images are called **map tiles**—they are pictures that can be stitched together to represent a geographic area. Map tiles are usually downloaded from a remote server, and then combined to display the complete map. The **ggmap**[15] package provides a nice extension to ggplot2 for both downloading map tiles and rendering them in R. Map tiles are also used with the Leaflet package, described in Chapter 17.

16.5 `ggplot2` in Action: Mapping Evictions in San Francisco

To demonstrate the power of `ggplot2` as a visualization tool for understanding pertinent social issues, this section visualizes eviction notices filed in San Francisco in 2017.[16] The complete code for this analysis is also available online in the book code repository.[17]

Before mapping this data, a minor amount of formatting needs to be done on the raw data set (shown in Figure 16.18):

```r
# Load and format eviction notices data
# Data downloaded from https://catalog.data.gov/dataset/eviction-notices

# Load packages for data wrangling and visualization
library("dplyr")
library("tidyr")
```

---

[15] ggmap repository on GitHub: https://github.com/dkahle/ggmap
[16] **data.gov: Eviction Notices:** https://catalog.data.gov/dataset/eviction-notices
[17] **ggplot2 in Action:** https://github.com/programming-for-data-science/in-action/tree/master/ggplot2

```
# Load .csv file of notices
notices <- read.csv("data/Eviction_Notices.csv", stringsAsFactors = F)

# Data wrangling: format dates, filter to 2017 notices, extract lat/long data
notices <- notices %>%
  mutate(date = as.Date(File.Date, format="%m/%d/%y")) %>%
  filter(format(date, "%Y") == "2017") %>%
  separate(Location, c("lat", "long"), ", ") %>% # split column at the comma
  mutate(
    lat = as.numeric(gsub("\\(", "", lat)), # remove starting parentheses
    long = as.numeric(gsub("\\)", "", long)) # remove closing parentheses
  )
```

	Eviction.ID	Address	City	State	Eviction.Notice.Source.Zipcode	File.Date
1	M172475	3400 Block Of Cabrillo Street	San Francisco	CA	94121	10/6/17
2	M172687	200 Block Of Lincoln Way	San Francisco	CA	94122	10/23/17
3	M172665	100 Block Of San Jose Avenue	San Francisco	CA	94110	10/27/17
4	M172474	1500 Block Of Gough Street	San Francisco	CA	94109	10/6/17
5	M172571	900 Block Of Larkin Street	San Francisco	CA	94109	10/16/17
6	M172642	2300 Block Of Mission Street	San Francisco	CA	94110	10/19/17
7	M172623	100 Block Of Charles Street	San Francisco	CA	94131	10/19/17
8	M172560	1200 Block Of 40th Avenue	San Francisco	CA	94122	10/13/17
9	M172484	1300 Block Of Clement Street	San Francisco	CA	94118	10/10/17
10	M172684	200 Block Of Lincoln Way	San Francisco	CA	94122	10/23/17

Figure 16.18   A subset of the eviction notices data downloaded from *data.gov*.

To create a background map of San Francisco, you can use the `qmplot()` function from the *development* version of ggmap package (see below). Because the ggmap package is built to work with ggplot2, you can then display points on top of the map as you normally would (using `geom_point()`). Figure 16.19 shows the location of each eviction notice filed in 2017, created using the following code:

> **Tip**: Installing the *development* version of a package using `devtools::install_github` (`"PACKAGE_NAME"`) provides you access to the most recent version of a package, including bug fixes and new—though not always fully tested—features.

```
# Create a map of San Francisco, with a point at each eviction notice address
# Use `install_github()` to install the newer version of `ggmap` on GitHub
# devtools::install_github("dkhale/ggmap") # once per machine
library("ggmap")
library("ggplot2")
```

```r
# Create the background of map tiles
base_plot <- qmplot(
    data = notices,              # name of the data frame
    x = long,                    # data feature for longitude
    y = lat,                     # data feature for latitude
    geom = "blank",              # don't display data points (yet)
    maptype = "toner-background", # map tiles to query
    darken = .7,                 # darken the map tiles
    legend = "topleft"           # location of legend on page
)

# Add the locations of evictions to the map
base_plot +
    geom_point(mapping = aes(x = long, y = lat), color = "red", alpha = .3) +
    labs(title = "Evictions in San Francisco, 2017") +
    theme(plot.margin = margin(.3, 0, 0, 0, "cm")) # adjust spacing around the map
```

Evictions in San Francisco, 2017

Figure 16.19 Location of each eviction notice in San Francisco in 2017. The image is generated by layering points on top of map tiles using the `ggplot2` package.

> **Tip:** You can store a plot returned by the ggplot() function in a variable (as in the preceding code)! This allows you to add different layers on top of a *base* plot, or to render the plot at chosen locations throughout a report (see Chapter 18).

While Figure 16.19 captures the gravity of the issue of evictions in the city, the overlapping nature of the points prevents ready identification of any patterns in the data. Using the geom_polygon() function, you can compute point density across two dimensions and display the computed values in *contours*, as shown in Figure 16.20.

```
# Draw a heatmap of eviction rates, computing the contours
base_plot +
  geom_polygon(
    stat = "density2d", # calculate two-dimensional density of points (contours)
    mapping = aes(fill = stat(level)), # use the computed density to set the fill
    alpha = .3 # Set the alpha (transparency)
  ) +
  scale_fill_gradient2(
    "# of Evictions",
    low = "white",
    mid = "yellow",
    high = "red"
  ) +
  labs(title="Number of Evictions in San Francisco, 2017") +
  theme(plot.margin = margin(.3, 0, 0, 0, "cm"))
```

This example of the geom_polygon() function uses the stat argument to automatically perform a **statistical transformation** (aggregation)—similar to what you could do using the dplyr functions group_by() and summarize()—that calculates the shape and color of each contour based on point density (a "density2d" aggregation). ggplot2 stores the result of this aggregation in an internal data frame in a column labeled level, which can be accessed using the stat() helper function to set the fill (that is, mapping = aes(fill = stat(level))).

> **Tip:** For more examples of producing maps with ggplot2, see this tutorial.[a]
> _____
> [a] http://eriqande.github.io/rep-res-web/lectures/making-maps-with-R.html

This chapter introduced the ggplot2 package for constructing precise data visualizations. While the intricacies of this package can be difficult to master, the investment is well worth the effort, as it enables you to control the granular details of your visualizations.

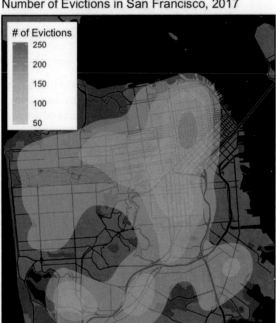

Figure 16.20 A *heatmap* of eviction notices in San Francisco. The image is created by aggregating eviction notices into 2D Contours with one of `ggplot2`'s *statistical transformations*.

> **Tip**: Similar to `dplyr` and many other packages, `ggplot2` has a large number of functions. A cheatsheet for the package is available through the RStudio menu: `Help > Cheatsheets`. In addition, this phenomenal cheatsheet[a] describes how to control the granular details of your `ggplot2` visualizations.
>
> ---
> [a]http://zevross.com/blog/2014/08/04/beautiful-plotting-in-r-a-ggplot2-cheatsheet-3/

For practice creating configurable visualizations with `ggplot2`, see the set of accompanying book exercises.[18]

---

[18]**ggplot2 exercises**: https://github.com/programming-for-data-science/chapter-16-exercises

Interactive Visualization in R

Adding interactivity to a visualization provides an additional mechanism through which data can be presented in an engaging, efficient, and communicative way. Interactions can allow users to effectively explore large data sets by *panning* and *zooming* through plots, or by *hovering* over specific plot geometry to gain additional details on demand.[1]

While ggplot2 is the definitive, leading package for making static plots in R, there is not a comparably popular package for creating *interactive* visualizations. Thus this chapter briefly introduces three different packages for building such visualizations. Instead of offering an in-depth description (as with ggplot2), this chapter provides a high-level "tour" of these packages. The first two (*Plotly* and *Bokeh*) are able to add basic interactions to the plots you might make with ggplot2, while the third (*Leaflet*) is used to create interactive map visualizations. Picking among these (and other) packages depends on the type of interactions you want your visualization to provide, the ease of use, the clarity of the package documentation, and your aesthetic preferences. And because these open source projects are constantly evolving, you will need to reference their documentation to make the best use of these packages. Indeed, exploring these packages further is great practice in learning to use new R packages!

The first two sections demonstrate creating interactive plots of the iris data set, a canonical data set in the machine learning and visualization world in which a flower's species is predicted using features of that flower. The data set is built into the R software, and is partially shown in Figure 17.1.

For example, you can use ggplot2 to create a static visualization of flower *species* in terms of the length of the petals and the sepals (the container for the buds), as shown in Figure 17.2:

```
# Create a static plot of the iris data set
ggplot(data = iris) +
  geom_point(mapping = aes(x = Sepal.Width, y = Petal.Width, color = Species))
```

The following sections show how to use the plotly and rbokeh packages to make this plot interactive. The third section of the chapter then explores interactive mapping with the leaflet package.

---

[1]Shneiderman, B. (1996). The eyes have it: A task by data type taxonomy for information visualizations. *Proceedings of the. 1996 IEEE Symposium on Visual Languages* (pp. 336–). Washington, DC: IEEE Computer Society. http://dl.acm.org/citation.cfm?id=832277.834354

	Sepal.Length	Sepal.Width	Petal.Length	Petal.Width	Species
1	5.1	3.5	1.4	0.2	setosa
2	4.9	3.0	1.4	0.2	setosa
3	4.7	3.2	1.3	0.2	setosa
4	4.6	3.1	1.5	0.2	setosa
5	5.0	3.6	1.4	0.2	setosa
6	5.4	3.9	1.7	0.4	setosa
7	4.6	3.4	1.4	0.3	setosa
8	5.0	3.4	1.5	0.2	setosa

Figure 17.1 A subset of the `iris` data set, in which each observation (row) represents the physical measurements of a flower. This canonical data set is used to practice the machine learning task of *classification*—the challenge is to predict (*classify*) each flower's `Species` based on the other features.

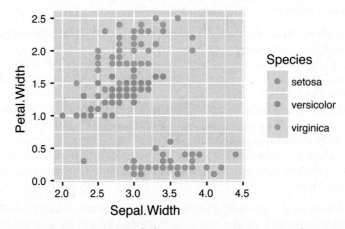

Figure 17.2 A static visualization of the `iris` data set, created using `ggplot2`.

17.1 The `plotly` Package

Plotly[2] is a piece of visualization software that provides open source APIs (programming libraries) for creating interactive visualizations in a wide variety of languages, including R, Python, Matlab, and JavaScript. By default, Plotly charts support a wide range of user interactions, including tooltips on hover, panning, and zooming in on selected regions.

Plotly is an external package (like `dplyr` or `ggplot2`), so you will need to install and load the package before you can use it:

```
install.packages("plotly") # once per machine
library("plotly")           # in each relevant script
```

---

[2] **Plotly:** https://plot.ly/r/

This will make all of the plotting functions you will need available.

With the package loaded, there are two main ways to create interactive plots. First, you can take any plot created using ggplot2 and "wrap" it in a Plotly plot,[3] thereby adding interactions to it. You do this by taking the plot returned by the ggplot() function and passing it into the **ggplotly()** function provided by the plotly package:

```
# Create (and store) a scatterplot of the `iris` data set using ggplot2
flower_plot <- ggplot(data = iris) +
  geom_point(mapping = aes(x = Sepal.Width, y = Petal.Width, color = Species))

# Make the plot interactive by passing it to Plotly's `ggplotly()` function
ggplotly(flower_plot)
```

This will render an interactive version of the iris plot! You can hover the mouse over any geometry element to see details about that data point, or you can click and drag in the plot area to zoom in on a cluster of points (see Figure 17.3).

When you move the mouse over a Plotly chart, you can see the suite of interaction types built into it through the menu that appears (see Figure 17.3). You can use these options to navigate and zoom into the data to explore it.

In addition to making ggplot plots interactive, you can use the Plotly API itself (e.g., calling its own functions) to build interactive graphics. For example, the following code will create an equivalent plot of the iris data set:

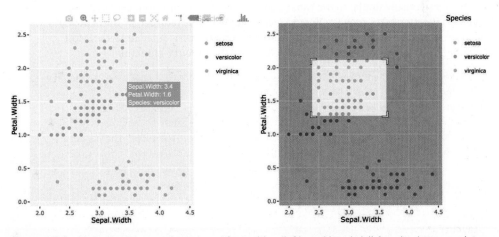

Figure 17.3   Plotly chart interactions: hover for tooltips (left), and brush (click + drag) to zoom into a region (right). More interactions, such as panning, are provided via the interaction menu at the top of the left-hand chart.

---

[3]**Plotly ggplot2 library**: https://plot.ly/ggplot2/ (be sure to check the navigation links in the menu on the left).

```
# Create an interactive plot of the iris data set using Plotly
plot_ly(
  data = iris,      # pass in the data to be visualized
  x = ~Sepal.Width, # use a formula to specify the column for the x-axis
  y = ~Petal.Width, # use a formula to specify the column for the y-axis
  color = ~Species, # use a formula to specify the color encoding
  type = "scatter", # specify the type of plot to create
  mode = "markers"  # determine the "drawing mode" for the scatter (points)
)
```

Plotly plots are created using the **plot_ly()** function, which is a sort of corollary to the ggplot()
function. The plot_ly() function takes as arguments details about how the chart should be
rendered. For example, in the preceding code, arguments are used to specify the data, the aesthetic
mappings, and the plot type (that is, geometry). Aesthetic mappings are specified as *formulas* (using
a tilde ~), indicating that the visual channel is a "function of" the data column. Also note that
Plotly will try to "guess" values such as type and mode if they are left unspecified (and in which
case it will print out a warning in the console).

For a complete list of options available to the plot_ly() function, see the official documentation.[4]
It's often easiest to learn to make Plotly charts by working from one of the many examples.[5] We
suggest that you find an example that is close to what you want to produce, and then read that code
and modify it to fit your particular use case.

In addition to using the plot_ly() function to specify how the data will be rendered, you can add
other chart options, such as titles and axes labels. These are specified using the **layout()** function,
which is conceptually similar to the labs() and theme() functions from ggplot2. Plotly's
layout() function takes as an argument a *Plotly chart* (e.g., one returned by the plot_ly()
function), and then modifies that object to produce a chart with a different layout. Most
commonly, this is done by *piping* the Plotly chart into the layout() function:

```
# Create a plot, then pipe that plot into the `layout()` function to modify it
# (Example adapted from the Plotly documentation)
plot_ly(
  data = iris,      # pass in the data to be visualized
  x = ~Sepal.Width, # use a formula to specify the column for the x-axis
  y = ~Petal.Width, # use a formula to specify the column for the y-axis
  color = ~Species, # use a formula to specify the color encoding
  type = "scatter", # specify the type of plot to create
  mode = "markers"  # determine the "drawing mode" for the scatter (points)
) %>%
  layout(
    title = "Iris Data Set Visualization",          # plot title
    xaxis = list(title = "Sepal Width", ticksuffix = "cm"), # axis label + format
    yaxis = list(title = "Petal Width", ticksuffix = "cm")  # axis label + format
  )
```

---

[4] **Plotly: R Figure Reference**: https://plot.ly/r/reference/
[5] **Plotly: Basic Charts** example gallery: https://plot.ly/r/#basic-charts

The chart created by this code is shown in Figure 17.4. The xaxis and yaxis arguments expect *lists* of axis properties, allowing you to control many aspects of each axis (such as the title and the ticksuffix to put after each numeric value in the axis). You can read about the structure and options to the other arguments in the API documentation.[6]

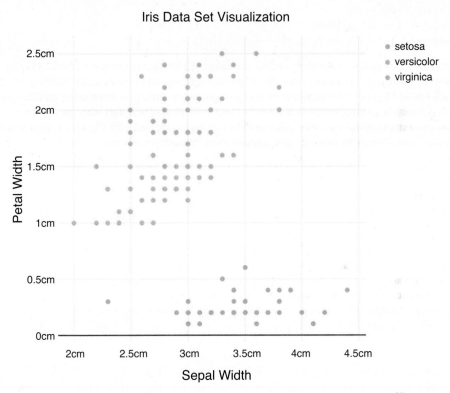

Figure 17.4   A Plotly chart with informative labels and axes added using the layout() function.

17.2   The **rbokeh** Package

Bokeh[7] is a visualization package that provides a similar set of interactive features as Plotly (including hover tooltips, drag-to-pan, and box zoom effects). Originally developed for the Python programming language, Bokeh can be used in R through the **rbokeh** package.[8] While not as popular as Plotly, Bokeh's API and documentation can be more approachable than Plotly's examples.

---

[6] **Plotly layout**: https://plot.ly/r/reference/#layout
[7] **Bokeh**: http://bokeh.pydata.org
[8] **rbokeh, R Interface for Bokeh**: http://hafen.github.io/rbokeh/

As with other packages, you will need to install and load the rbokeh package before you can use it. At the time of this writing, the version of rbokeh on CRAN (what is installed with install.packages()) gives warnings—but not errors!—for R version 3.4; installing a development version from the package's maintainer Ryan Hafen fixes this problem.

```
# Use `install_github()` to install the version of a package on GitHub
# (often newer)
devtools::install_github("hafen/rbokeh") # once per machine
library("rbokeh") # in each relevant script
```

You create a new plot with Bokeh by calling the **figure()** function (which is a corollary to the ggplot() and plot_ly() functions). The figure() function will create a new plotting area, to which you add layers of plot elements such as plot geometry. Similar to when using geometries in ggplot2, each layer is created with a different function—all of which start with the ly_ prefix. These layer functions take as a first argument the plot region created with figure(), so in practice they are "added" to a plot through piping rather than through the addition operator.

For example, the following code shows how to recreate the iris visualization using Bokeh (shown in Figure 17.5):

```
# Create an interactive plot of the iris data set using Bokeh
figure(
  data = iris,                        # data for the figure
  title = "Iris Data Set Visualization" # title for the figure
) %>%
  ly_points(
    Sepal.Width,    # column for the x-axis (without quotes!)
    Petal.Width,    # column for the y-axis (without quotes!)
    color = Species # column for the color encoding (without quotes!)
  ) %>%
  x_axis(
    label = "Sepal Width",          # label for the axis
    number_formatter = "printf",    # formatter for each axis tick
    format = "%s cm",               # specify the desired tick labeling
  ) %>%
  y_axis(
    label = "Petal Width",          # label for the axis
    number_formatter = "printf",    # formatter for each axis tick
    format = "%s cm",               # specify the desired tick labeling
  )
```

The code for adding layers is reminiscent of how geometries act as layers in ggplot2. Bokeh even supports non-standard evaluation (referring to column names without quotes) just like ggplot2—as opposed to Plotly's reliance on formulas. However, formatting the axis tick marks is more verbose with Bokeh (and is not particularly clear in the documentation).

The plot that is generated by Bokeh (Figure 17.5) is quite similar to the version generated by Plotly (Figure 17.4) in terms of general layout, and offers a comparable set of interaction utilities through a

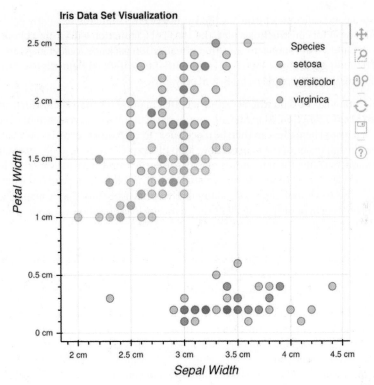

Figure 17.5   A Bokeh chart with styled axes. Note the interaction menu to the right of the chart.

toolbar to the right of the chart. Thus you might choose between these packages based on which coding style you prefer, as well as any other aesthetic or interactive design choices of the packages.

17.3   The `leaflet` Package

Leaflet[9] is an open source JavaScript library for building interactive maps, which you can use in R through the **leaflet** package.[10] Maps built with Leaflet have rich interactivity by default, including the ability to pan, zoom, hover, and click on map elements and markers. They can also be customized to support formatted labels or respond to particular actions. Indeed, many of the interactive maps you see accompanying online news articles are created using Leaflet.

As with other packages, you will need to install and load the `leaflet` package before you can use it:

```
install.packages("leaflet") # once per machine
library("leaflet")          # in each relevant script
```

---

[9]**Leaflet**: https://leafletjs.com
[10]**Leaflet for R**: https://rstudio.github.io/leaflet/

You can create a new Leaflet map by calling the **leaflet()** function. Just as calling ggplot() will create a blank canvas for constructing a plot, the leaflet() function will create a blank canvas on which you can build a map. Similar to the other visualization packages, Leaflet maps are then constructed by adding (via pipes) a series of layers with different visual elements to constitute the image—including map tiles, markers, lines, and polygons.

The most important layer to add when creating a Leaflet map are the **map tiles**, which are added with the **addTiles()** function. Map tiles are a series of small square images, each of which shows a single piece of a map. These tiles can then be placed next to each other (like tiles on a bathroom floor) to form the full image of the map to show. Map tiles power mapping applications like Leaflet and Google Maps, enabling them to show a map of the entire world at a wide variety of levels of zoom (from street level to continent level); which tiles will be rendered depends on what region and zoom level the user is looking at. As you interactively navigate through the map (e.g., panning to the side or zooming in or out), Leaflet will automatically load and show the appropriate tiles to display the desired map!

> **Fun Fact**: It takes 366,503,875,925 tiles (each 256×256 pixels) to map the entire globe for the (standard) 20 different zoom levels!

There are many different sources of map tiles that you can use in your maps, each of which has its own appearance and included information (e.g., rivers, streets, and buildings). By default, Leaflet will use tiles from *OpenStreetMap*,[11] an open source set of map tiles. OpenStreetMap provides a number of different tile sets; you can choose which to use by passing in the name of the tile set (or a URL schema for the tiles) to the addTiles() function. But you can also choose to use another map tile provider[12] depending on your aesthetic preferences and desired information. You do this by instead using the addProviderTiles() function (again passing in the name of the tile set). For example, the following code creates a basic map (Figure 17.6) using map tiles from the *Carto*[13] service. Note the use of the setView() function to specify where to center the map (including the "zoom level").

```
# Create a new map and add a layer of map tiles from CartoDB
leaflet() %>%
  addProviderTiles("CartoDB.Positron") %>%
  setView(lng = -122.3321, lat = 47.6062, zoom = 10) # center the map on Seattle
```

The rendered map will be *interactive* in the sense that you can drag and scroll to pan and zoom—just as with other online mapping services!

After rendering a basic map with a chosen set of map tiles, you can add further layers to the map to show more information. For instance, you can add a layer of **shapes** or **markers** to help answer questions about events that occur at specific geographic locations. To do this, you will need to pass the data to map into the leaflet() function call as the data argument (i.e., leaflet(data = SOME_DATA_FRAME)). You can then use the addCircles() function to add a layer of circles to the

---

[11] **OpenStreetMap** map data service: https://www.openstreetmap.org
[12] **Leaflet-providers preview** http://leaflet-extras.github.io/leaflet-providers/preview/
[13] **Carto** map data service: https://carto.com

Figure 17.6 A map of Seattle, created using the `leaflet` package. The image is constructed by stitching together a layer of *map tiles*, provided by the Carto service.

map (similar to adding a geometry in `ggplot2`). This function will take as arguments the data columns to map to the circle's location aesthetics, specified as formulas (with a ~).

```
# Create a data frame of locations to add as a layer of circles to the map
locations <- data.frame(
  label = c("University of Washington", "Seattle Central College"),
  latitude = c(47.6553, 47.6163),
  longitude = c(-122.3035, -122.3216)
)

# Create the map of Seattle, specifying the data to use and a layer of circles
leaflet(data = locations) %>% # specify the data you want to add as a layer
  addProviderTiles("CartoDB.Positron") %>%
  setView(lng = -122.3321, lat = 47.6062, zoom = 11) %>% # focus on Seattle
  addCircles(
    lat = ~latitude,    # a formula specifying the column to use for latitude
    lng = ~ longitude,  # a formula specifying the column to use for longitude
    popup = ~label,     # a formula specifying the information to pop up
    radius = 500,       # radius for the circles, in meters
    stroke = FALSE      # remove the outline from each circle
  )
```

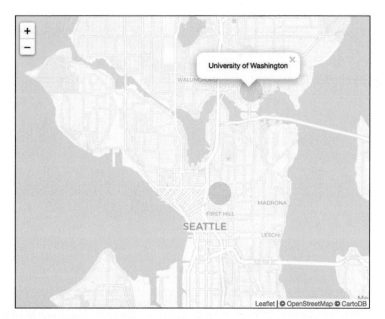

Figure 17.7   A map showing two universities in Seattle, created by adding a layer of markers (addCircles()) on top of a layer of map tiles.

> **Caution**: Interactive visualization packages such as plotly and leaflet are limited in the number of markers they can display. Because they render *scalable vector graphics* (SVGs) rather than raster images, they actually add a new visual element for each marker. As a result they are often unable to handle more than a few thousand points (something that isn't an issue with ggplot2).

The preceding code also adds interactivity to the map by providing *popups*—information that pops up *on click* and remains displayed—as shown in Figure 17.7. Because these popups appear when users are interacting with the circle elements you created, they are specified as another argument to the addCircles() function—that is, as a value of the formula for which column to map to the popup. Alternatively, you can cause labels to appear *on hover* by passing in the label argument instead of popup.

17.4   Interactive Visualization in Action: Exploring Changes to the City of Seattle

This section demonstrates using an interactive visualization in an attempt to evaluate the claim that *"The City of Seattle is changing"* (in large part due to the growing technology industry) by analyzing construction projects as documented through building permit data[14] downloaded from

---

[14]**City of Seattle Land use permits**: https://data.seattle.gov/Permitting/Building-Permits/76t5-zqzr

	PermitNum	PermitClass	PermitClassMapped	PermitTypeMapped	PermitTypeDesc	Description
1	6243602-CN	Commercial	Non-Residential	Building	New	Install portable office building (unit...
2	6408217-CN	Single Family/Duplex	Residential	Building	New	Establish use as and Construct new...
3	6285442-CN	Single Family/Duplex	Residential	Building	New	Establish use as and construct new ...
4	6343245-CN	Multifamily	Residential	Building	New	Establish use as and construct six-...
5	6547255-CN	Single Family/Duplex	Residential	Building	New	Establish use as and construct new ...
6	6271097-CN	Commercial	Non-Residential	Building	New	Establish use as car wash and mino...
7	6454733-CN	Single Family/Duplex	Residential	Building	New	Establish use as and construct new ...
8	6213875-PH	Multifamily	Residential	Building	New	Phased project: Construction of a r...
9	6583694-CN	Single Family/Duplex	Residential	Building	New	Construct East duplex, per plan (Es...
10	6312868-CN	Single Family/Duplex	Residential	Building	New	Establish use as and construct new ...
11	6464007-CN	Single Family/Duplex	Residential	Building	New	Establish use as and Construct new...
12	6363269-CN	Single Family/Duplex	Residential	Building	New	Construct CNTR single family resid...

Figure 17.8 City of Seattle data on permits for buildings in Seattle, showing the subset of new permits since 2010.

the City of Seattle's open data program. A subset of this data is shown in Figure 17.8. The complete code for this analysis is also available online in the book code repository.[15]

First, the data needs to be loaded into R and filtered down to the subset of data of interest (new buildings since 2010):

```
# Load data downloaded from
# https://data.seattle.gov/Permitting/Building-Permits/76t5-zqzr
all_permits <- read.csv("data/Building_Permits.csv", stringsAsFactors = FALSE)

# Filter for permits for new buildings issued in 2010 or later
new_buildings <- all_permits %>%
  filter(
    PermitTypeDesc == "New",
    PermitClass != "N/A",
    as.Date(all_permits$IssuedDate) >= as.Date("2010-01-01") # filter by date
  )
```

Before mapping these points, you may want to get a higher-level view of the data. For example, you could aggregate the data to show the number of permits issued per year. This will again involve a bit of data wrangling, which is often the most time-consuming part of visualization:

```
# Create a new column storing the year the permit was issued
new_buildings <- new_buildings %>%
  mutate(year = substr(IssuedDate, 1, 4)) # extract the year

# Calculate the number of permits issued by year
by_year <- new_buildings %>%
  group_by(year) %>%
  count()
```

---

[15]**Interactive visualization in action**: https://github.com/programming-for-data-science/in-action/tree/master/interactive-vis

```
# Use plotly to create an interactive visualization of the data
plot_ly(
  data = by_year, # data frame to show
  x = ~year,       # variable for the x-axis, specified as a formula
  y = ~n,          # variable for the y-axis, specified as a formula
  type = "bar",    # create a chart of type "bar" -- a bar chart
  alpha = .7,      # adjust the opacity of the bars
  hovertext = "y" # show the y-value when hovering over a bar
) %>%
  layout(
    title = "Number of new building permits per year in Seattle",
    xaxis = list(title = "Year"),
    yaxis = list(title = "Number of Permits")
  )
```

The preceding code produces the bar chart shown in Figure 17.9. Keep in mind that the data was downloaded before the summer of 2018, so the observed downward trend is an artifact of when the visualization was created!

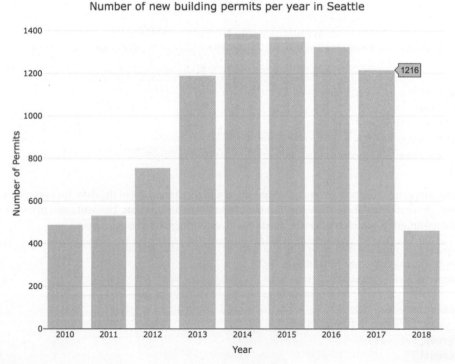

Figure 17.9 The number of permits issued for new buildings in Seattle since 2010. The chart was built before the summer of 2018.

After understanding this high-level view of the data, you likely want to know *where* buildings are being constructed. To do so, you can take the previous map of Seattle and add an additional layer of circles on top of the tiles (one for each building constructed) using the `addCircles()` function:

```
# Create a Leaflet map, adding map tiles and circle markers
leaflet(data = new_buildings) %>%
  addProviderTiles("CartoDB.Positron") %>%
  setView(lng = -122.3321, lat = 47.6062, zoom = 10) %>%
  addCircles(
    lat = ~Latitude,      # specify the column for `lat` as a formula
    lng = ~Longitude,     # specify the column for `lng` as a formula
    stroke = FALSE,       # remove border from each circle
    popup = ~Description  # show the description in a popup
  )
```

The results of this code are shown in Figure 17.10—it's a lot of new buildings. And because the map is interactive, you can click on each one to get more details!

While this visualization shows all of the new construction, it leaves unanswered the question of *who benefits* and *who suffers* as a result of this change. You would need to do further research into the number of affordable housing units being built, and the impact on low-income and homeless communities. As you may discover, building at such a rapid pace often has a detrimental effect on housing security in a city.

As with `ggplot2`, the visual attributes of each shape or marker (such as the size or color) can also be driven by data. For example, you could use information about the permit classification (i.e., if the

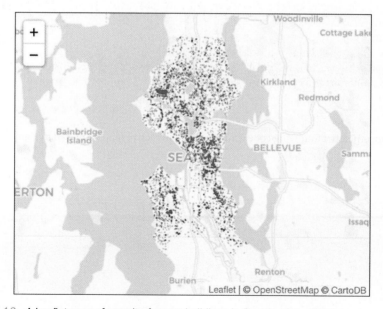

Figure 17.10   A Leaflet map of permits for new buildings in Seattle since 2010.

permit is for a home versus a commercial building) to color the individual circles. To effectively map this (categorical) data to a set of colors in Leaflet, you can use the **colorFactor( )** function. This function is a lot like a scale in `ggplot2`, in that it returns a specific mapping to use:

```
# Construct a function that returns a color based on the PermitClass column
# Colors are taken from the ColorBrewer Set3 palette
palette_fn <- colorFactor(palette = "Set3", domain = new_buildings$PermitClass)
```

The `colorFactor( )` function returns a new function (here called `palette_fn( )`) that maps from a set of data values (here the unique values from the `PermitClass` column) to a set of colors—it performs an aesthetic mapping. You can use this function to specify how the circles on the map should be rendered (as with `ggplot2` geometries, further arguments can be used to customize the shape rendering):

```
# Modify the `addCircles()` method to specify color using `palette_fn()`
addCircles(
  lat = ~Latitude, # specify the column for `lat` as a formula
  lng = ~Longitude, # specify the column for `lng` as a formula
  stroke = FALSE, # remove border from each circle
  popup = ~Description, # show the description in a popup
  color = ~palette_fn(PermitClass) # a "function of" the palette mapping
)
```

To make these colors meaningful, you will need to add a legend to your map. As you might have expected, you can do this by adding another layer with a legend in it, specifying the color scale, values, and other attributes:

```
# Add a legend layer in the "bottomright" of the map
addLegend(
  position = "bottomright",
  title = "New Buildings in Seattle",
  pal = palette_fn, # the color palette described by the legend
  values = ~PermitClass, # the data values described by the legend
  opacity = 1
)
```

Putting it together, the following code generates the interactive map displayed in Figure 17.11.

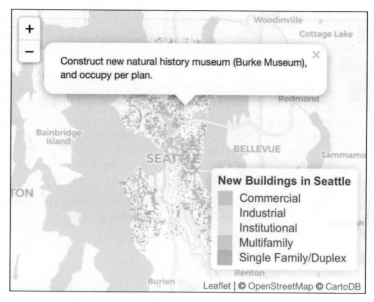

Figure 17.11   A Leaflet map of permits for new buildings in Seattle since 2010, colored by construction category.

```
# Create a Leaflet map of new building construction by category
leaflet(data = new_buildings) %>%
  addProviderTiles("CartoDB.Positron") %>%
  setView(lng = -122.3321, lat = 47.6062, zoom = 10) %>%
  addCircles(
    lat = ~Latitude, # specify the column for `lat` as a formula
    lng = ~Longitude, # specify the column for `lng` as a formula
    stroke = FALSE, # remove border from each circle
    popup = ~Description, # show the description in a popup
    color = ~palette_fn(PermitClass), # a "function of" the palette mapping
    radius = 20,
    fillOpacity = 0.5
  ) %>%
  addLegend(
    position = "bottomright",
    title = "New Buildings in Seattle",
    pal = palette_fn, # the palette to label
    values = ~PermitClass, # the values to label
    opacity = 1
  )
```

In summary, packages for developing interactive visualizations (whether plots or maps) use the same general concepts as ggplot2, but with their own preferred syntax for specifying plot options and customizations. As you choose among these (and other) packages for making visualizations, consider the style of code you prefer to use, the trade-off of customizability versus ease of use, and the visual design choices of each package. There are dozens (if not hundreds) of other packages available and more created every day; exploring and learning these packages is an excellent way to expand your programming and data science skills.

That said, when you are exploring new packages, be careful about using code that is poorly documented or not widely used—such packages may have internal errors, memory leaks, or even security flaws that haven't been noticed or addressed yet. It's a good idea to view the package code *on GitHub*, where you can check the popularity by looking at the number of *stars* (similar to "likes") and *forks* for the project, as well as how actively and recently new commits have been made to the code. Such research and consideration are vital when choosing one of the many packages for building interactive visualizations—or doing any other kind of work—with R.

For practice building interactive visualizations, see the set of accompanying book exercises.[16]

---

[16]**Interactive visualization exercises:** https://github.com/programming-for-data-science/chapter-17-exercises

VI

Building and Sharing Applications

The final part of this book focuses on the technologies that allow you to collaborate with others and share your work with the world. It walks through multiple approaches to building interactive web applications (Chapter 18, Chapter 19), and explains how to leverage git and GitHub when working as a member of a team (Chapter 20).

Dynamic Reports with R Markdown

The insights you discover through your analysis are only valuable if you can share them with others. To do this, it's important to have a simple, repeatable process for combining the set of charts, tables, and statistics you generate into an easily presentable format.

This chapter introduces **R Markdown**[1] as a tool for compiling and sharing your results. R Markdown is a development framework that supports using R to dynamically create documents, such as websites (`.html` files), reports (`.pdf` files), and even slideshows (using `ioslides` or `slidy`).

As you may have guessed, R Markdown does this by providing the ability to blend Markdown syntax and R code so that, when compiled and executed, the results from your code will be automatically injected into a formatted document. The ability to automatically generate reports and documents from a computer script eliminates the need to manually update the results of a data analysis project, enabling you to more effectively share the information that you've produced from your data. In this chapter, you will learn the fundamentals of the R Markdown package so that you can create well-formatted documents that combine analysis and reporting.

> **Fun Fact**: This book was written using R Markdown!

18.1 Setting Up a Report

R Markdown documents are created from a combination of two packages: **rmarkdown** (which processes the markdown and generates the output) and **knitr**[2] (which runs R code and produces Markdown-like output). These packages are produced by and already included in RStudio, which provides direct support for creating and viewing R Markdown documents.

---

[1] **R Markdown**: https://rmarkdown.rstudio.com
[2] **knitr** package: https://yihui.name/knitr/

18.1.1 Creating .Rmd Files

The easiest way to create a new R Markdown document in RStudio is to use the `File > New File > R Markdown` menu option (see Figure 18.1), which opens a document creation wizard.

RStudio will then prompt you to provide some additional details about what kind of R Markdown document you want to create (shown in Figure 18.2). In particular, you will need to choose a default document type and output format. You can also provide a title and author information that will be

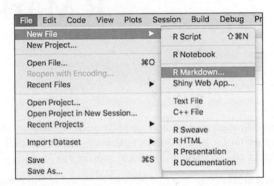

Figure 18.1 Create a new R Markdown document in RStudio via the dropdown menu (`File > New File > R Markdown`).

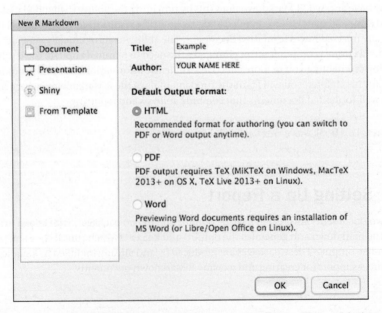

Figure 18.2 RStudio wizard for creating R Markdown documents. Enter a Title and Author, and select the document output format (we suggest HTML to start).

included in the document. This chapter focuses on creating HTML documents (websites, the default format); other formats require the installation of additional software.

Once you've chosen your desired document type and output format, RStudio will open up a new script file for you. You should save this file with the extension `.Rmd` (for "R Markdown"), which tells the computer and RStudio that the document contains Markdown content with embedded R code. If you use a different extension, RStudio won't know how to interpret the code and render the output!

The wizard-generated file contains some example code demonstrating how to write an R Markdown document. Understanding the basic structure of this file will enable you to insert your own content into this structure.

A `.Rmd` file has three major types of content: the header, the Markdown content, and R code chunks.

- The **header** is found at the top of the file, and includes text with the following format:

```
---
title: "EXAMPLE_TITLE"
author: "YOUR_NAME"
date: "2/01/2018"
output: html_document
---
```

  This header is written in *YAML*[3] format, which is *yet another* way of formatting structured data, similar to CSV or JSON. In fact, YAML is a superset of JSON and can represent the same data structures, just using indentation and dashes instead of braces and commas.

  The header contains **meta-data**, or information about the file and how it should be processed and rendered. For example, the title, author, and date will be automatically included and displayed at the top of your generated document. You can include additional information and configuration options as well, such as whether there should be a table of contents. See the R Markdown documentation[4] for further details.

- Everything below the header is the content that will be included in your report, and is primarily made up of **Markdown content**. This is normal Markdown text like that described in Chapter 4. For example, you could include the following markdown code in your `.Rmd` file:

```
## Second Level Header
This is just plain markdown that can contain **bold** or _italics_.
```

  R Markdown also provides the ability to render code content *inline* with the Markdown content, as described later in this chapter.

- **R code chunks** can be included in the middle of the regular Markdown content. These segments (chunks) of R code look like normal code block elements (using three backticks ` ``` `), but with an extra `{r}` immediately after the opening set of backticks. Inside these code chunks you include regular R code, which will be evaluated and then rendered

---

[3] **YAML**: http://yaml.org
[4] **R Markdown HTML Documents**: http://rmarkdown.rstudio.com/html_document_format.html

into the document. Section 18.2 provides more details about the format and process used by these chunks.

```{r}
# R code chunk in an R Markdown file
some_variable <- 100
```

Combining these content types (header, markdown, and code chunks), you will be able to reproducibly create documents to share your insights.

18.1.2   Knitting Documents

RStudio provides a direct interface to compile your .Rmd source code into an actual document (a process called **knitting**, performed by the knitr package). To do so, click the **Knit** button at the top of the script panel, shown in Figure 18.3. This button will compile the code and generate the document (into the same directory as your saved .Rmd file), as well as open up a preview window in RStudio.

While it is straightforward to generate such documents, the knitting process can make it hard to debug errors in your R code (whether syntax or logical), in part because the output may or may not show up in the document! We suggest that you write complex R code in another script and then use the source() function to insert that script into your .Rmd file and use calculated variables in your output (see Chapter 14 for details and examples of the source() function). This makes it possible to test your data processing work outside of the knitted document. It also **separates the concerns** of the data and its representation—which is good programming practice.

Nevertheless, you should be sure to knit your document frequently, paying close attention to any errors that appear in the console.

> **Tip:** If you're having trouble finding your error, a good strategy is to systematically remove ("comment out") segments of your code and attempt to re-knit the document. This will help you identify the problematic syntax.

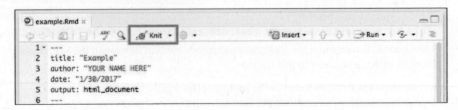

Figure 18.3   Click on RStudio's Knit button to compile your code to the desired document type (e.g., HTML).

18.2   Integrating Markdown and **R** Code

What makes R Markdown distinct from simple Markdown code is the ability to actually execute your R code and include the output directly in the document. R code can be executed and included in the document in blocks of code, or even inline with other content!

18.2.1   R Code Chunks

Code that is to be executed (rather than just displayed as formatted text) is called a **code chunk**. To specify a code chunk, you need to include **{r}** immediately after the backticks that start the code block (the ```` ``` ````). You can type this out yourself, or use the keyboard shortcut (cmd+alt+i) to create one. For example:

```
Write normal **markdown** out here, then create a code block:

```{r}
# Execute R code in here
course_number <- 201
```

Back to writing _markdown_ out here.
```

By default, the code chunk will execute the R code listed, and then render both the code that was executed and the result of the last statement into the Markdown—similar to what would be returned by a function. Indeed, you can think of code chunks as functions that calculate and return a value that will be included in the rendered report. If your code chunk doesn't return a particular expression (e.g., the last line is just an assignment), then no returned output will be rendered, although R Markdown will still render the code that was executed.

It is also possible to specify additional **configuration options** by including a comma-separated list of named arguments (as you've done with lists and functions) inside the curly braces following the r:

```
```{r options_example, echo = FALSE, message = TRUE)
# A code chunk named "options_example", with argument `echo` assigned FALSE
# and argument `message` assigned TRUE

# Would execute R code in here
```
```

The first "argument" (options_example) is a "name" or label for the chunk; it is followed by named arguments (written in option = VALUE format) for the options. While including chunk names is technically optional, this practice will help you create well-documented code and reference results in the text. It will also help in the debugging process, as it will allow RStudio to produce more detailed error messages.

There are many options[5] you can use when creating code chunks. Some of the most useful ones have to do with how the executed code is output in the document:

- **echo** indicates whether you want the *R code itself* to be displayed in the document (i.e., if you want readers to be able to see your work and reproduce your calculations and analysis). The value is either TRUE (do display; the default) or FALSE (do not display).

- **message** indicates whether you want any messages generated by the code to be displayed. *This includes print statements!* The value is either TRUE (do display; the default) or FALSE (do not display).

- **include** indicates if *any* results of the code should be output in the report. Note that any code in this chunk will still be executed—it just won't be included in the output. It is extremely common and best practice to have a "setup" code chunk at the beginning of your report that has the include = FALSE option and is used to do initial processing work—such as library() packages, source() analysis code, or perform some other data wrangling. The R Markdown reports produced by RStudio's wizard include a code chunk like this.

If you want to *show* your R code but not *evaluate* it, you can use a standard Markdown code block that indicates the r language (```r instead of ```{r}), or set the eval option to FALSE.

18.2.2 Inline Code

In addition to creating distinct code blocks, you will commonly want to execute R code inline with the rest of your text. This empowers you to reference a variable defined in a code chunk in a section of Markdown—injecting the value stored in a variable into the text you have written. Using this technique, you can include a specific result inside a paragraph of text; if the computation changes, re-knitting your document will update the values inside the text without any further work needed.

Recall that a single backtick (` ) is the Markdown syntax for making text display as `code`. You can make R Markdown *evaluate*—rather than display—inline code by adding the letter **r** and a space immediately after the first backtick. For example:

```
To calculate 3 + 4 inside some text, you can use `r 3 + 4` right in the _middle_.
```

When you knit this text, `r 3 + 4` would be replaced with the number 7 (what 3 + 4 evaluates to).

You can also reference values computed in any code chunks that precede the inline code. For example, `r SOME_VARIABLE` would include the value of SOME_VARIABLE inline with the paragraph. In fact, it is best practice to do your calculations in a code block (with the echo = FALSE option), save the result in a variable, and then inline that variable to display it.

> **Tip:** To quickly access the R Markdown Cheatsheet and Reference, use the RStudio menu: Help > Cheatsheets.

---

[5]**knitr** Chunk options and package options: https://yihui.name/knitr/options/

18.3   Rendering Data and Visualizations in Reports

R Markdown's code chunks let you perform data analysis directly in your document, but you will often want to include more complex data output than just the resulting numbers. This section discusses a few tips for specifying dynamic, complex output to render using R Markdown.

18.3.1   Rendering Strings

If you experiment with knitting R Markdown, you will quickly notice that using `print()` will generate content that looks like a printed vector (e.g., what you see in the console in RStudio). For example:

````
```{r raw_print_example, echo = FALSE}
print("Hello world")
```
````

will produce:

```
## [1] "Hello world"
```

For this reason, you usually want to have the code block generate a string that you save in a variable, which you can then display with an inline expression (e.g., on its own line):

````
```{r stored_print_example, echo = FALSE}
msg <- "**Hello world**"
```
````

```
Below is the message to see:
```

`` `r msg` ``

When `knit`, this code produces the text shown in Figure 18.4. Note that the Markdown syntax included in the variable is rendered as well: `` `r msg` `` is replaced by the value of the expression just as if you had typed that Markdown in directly. This allows you to even include dynamic styling if you construct a "Markdown string" (i.e., containing Markdown syntax) from your data.

Figure 18.4   A preview of the `.html` file that is created by knitting an R Markdown document containing a chunk that stores a message in a variable and an inline expression of that message.

Alternatively, you can give your chunk a `results` option[6] with a value `"asis"`, which will cause the output to be rendered directly into the Markdown. When combined with the base R function **cat( )** (which concatenates content without specifying additional information such as vector position), you can make a code chunk effectively render a specific string:

```
```{r asis_example, results = "asis", echo = FALSE}
cat("**Hello world**")
```
```

18.3.2   Rendering Markdown Lists

Because output strings render any Markdown they contain, it's possible to construct these Markdown strings so that they contain more complex structures such as **unordered lists**. To do this, you specify the string to include the - symbols used to indicate a Markdown list (with each item in the list separated by a line break or a \n character):

```
```{r list_example, echo = FALSE}
markdown_list <- "
- Lions
- Tigers
- Bears
- Oh mys
"
```


`r markdown_list`
```

This code outputs a list that looks like this:

- Lions

- Tigers

- Bears

- Oh mys

When this approach is combined with the vectorized `paste( )` function and its `collapse` argument, it becomes possible to convert vectors into Markdown lists that can be rendered:

```
```{r pasted_list_example, echo = FALSE}
# Create a vector of animals
animals <- c("Lions", "Tigers", "Bears", "Oh mys")
```
```

---

[6]**knitr** text result options: https://yihui.name/knitr/options/#text-results

```
# Paste `-` in front of each animal and join the items together with
# newlines between
markdown_list <- paste("-", animals, collapse = "\n")
```

`r markdown_list`

Of course, the contents of the vector (e.g., the text `"Lions"`) could include additional Markdown syntax to make it bold, italic, or hyperlinked text.

> **Tip**: Creating a "helper function" to help with formatting your output is a great approach. For some other work in this area, see the **pander**[a] package.
>
> ─────────────────
> [a] http://rapporter.github.io/pander/

18.3.3 Rendering Tables

Because data frames are so central to programming with R, R Markdown includes capabilities that enable you to render data frames as Markdown *tables* via the knitr package's **kable()** function. This function takes as an argument the data frame you wish to render, and it will automatically convert that value into a string of text representing a Markdown table:

```
```{r kable_example, echo = FALSE}
library("knitr") # make sure you load the package (once per document)

# Make a data frame
letters <- c("a", "b", "c", "d")
numbers <- 1:4
df <- data.frame(letters = letters, numbers = numbers)

# "Return" the table to render it
kable(df)
```
```

Figure 18.5 compares the rendered R Markdown results with and without the kable() function. The kable() function supports a number of other arguments that can be used to customize how it outputs a table; see the documentation for details. Again, if the values in the data frame are strings that contain Markdown syntax (e.g., bold, italics, or hyperlinks), they will be rendered as such in the table!

> **Going Further**: Tables generated with the kable() function can be further customized using additional packages, such as **kableExtra**.[a] This package allows you to add more layers and styling to a table using a format similar to how you add labels and themes with ggplot2.
>
> ─────────────────
> [a] http://haozhu233.github.io/kableExtra/

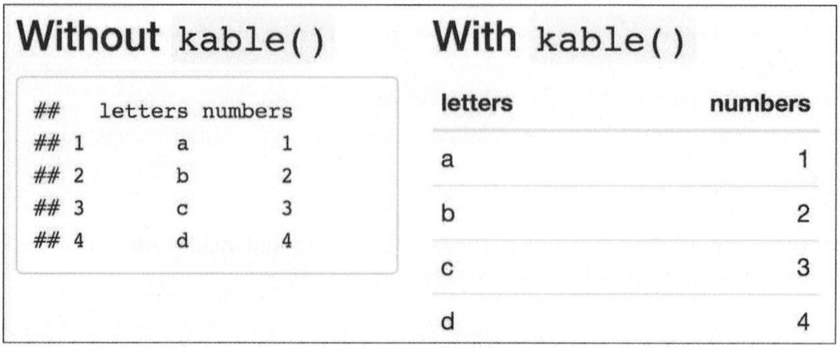

Without kable()			With kable()	
## letters numbers			letters	numbers
## 1 a 1			a	1
## 2 b 2				
## 3 c 3			b	2
## 4 d 4			c	3
			d	4

Figure 18.5   R Markdown rendering a data frame with and without the kable() function.

So while you may need to do a little bit of work to manually generate the Markdown syntax, R Markdown makes it is possible to dynamically produce complex documents based on dynamic data sources.

18.3.4   Rendering Plots

You can also include visualizations created by R in your rendered reports! To do so, you have the code chunk "return" the plot you wish to render:

```
```{r plot_example, echo = FALSE}
library("ggplot2") # make sure you load the package (once per document)

# Plot of college education vs. poverty rates in the Midwest
ggplot(data = midwest) +
  geom_point(
    mapping = aes(x = percollege, y = percadultpoverty, color = state)
  ) +
  scale_color_brewer(palette = "Set3")
```
```

When knit, the document generated that includes this code would include the ggplot2 chart. Moreover, RStudio allows you to *preview* each code chunk before knitting—just click the green play button icon above each chunk, as shown in Figure 18.6. While this can help you debug individual chunks, it may be tedious to do in longer scripts, especially if variables in one code chunk rely on an earlier chunk.

It is best practice to do any data wrangling necessary to prepare the data for your plot in a separate .R file, which you can then source() into the R Markdown (in an initial setup code chunk with the include = FALSE option). See Section 18.5 for an example of this organization.

18.4   Sharing Reports as Websites

The default output format for new R Markdown scripts created with RStudio is **HTML** (with the content saved in a .html file). HTML stands for *HyperText Markup Language* and, like the Markdown

```
```{r stored_print_example, echo=FALSE}
library("ggplot2")  # make sure you load the package (once per document)

# Plot of college education vs. poverty rates in the midwest
ggplot(data = midwest) +
  geom_point(mapping = aes(x = percollege, y = percadultpoverty, color = state)) +
  scale_color_brewer(palette = "Set3")
```
```

Click to produce
preview below.

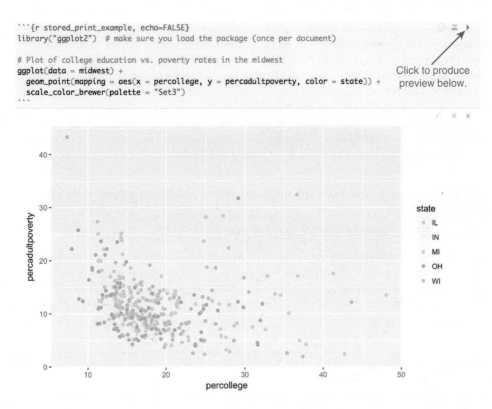

Figure 18.6 A preview of the content generated by `knitr` is displayed when you click the green *play button* icon (very helpful for debugging `.Rmd` files!).

language, is a syntax for describing the structure and formatting of content (though HTML is far more extensive and detailed). In particular, HTML is a markup language that can be automatically rendered by web browsers, so it is the language used to create webpages. In fact, you can open up `.html` files generated by RStudio in any web browser to see the content. Additionally, this means that the `.html` files you create with R Markdown can be put online as webpages for others to view!

As it turns out, you can use GitHub not only to host versions of your code repository, but also to serve (display) `.html` files—including ones generated from R Markdown. Github will **host** webpages on a publicly accessible web server that can "serve" the page to anyone who requests it (at a particular URL on the `github.io` domain). This feature is known as *GitHub Pages*.[7]

Using GitHub Pages involves a few steps. First, you need to knit your document into a `.html` file with the name **index.html**—this is the traditional name for a website's homepage (and the file that will be served at a particular URL by default). You will need to have pushed this file to a GitHub repository; the `index.html` file will need to be in the **root folder** of the repo.

---

[7] **What Is GitHub Pages**: https://help.github.com/articles/what-is-github-pages/

Next, you need to configure that GitHub repository to enable GitHub Pages. On the web portal page for your repo, click on the *"Settings"* tab, and scroll down to the section labeled *"GitHub Pages."* From there, you need to specify the "Source" of the .html file that Github Pages should serve. Select the *"master branch"* option to enable GitHub Pages and have it serve the "master" version of your index.html file (see Figure 18.7).

> **Going Further**: If you push code to a different branch on GitHub with the name gh-pages, GitHub Pages will automatically be enabled—serving the files on that branch—without any need to adjust the repository settings. See Section 20.1 for details on working with branches.

Once you've enabled GitHub Pages, you will be able to view your hosted webpage at the URL:

```
# The URL for a website hosted with GitHub Pages
https://GITHUB_USERNAME.github.io/REPO_NAME
```

Replace GITHUB_USERNAME with the username of *the account hosting the repo*, and REPO_NAME with your repository name. Thus, if you pushed your code to the mkfreeman/report repo on GitHub (stored online at https://github.com/mkfreeman/report), the webpage would be available at https://mkfreeman.github.io/report. See the official documentation[8] for more details and options.

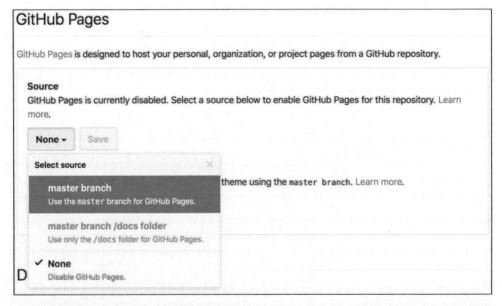

Figure 18.7 Enable hosting via GitHub Pages for a repository by navigating to the *Settings* tab on a repository and scrolling down to the *GitHub Pages* section. Set the "source" as the master branch to host your compiled index.html file as a website!

---

[8]**Documentation for GitHub Pages**: https://help.github.com/ articles/user-organization-and-project-pages/

18.5 R Markdown in Action: Reporting on Life Expectancy

To demonstrate the power of using R Markdown as a tool to generate dynamic reports, this section walks through authoring a report about the life expectancy in each country from 1960 to 2015. The data for the example can be downloaded from the World Bank.[9] The complete code for this analysis is also available online in the book code repository.[10] A subset of the data is shown in Figure 18.8.

To keep the code organized, the report will be written in two separate files:

- `analysis.R`, which will contain the analysis and save important values in variables

- `index.Rmd`, which will `source()` the `analysis.R` script, and generate the report (the file is named so that it can be hosted on GitHub Pages when rendered)

The `analysis.R` file will need to complete the following tasks:

- Load the data.

- Compute metrics of interest.

- Generate data visualizations to display.

As each step is completed in this file, key reporting values and charts are saved to variables so that they can be referenced in the `index.Rmd` file.

To reference these variables, you load the `analysis.R` script (with `source()`) in a "setup" block of the `index.Rmd` file, enabling its data to be referenced within the Markdown. The `include = FALSE` code chunk option means that the block will be evaluated, but not rendered in the document.

```
```{r setup, include = FALSE}
# Load results from the analysis
# Errors and messages will not be printed because `include` is set to FALSE
source("analysis.R")
```
```

	Country.Name	Country.Code	Indicator.Name	Indicator.Code	X1960	X1961
1	Aruba	ABW	Life expectancy at birth, total (years)	SP.DYN.LE00.IN	65.56937	65.98802
2	Afghanistan	AFG	Life expectancy at birth, total (years)	SP.DYN.LE00.IN	32.33756	32.78698
3	Angola	AGO	Life expectancy at birth, total (years)	SP.DYN.LE00.IN	33.22602	33.54776
4	Albania	ALB	Life expectancy at birth, total (years)	SP.DYN.LE00.IN	62.25437	63.27346
5	Andorra	AND	Life expectancy at birth, total (years)	SP.DYN.LE00.IN	NA	NA
6	Arab World	ARB	Life expectancy at birth, total (years)	SP.DYN.LE00.IN	46.81505	47.39723
7	United Arab Emirates	ARE	Life expectancy at birth, total (years)	SP.DYN.LE00.IN	52.28871	53.33405
8	Argentina	ARG	Life expectancy at birth, total (years)	SP.DYN.LE00.IN	65.21554	65.33851
9	Armenia	ARM	Life expectancy at birth, total (years)	SP.DYN.LE00.IN	65.86346	66.28439
10	American Samoa	ASM	Life expectancy at birth, total (years)	SP.DYN.LE00.IN	NA	NA

Figure 18.8 A subset of the World Bank data on the life expectancy in each country from 1960 to 2015.

[9] **World Bank: life expectancy at birth data**: https://data.worldbank.org/indicator/SP.DYN.LE00.IN
[10] **R Markdown in Action**: https://github.com/programming-for-data-science/in-action/tree/master/r-markdown

> **Remember**: All "algorithmic" work should be done in the separate `analysis.R` file, allowing you to more easily debug and iterate your analysis. Since visualizations are part of the "presented" information, they could instead be generated directly in the R Markdown, though the data to be visualized should be preprocessed in the `analysis.R` file.

To compute the metrics of interest in your `analysis.R` file, you can use `dplyr` functions to ask questions of the data set. For example:

```
# Load the data, skipping unnecessary rows
life_exp <- read.csv(
  "data/API_SP.DYN.LE00.IN_DS2_en_csv_v2.csv",
  skip = 4,
  stringsAsFactors = FALSE
)


# Which country had the longest life expectancy in 2015?
longest_le <- life_exp %>%
  filter(X2015 == max(X2015, na.rm = T)) %>%
  select(Country.Name, X2015) %>%
  mutate(expectancy = round(X2015, 1)) # rename and format column
```

In this example, the data frame `longest_le` stores an answer to the question *Which country had the longest life expectancy in 2015?* This data frame could be included directly as content of the `index.Rmd` file. You will be able to reference values from this data frame inline to ensure the report contains the most up-to-date information, even if the data in your analysis changes:

```
The data revealed that the country with the longest life expectancy is
`r longest_le$Country.Name`, with a life expectancy of
`r longest_le$expectancy`.
```

When rendered, this code snippet would replace `` `r longest_le$Country.Name` `` with the value of that variable. Similarly, if you want to show a table as part of your report, you can construct a data frame with the desired information in your `analysis.R` script, and render it in your `index.Rmd` file using the `kable()` function:

```
# What are the 10 countries that experienced the greatest gain in
# life expectancy?
top_10_gain <- life_exp %>%
  mutate(gain = X2015 - X1960) %>%
  top_n(10, wt = gain) %>% # a handy dplyr function!
  arrange(-gain) %>%
  mutate(gain_str = paste(format(round(gain, 1), nsmall = 1),"years")) %>%
  select(Country.Name, gain_formatted)
```

Once you have stored the desired information in the `top_10_gain` data frame in your `analysis.R` script, you can display that information in your `index.Rmd` file using the following syntax:

```
```{r top_10_gain, echo = FALSE}
# Show the top 10 table (specifying the column names to display)
kable(top_10_gain, col.names = c("Country", "Change in Life Expectancy"))
```
```

Figure 18.9 shows the entire report; the complete analysis and R Markdown code to generate this report follows. Note that the report uses a package called `rworldmap` to quickly generate a simple, static world map (as an alternative to mapping with `ggplot2`).

```r
# analysis.R script

# Load required libraries
library(dplyr)
library(rworldmap) # for easy mapping
library(RColorBrewer) # for selecting a color palette

# Load the data, skipping unnecessary rows
life_exp <- read.csv(
  "data/API_SP.DYN.LE00.IN_DS2_en_csv_v2.csv",
  skip = 4,
  stringsAsFactors = FALSE
)

# Notice that R puts the letter "X" in front of each year column,
# as column names can't begin with numbers

# Which country had the longest life expectancy in 2015?
longest_le <- life_exp %>%
  filter(X2015 == max(X2015, na.rm = T)) %>%
  select(Country.Name, X2015) %>%
  mutate(expectancy = round(X2015, 1)) # rename and format column

# Which country had the shortest life expectancy in 2015?
shortest_le <- life_exp %>%
  filter(X2015 == min(X2015, na.rm = T)) %>%
  select(Country.Name, X2015) %>%
  mutate(expectancy = round(X2015, 1)) # rename and format column

# Calculate range in life expectancies
le_difference <- longest_le$expectancy - shortest_le$expectancy
```

```
# What 10 countries experienced the greatest gain in life expectancy?
top_10_gain <- life_exp %>%
  mutate(gain = X2015 - X1960) %>%
  top_n(10, wt = gain) %>% # a handy dplyr function!
  arrange(-gain) %>%
  mutate(gain_str = paste(format(round(gain, 1), nsmall = 1), "years")) %>%
  select(Country.Name, gain_str)
```

Life Expectancy Report

Overview

This is a brief report regarding life expectancy for each country from 1960 to 2015 (source). The data reveals that the country with the longest life expectancy was Hong Kong SAR, China, with a life expectancy of 84.3. That life expectancy was 32.9 years longer than the life expectancy in Central African Republic.

Here are the countries whose life expectancy **improved the most** since 1960.

Country	Change in Life Expectancy
Maldives	39.8 years
Bhutan	35.3 years
Timor-Leste	34.8 years
Nepal	34.7 years
Oman	34.5 years
Tunisia	33.5 years
China	32.3 years
Afghanistan	31.0 years
Iran, Islamic Rep.	30.8 years
Yemen, Rep.	30.4 years

Life Expectancy in 2015

To identify geographic variation in life expectancy, here is a choropleth map of life expectancy in 2015:

Life Expectancy in 2015

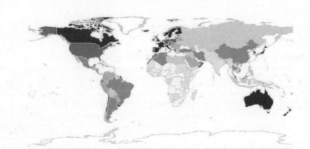

Figure 18.9   A report on life expectancy generated with R Markdown.

```r
# Join this data frame to a shapefile that describes how to draw each country
# The `rworldmap` package provides a helpful function for doing this
mapped_data <- joinCountryData2Map(
  life_exp,
  joinCode = "ISO3",
  nameJoinColumn = "Country.Code",
  mapResolution = "high"
)
```

The following index.Rmd file renders the report using the preceding analysis.R script:

```
---
title: "Life Expectancy Report"
output: html_document
---


```{r setup, include = FALSE}
# Load results from the analysis
# errors and messages will not be printed given the `include = FALSE` option
source("analysis.R")

# Also load additional libraries that may be needed for output
library("knitr")
```


Overview
This is a brief report regarding life expectancy for each country from
1960 to 2015 ([source](https://data.worldbank.org/indicator/SP.DYN.LE00.IN)).
The data reveals that the country with the longest life expectancy was
`r longest_le$Country.Name`, with a life expectancy of
`r longest_le$expectancy`. That life expectancy was `r le_difference`
years longer than the life expectancy in `r shortest_le$Country.Name`.

Here are the countries whose life expectancy **improved the most** since 1960.

```{r top_10_gain, echo = FALSE}
# Show the top 10 table (specifying the column names to display)
kable(top_10_gain, col.names = c("Country", "Change in Life Expectancy"))
```


Life Expectancy in 2015
To identify geographic variations in life expectancy,
here is a choropleth map of life expectancy in 2015:
```

```r
```{r le_map, echo = FALSE}
# Create and render a world map using the `rworldmap` package
mapCountryData(
  mapped_data, # indicate the data to map
  mapTitle = "Life Expectancy in 2015",
  nameColumnToPlot = "X2015",
  addLegend = F, # exclude the legend
  colourPalette = brewer.pal(7, "Blues") # set the color palette
)
```
```

For practice creating reports with R Markdown, see the set of accompanying book exercises.[11]

---

[11] **R Markdown exercises:** https://github.com/programming-for-data-science/chapter-18-exercises

19

Building Interactive Web Applications with Shiny

Adding interactivity to a data report is a highly effective way of communicating information and enabling users to explore a data set. This chapter describes the **Shiny**[1] framework for building interactive applications using R. This will allow you to create dynamic systems in which users can choose what information they want to see, and how they want to see it.

Shiny provides a structure for communicating between a user interface (i.e., a web browser) and a data server (i.e., an R session), allowing users to interactively change the "code" that is run and the data that are output. This not only enables developers to create interactive data presentations, but provides a way for users to interact directly with an R session (without requiring them to write any code).

19.1 The Shiny Framework

Shiny is a **web application framework** for R. As opposed to a simple (static) webpage as you would create with R Markdown, a **web application** is an interactive, dynamic webpage—the user can click on buttons, check boxes, or input text to change how and what data is presented. Shiny is a **framework** in that it provides the "code" for producing and enabling this interaction, while you as the developer "fill in the blanks" by providing variables or functions that the provided code will use to create the interactive page.

Sharing data with others requires your code to perform two different tasks: it needs to *process and analyze information*, and then *present that information* for the user to see. Moreover, with an interactive application, the user is able to *interact* with the presented data (e.g., click on a button or enter a search term into a form). That user **input** then needs to be used to *re*-process the information, and then *re*-present the **output** results.

The Shiny framework provides a structure for applications to perform this exchange: it enables you to write R functions that are able to output (*serve*) results to a web browser, as well as an *interface* for showing those outputs in the browser. Users can interact with this interface to send information to

[1]**Shiny:** http://shiny.rstudio.com

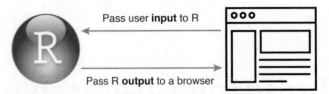

Figure 19.1   Passing content between an R session and a web browser.

the server, which will then output new content for the user. Passing these inputs and outputs back and forth (as illustrated in Figure 19.1) allows Shiny to provide a dynamic and interactive user experience!

> **Fun Fact**: Because Shiny is rendering a user interface for a web browser, it actually generates a *website*. That is, the framework will create all of the necessary components (HTML elements), their styles (CSS rules), and the scripts (JavaScript code) to enable interactivity. But don't worry: you don't need to know anything about these languages; Shiny code is written entirely in R. However, if you already know a few things about web development, you can augment the Shiny-generated elements and interactivity to really make your application shine.

19.1.1 Shiny Core Concepts

The Shiny framework involves a number of different components; you will need to be familiar with and distinguish between these terms to understand how to implement Shiny apps.

- **User interface (UI):** The UI of a Shiny app defines how the application is displayed in the browser. The UI provides a webpage that renders R content such as text or graphics (just like a knitted R Markdown document). Moreover, a Shiny UI supports interactivity through **control widgets**, which are interactive controls for the application (think: buttons or sliders). The UI can specify a **layout** for these components, allowing you to organize your content in side-by-side panels, or across multiple tabs.

- **Server:** The server of a Shiny app defines and processes the data that will be displayed by the UI. Generally speaking, a server is a program running on a computer (often remotely) that receives requests and provides ("serves") content based on the request. For example, when you request information from a web API, you submit a request to a server that processes the request and returns the desired information. In a Shiny application, you can think of the server as an interactive R session that the user will use to "run" data processing functions by interacting with the UI in the web browser (*not* in RStudio). The server takes in inputs from the user (based on their interactions) and runs functions that provide outputs (e.g., text or charts) for the UI to display. These data processing functions are **reactive**, which means they are automatically rerun whenever the input changes (they "react" to it). This allows the output to be dynamic and interactive.

- **Control widget:** An element in the UI that allows the user to provide input to the server—for example, a text input box, a dropdown menu, or a slider. Control widgets store input values, which are automatically updated as the user interacts with the widget. Updates to the value stored by the widget are sent from the UI to the server, which will react to those changes to generate new content to display.

- **Reactive output:** An element in the UI that displays dynamic (changing) content produced by the server—for example, a chart that dynamically updates when the user selects different data to display, or a table that responds to a search query. A reactive output will automatically update whenever the server sends it a new value to display.

- **Render function:** Functions in the server that produce output that can be understood and displayed by the UI's reactive outputs. A render function will automatically "re-execute" whenever a related control widget changes, producing an updated value that will be read and displayed by a reactive output.

- **Reactivity:** Shiny apps are designed around reactivity: updating some components in the UI (e.g., the control widgets) will cause other components (e.g., the render functions in the server) to "react" to that change and automatically re-execute. This is similar to how equations in a spreadsheet program like Microsoft Excel work: when you change the value in one cell, any others that reference it "react" and change as well.

19.1.2 Application Structure

A Shiny application is written in a script file named **app.R** (it *must* to have that exact name so that RStudio will handle the file correctly). This file should be saved in the root directory of a project (i.e., the root of a `git` repository). You can create this file and folder yourself, or alternatively you can create a new Shiny project through the RStudio interface (via `File > New File > Shiny Web App...`).

Shiny is made available through the `shiny` package—another external package (like `dplyr` and `ggplot2`) that you will need to install and load before you can use it:

```
install.packages("shiny") # once per machine
library("shiny")          # in each relevant script
```

This will make all of the framework functions and variables you will need to work with available.

Mirroring Figure 19.1, Shiny applications are separated into two components (parts of the application): the UI and the server.

1. The UI defines how the application is displayed in the browser. The UI for a Shiny application is defined as a *value*, almost always one returned from calling one of Shiny's **layout functions**.

   The following example UI defines a `fluidPage()` (where the content will "fluidly" flow down the page based on the browser size) that contains three content elements: static text content for the page heading, a text input box where the user can type a name, and the output text of a calculated `message` value (which is defined by the server). These functions and their usage are described in more detail in Section 19.2.

```
# The UI is the result of calling the `fluidPage()` layout function
my_ui <- fluidPage(
    # A static content element: a 2nd level header that displays text
    h2("Greetings from Shiny"),

    # A widget: a text input box (save input in the `username` key)
    textInput(inputId = "username", label = "What is your name?"),

    # An output element: a text output (for the `message` key)
    textOutput(outputId = "message")
)
```

2. The server defines and processes the data that will be displayed by the UI. The server for a Shiny application is defined as a *function* (in contrast, the UI is a *value*). This function needs to take in two lists as arguments, conventionally called input and output. The values in the input list are received from the user interface (e.g., web browser), and are used to create content (e.g., calculate information or make graphics). This content is then saved in the output list so that it can be sent back to the UI to be rendered in the browser. The server uses **render functions** to assign these values to output so that the content will automatically be recalculated whenever the input list changes. For example:

```
# The server is a function that takes `input` and `output` arguments
my_server <- function(input, output) {
    # Assign a value to the `message` key in the `output` list using
    # the renderText() method, creating a value the UI can display
    output$message <- renderText({
        # This block is like a function that will automatically rerun
        # when a referenced `input` value changes

        # Use the `username` key from `input` to create a value
        message_str <- paste0("Hello ", input$username, "!")

        # Return the value to be rendered by the UI
        message_str
    })
}
```

(The specifics of the server and its functions are detailed in Section 19.3.)

The UI and the server are both written in the app.R file. They are combined by calling the **shinyApp()** function, which takes a UI value and a server function as arguments. For example:

```
# To start running your app, pass the variables defined in previous
# code snippets into the `shinyApp()` function
shinyApp(ui = my_ui, server = my_server)
```

Executing the shinyApp( ) function will start the app. Alternatively, you can launch a Shiny app using the "Run App" button at the top of RStudio (see Figure 19.2). This will launch a viewer window presenting your app (Figure 19.3); you can also click the "Open in Browser" button at the top to show the app running in your computer's web browser. Note that if you need to stop the app, you can close the window or click the "Stop Sign" icon that appears on the RStudio console.

> **Tip:** If you change the UI or the server, you generally do not need to stop and start the app. Instead, you can refresh the browser or viewer window, and it will reload with the new UI and server.

When this example application is run, Shiny will combine the UI and server components into a webpage that allows the user to type a name into an input box; the page will then say "Hello" to whatever name is typed in (as shown in Figure 19.3). As the user types into the input box (created by the textInput( ) function), the UI sends an updated username value to the server; this value is stored in the input argument list as input$username. The renderText( ) function in the server then reacts to the change to the input$username value, and automatically re-executes to calculate a new renderable value that is stored in output$message and sent back to the UI (illustrated in Figure 19.4). Through this process, the app provides a dynamic experience in which the user types into a box and sees the message change in response. While this is a simple example, the same structure can be used to create searchable data tables, change the content of interactive graphics, or even specify the parameters of a machine learning model!

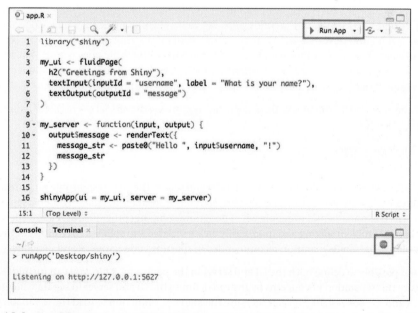

```r
library("shiny")

my_ui <- fluidPage(
  h2("Greetings from Shiny"),
  textInput(inputId = "username", label = "What is your name?"),
  textOutput(outputId = "message")
)

my_server <- function(input, output) {
  output$message <- renderText({
    message_str <- paste0("Hello ", input$username, "!")
    message_str
  })
}

shinyApp(ui = my_ui, server = my_server)
```

```
> runApp('Desktop/shiny')

Listening on http://127.0.0.1:5627
```

Figure 19.2 Use RStudio to run a Shiny app. The *"Run App"* button starts the application, while the *"Stop Sign"* icon in the console stops it.

Figure 19.3   A Shiny application that greets a user based on an input name, running in the RStudio viewer. Note the "Open in Browser" and Refresh buttons at the top.

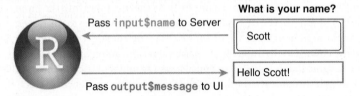

Figure 19.4   Variables passing between a UI and a server. The server function accepts inputs from the UI and generates a set of outputs that are passed back to the UI to be rendered.

> **Tip:** The reactivity involved in Shiny apps can make them difficult to debug. Code statements don't flow directly from top to bottom as with most scripts, and Shiny may produce somewhat obscure error messages in the console when something goes wrong. As with R Markdown, a good strategy for identifying problematic code is to systematically remove ("comment out") segments of your project and attempt to rerun your application.
>
> For additional advice on how to fix issues in Shiny apps, see the official *Debugging Shiny applications*[a] guide.
>
> ――――――――――――――――――
> [a] https://shiny.rstudio.com/articles/debugging.html

A Shiny app divides responsibilities between its UI and server: the UI is responsible for *presenting* information, while the server is responsible for *processing* information. Enabling such a **separation of concerns** is a fundamental principle when designing computer programs, as it allows developers to isolate their problem solving and more easily create scalable and collaborative projects. Indeed, this division is the same separation recommended in splitting code across .R and .Rmd files.

While it is possible to define both the UI and server in the same app.R file, you can further emphasize this separation of concerns by instead defining the UI and server in separate files (e.g., my_ui.R and my_server.R). You can then use the source() function to load those variables into the app.R script for combining. Such a division can help keep your code more organized and understandable, particularly as your apps grow larger.

If you name the separate files exactly **ui.R** and **server.R** (and have the last value returned in each script be the UI value and the server function, respectively), RStudio will be able to launch your Shiny application without having a unified app.R file. Even so, it is better practice to use a single app.R script to run your Shiny app, and then source() in the UI and server to keep them separated.

> **Caution:** Avoid creating both an app.R *and* files named exactly ui.R and server.R in your project. This can confuse RStudio and cause your application not to run. Pick one approach or the other!

> **Going Further**: You can use the Shiny framework to add interactive widgets to HTML documents created using R Markdown! See the *Introduction to Interactive Documents* article.[a] Note that the webpage will still need to be hosted somewhere that supports a Shiny server (such as *shinyapps.io*, described in Section 19.4).
>
> ――――――――――――――――
> [a]https://shiny.rstudio.com/articles/interactive-docs.html

19.2 Designing User Interfaces

To enable the rapid discovery of information, you will want to create interfaces that prioritize pertinent information in a clear and organized fashion. The Shiny framework provides structural elements that you can use to construct such a well-organized page.

When you write code defining a UI, you are defining how the app will be displayed in the browser. You create a UI by calling a layout function such as **fluidPage()**, which will return a UI definition that can be used by the shinyApp() function. Layout functions take as arguments the **content elements** (pieces of content) that you want the layout to contain (and thus will be shown in the app's UI):

```
# A "pseudocode" example rendering 3 UI elements in a fluid layout
ui <- fluidPage(element1, element2, element3)
```

A layout function can take as many content elements as needed, each as an additional argument (often placed onto separate lines for readability). For example, the UI shown in Figure 19.2 has three content elements: one produced by the h2() function, one produced by the textInput() function, and one produced by the textOutput() function.

Many different types of content elements can be passed to a layout function, as described in the following sections.

> **Tip:** You can initially implement your app with an "empty" server function as a way to design and test your UI—a UI does not require any actual content in the server! See Section 19.3 for an example of an empty server function.

19.2.1 Static Content

The simplest type of content element a UI can include is a **static content** element. These elements specify content that will *not* change as the user interacts with the page. They are generally used to provide further explanatory information about what the user is looking at—similar to the Markdown portion of an R Markdown document.

Content elements are created by calling specific functions that create them. For example, the h1( ) function will create an element that has a first-level heading (similar to using a # in Markdown). These functions are passed arguments that are the content (usually strings) that should be shown:

```
# A UI whose layout contains a single static content element
ui <- fluidPage(
  h1("My Static App")
)
```

Static content functions can alternatively be referenced as elements of the tags list (e.g., tags$h1( )), so they are also known as "tag functions." This is because static content functions are used to produce **HTML**,[2] the language used to specify the content of webpages (recall that a Shiny app is an interactive webpage). As such, static content functions are all named after HTML tags. But since Markdown is also compiled into HTML tags (as when you knit an R Markdown document), many static content functions correspond to Markdown syntax, such as those described in Table 19.1. See the *HTML Tags Glossary*[3] for more information about the meaning of individual functions and their common arguments.

Static content functions can be passed multiple unnamed arguments (i.e., multiple strings), all of which are included as that kind of static content. You can even pass other content elements as arguments to a tag function, allowing you to "nest" formatted content:

```
# Create a UI using multiple nested content elements
ui <- fluidPage(
  # An `h1()` content element that contains an `em()` content element
  # This will render like the Markdown content `# My _Awesome_ App`
  h1("My", em("Awesome"), "App"),

  # Passing multiple string arguments will cause them to be concatenated (within
  # the same paragraph)
  p("My app is really cool.", "It's the coolest thing ever!"),
)
```

It is common practice to include a number of static elements (often with such nesting) to describe your application—similar to how you would include static Markdown content in an R Markdown document. In particular, almost all Shiny apps include a **titlePanel( )** content element, which provides both a second-level heading (h2( )) element for the page title and specifies the title shown in the tab of a web browser.

---

[2] **HTML** tutorials and reference from Mozilla developer network: https://developer.mozilla.org/en-US/docs/Web/HTML
[3] **Shiny HTML Tags Glossary:** https://shiny.rstudio.com/articles/tag-glossary.html

Table 19.1   **Some example static content functions and their Markdown equivalents**

Static Content Function	Markdown Equivalent	Description
h1("Heading 1")	# Heading 1	A first-level heading
h2("Heading 2")	## Heading 2	A second-level heading
p("some text")	some text (on own line)	A paragraph (of plain text)
em("some text")	_some text_	*Emphasized* (italic) text
strong("some text")	**some text**	**Strong** (bold) text
a("some text", href = "url")	[some text](url)	A hyperlink (**a**nchor)
img("description", src = "path")		An image

Going Further: If you are familiar with HTML syntax, you can write such content directly using the HTML() function, passing in a string of the HTML you want to include. Similarly, if you are familiar with CSS, you can include stylesheets using the includeCSS() content function. See the article *Style Your Apps with CSS*[a] for other options and details.

[a]https://shiny.rstudio.com/articles/css.html

19.2.2 Dynamic Inputs

While your application will include many static elements, the true power and purpose of Shiny come from its support for user interactions. In a Shiny app, users interact with content elements called control widgets. These elements allow users to provide input to the server, and include elements such as text input boxes, dropdown menus, and sliders. Figure 19.5 shows a sample of widgets available in the Shiny package.

Each widget handles user input by storing a value that the user has entered—whether by typing into a box, moving a slider, or clicking a button. When the user interacts with the widget and changes the input, the stored value automatically changes as well. Thus you can almost think of each widget's value as a "variable" that the user is able to modify by interacting with the web browser. Updates to the value stored by the widget are sent to the server, which will react to those changes to generate new content to display.

Like static content elements, control widgets are created by calling an appropriate function—most of which include the word "input" in the name. For example:

- **textInput()** creates a box in which the user can enter text. The "Greeting" app described previously includes a textInput().

- **sliderInput()** creates a slider that the user can drag to choose a value (or range of values).

- **selectInput()** creates a dropdown menu the user can choose from.

- **checkboxInput()** creates a box the user can check (using checkboxGroupInput() to group them).

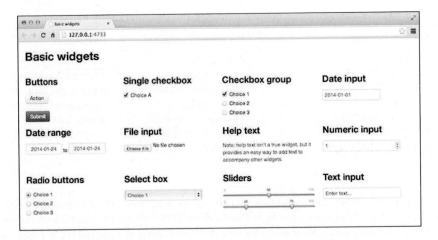

Figure 19.5   Examples of control widgets that can be included in the UI of a Shiny application (image from shiny.rstudio.com).

- **`radioButtons()`** creates "radio" buttons (the user can select only one of these buttons at a time, just like selecting the station on a radio).

See the documentation[4] for a complete list of available control widgets, and the widgets gallery[5] for examples.

All widget functions take at least two arguments:

- An **`inputId`** (a string) or "name" for the widget's value. This is the "key" that allows the server to access that widget's value (literally, it is the key for that value in the `input` list argument).

- A **`label`** (as a string or static content element) that will be shown alongside the widget and tell the user what the value represents. The label can be an empty string (`" "`) if you don't want to show anything.

Other arguments may be required by a particular widget. For example, a slider widget requires a `min`, `max`, and (starting) `value`, as in the code below.

Control widgets are used to solicit input values from the user, which are then sent to the server for processing. See Section 19.3 for details on how to use these input values.

---

[4]**Shiny reference:** http://shiny.rstudio.com/reference/shiny/latest/
[5]**Shiny Widgets Gallery:** http://shiny.rstudio.com/gallery/widget-gallery.html

```
# A UI containing a single slider
ui <- fluidPage(
  sliderInput(
    inputId = "age",          # key this value will be assigned to
    label = "Age of subjects", # label to display alongside the slider
    min = 18,                 # minimum slider value
    max = 80,                 # maximum slider value
    value = 42                # starting value for the slider
  )
)
```

19.2.3   Dynamic Outputs

To display output values from the server, a UI uses **reactive output** elements. This kind of content element is similar to a static content element, but instead of displaying unchanging content, it displays dynamic (changing) content produced by the server—for example, a chart that dynamically updates when a user selects different data to display, or a table that responds to a search query. A reactive output will automatically update whenever the server sends it a new value to display.

As with other content elements, reactive outputs are created by calling an appropriate function, most of which include the word "output" in the name. For example:

- **textOutput()** displays output as plain text; use htmlOutput() if you want to render HTML content.

- **tableOutput()** displays output as a data table (similar to kable() in R Markdown). Note that the dataTableOutput() function from the DT package will display an interactive table.

- **plotOutput()** displays a graphical plot, such as one created with the ggplot2 package. The plotlyOutput() function from the plotly package can be used to render an interactive plot, or you can make a ggplot2 plot interactive.[6]

- **verbatimTextOutput()** displays content as a formatted code block, such as if you wanted to print a non-string variable like a vector or data frame.

Each of these functions takes as an argument the **outputId** (a string) or "name" for the value that will be displayed. The function uses this **"key"** to access the value that is output by the server. For example, you could show the following information generated by your server:

```
# A UI containing different reactive outputs
ui <- fluidPage(
  textOutput(outputId = "mean_value"),  # display text stored in `output$mean_value`
  tableOutput(outputId = "table_data"), # display table stored in `output$table_data`
  plotOutput(outputId = "my_chart")     # display plot stored in `output$my_chart`
)
```

Note that each function may support additional arguments as well (e.g., to specify the size of a plot). See the documentation for details on individual functions.

---

[6]**Interactive Plots:** http://shiny.rstudio.com/articles/plot-interaction.html

> **Caution:** Each page can show a single output value just once (because it needs to be given a unique id in the generated HTML). For example, you can't include `textOutput(outputId = "mean_value")` twice in the same UI.

> **Remember:** As you build your application's UI, be careful to keep track of the names (`inputId` and `outputId`) you give to each control widget and reactive output; you will need these to match with the values referenced by the server!

19.2.4 Layouts

You can specify how content is organized on the page by using different **layout** content elements. Layout elements are similar to other content elements, but are used to specify the position of different pieces of content on the page—for example, organizing content into columns or grids, or breaking up a webpage into tabs.

Layout content elements are also created by calling associated functions; see the Shiny documentation or the Layout Guide[7] for a complete list. Layout functions all take as arguments a sequence of other content elements (created by calling other functions) that will be shown on the page following the specified layout. For example, the previous examples use a `fluidPage()` layout to position content from top to bottom in a way that responds to the size of the browser window.

Because layouts themselves are content elements, it's also possible to pass the result of calling one layout function as an argument to another. This allows you to specify some content that is laid out in "columns," and then have the "columns" be placed into a "row" of a grid. As an example, the commonly used **sidebarLayout()** function organizes content into two columns: a "sidebar" (shown in a gray box, often used for control widgets or related content) and a "main" section (often used for reactive outputs such as plots or tables). Thus `sidebarLayout()` needs to be passed two arguments: a `sidebarPanel()` layout element that contains the content for the sidebar, and a `mainPanel()` layout element that contains the content for the main section:

```
ui <- fluidPage(  # lay out the passed content fluidly
  sidebarLayout(  # lay out the passed content into two columns
    sidebarPanel( # lay out the passed content inside the "sidebar" column
      p("Sidebar panel content goes here")
    ),
    mainPanel(    # lay out the passed content inside the "main" column
      p("Main panel content goes here"),
      p("Layouts usually include multiple content elements")
    )
  )
)
```

An example of a sidebar layout is also shown in Figure 19.6.

---

[7]**Shiny Application Layout Guide:** http://shiny.rstudio.com/articles/layout-guide.html

> **Caution:** Because Shiny layouts are usually responsive to web browser size, on a small window (such as the default app viewer) the sidebar may be placed *above* the content—since there isn't room for it to fit nicely on the side!

Since a layout and its content elements are often *nested* (similar to some static content elements), you almost always want to use line breaks and indentation to make that nesting apparent in the code. With large applications or complex layouts, you may need to trace down the page to find the closing parenthesis **)** that indicates exactly where a particular layout's argument list (passed in content) ends.

Because layout functions can quickly become complex (with many other nested content functions), it is also useful to store the returned layouts in variables. These variables can then be passed into higher-level layout functions. The following example specifies multiple "tabs" of content (created using the `tabPanel()` layout function), which are then passed into a `navbarPage()` layout function to create a page with a "navigation bar" at the top to browse the different tabs. The result is shown in Figure 19.6.

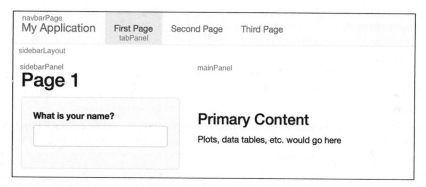

Figure 19.6   A "multi-page" application built with Shiny's layout functions, including `navbarPage()` and `sidebarLayout()`. Red notes are added.

```
# Define the first page content; uses `tabPanel()` and `sidebarLayout()`
# layout functions together (as an example)
page_one <- tabPanel(
  "First Page", # label for the tab in the navbar
  titlePanel("Page 1"), # show with a displayed title

  # This content uses a sidebar layout
  sidebarLayout(
    sidebarPanel(
      textInput(inputId = "username", label = "What is your name?")
    ),
    mainPanel(
      h3("Primary Content"),
```

```r
      p("Plots, data tables, etc. would go here")
    )
  )
)


# Define content for the second page
page_two <- tabPanel(
  "Second Page" # label for the tab in the navbar
  # ...more content would go here...
)


# Define content for the third page
page_three <- tabPanel(
  "Third Page" # label for the tab in the navbar
  # ...more content would go here...
)


# Pass each page to a multi-page layout (`navbarPage`)
ui <- navbarPage(
  "My Application", # application title
  page_one,         # include the first page content
  page_two,         # include the second page content
  page_three        # include the third page content
)
```

The Shiny framework can be used to develop highly complex layouts just by calling R functions. For more examples and details on how to achieve particular layout and UI effects, check the Shiny documentation and application gallery.

> **Fun Fact**: Much of Shiny's styling and layout structure is based on the Bootstrap[a] web framework, which is how it supports layouts that are *responsive* to window size. Note that Shiny uses Bootstrap 3, not the more recent Bootstrap 4.
>
> ───────────────
> [a]http://getbootstrap.com/docs/3.3/

19.3 Developing Application Servers

To generate dynamic data views that can be shown in the UI (as reactive outputs), you will need to specify how you want that data to be manipulated based on the user input (through control widgets). In the Shiny framework, you define this manipulation as the application's server.

You create a Shiny server by defining a function (rather than calling a provided one, as with a UI). The function must be defined to take at least two arguments: a list to hold the input values, and a list to hold the output values:

```
# Define a server function for a Shiny app
server <- function(input, output) {
  # assign values to `output` here
}
```

Note that a server function is just a normal function, albeit one that will be executed to "set up" the application's reactive data processing. Thus you can include any code statements that would normally go in a function—though that code will be run only once (when the application is first started) unless defined as part of a render function.

When the server function is called to set up the application, it will be passed the input and output list arguments. The first argument (input) will be a list containing any values stored by the control widgets in the UI: each inputId ("name") in a control widget will be a key in this list, whose value is the value currently stored by the widget. For example, the textInput() shown in Figure 19.2 has an inputId of username, so would cause the input list to have a username key (referenced as input$username inside of the server function). This allows the server to access any data that the user has input into the UI. Importantly, these lists are reactive, so the values inside of them will automatically change as the user interacts with the UI's control widgets.

The primary purpose of the server function is to assign new values to the output list (each with an appropriate key). These values will then be displayed by the reactive outputs defined in the UI. The output list is assigned values that are produced by render functions, which are able to produce output in a format that can be understood by the UI's outputs (reactive outputs can't just display plain strings). As with the UI's reactive output functions, the server uses different render functions for the different types of output it provides, as shown in Table 19.2.

The result of a render function *must* be assigned to a key in the output list argument that matches the outputId ("name") specified in the reactive output. For example, if the UI includes textOutput(outputId = "message"), then the value must be assigned to output$message. If the keys don't match, then the UI won't know what output to display! In addition, the *type* of render function must match the *type* of reactive output: you can't have the server provide a plot to render but have the UI try to output a table for that value! This usually means that the word after "render" in the render function needs to be the same as the word before "Output" in the reactive output function. Note that Shiny server functions will usually have multiple render functions assigning values to the output list—one for each associated reactive output in the UI.

Table 19.2 **Some example render functions and their associated reactive outputs**

Render Function (Server)	Reactive Output (UI)	Content Type
renderText()	textOutput()	Unformatted text (character strings)
renderTable()	tableOutput()	A simple data table
renderDataTable()	dataTableOutput()	An interactive data table (use the DT package)
renderPlot()	plotOutput()	A graphical plot (e.g., created with ggplot2)
renderPlotly()	plotlyOutput()	An interactive Plotly plot
renderLeaflet()	leafletOutput()	An interactive Leaflet map
renderPrint()	verbatimTextOutput()	Any output produced with print()

All render functions take as an argument a **reactive expression**. A reactive expression is a lot like a function: it is written as a block of code (in braces **{}**) that returns the value to be rendered. Indeed, the only difference between writing a function and writing a reactive expression is that you don't include the keyword function or a list of arguments—you just include the block (the braces and the code inside it).

```
# Create a server function that defines a `message` output based on a
# `username` input
server <- function(input, output) {
  # Define content to be displayed by the `message` reactive output
  # `renderText()` is passed a reactive expression
  output$message <- renderText({
    # A render function block can include any code used in a regular function
    my_greeting <- "Hello "

    # Because the render function references an `input` value, it will be
    # automatically rerun when that value changes, producing an updated output
    message_str <- paste0(my_greeting, input$username, "!")
    message_str # return the message to be output
  })
}
```

> **Going Further**: Reactive expressions technically define a *closure*, which is a programming concept used to encapsulate functions and the context for those functions.

What is significant about render functions is that they will automatically "rerun" their passed-in code block every time a value they reference in the input list changes. So if the user interacts with the username control widget in the UI (and thereby changes the input$username value), the function in the preceding example will be executed again—producing a new value that will be reassigned to output$message. And once output$message changes, any reactive output in the UI (e.g., a textOutput()) will update to show the latest value. This makes the app interactive!

> **Remember**: In effect, render functions are functions that will be *rerun automatically* when an input changes, without you having to call them explicitly! You can think of them as the functions you define for how the output should be determined—and those functions will be rerun when the input changes.

Thus your server defines a series of "functions" (render functions) that specify how the output should change based on changes to the input—when that input changes, the output changes along with it.

> **Tip:** Data values that are not reactive (that will not change based on user interaction) can be defined elsewhere in the server function, as normal. If you want a nonreactive data value to be available to the UI as well—such as one that contains configuration or static data range information—you should create it outside of the server function in the app.R file, or in a separate global.R file. See the *Scoping Rules for Shiny Apps* article[a] for details.
>
> _____
>
> [a]https://shiny.rstudio.com/articles/scoping.html

> **Going Further:** Understanding the flow of data in and between render functions and other reactive expressions is the key to developing complex Shiny applications. For more details on reactivity in Shiny, see RStudio's articles on reactivity,[a] particularly *Reactivity: An Overview*[b] and *How to Understand Reactivity in R.*[c]
>
> _____
>
> [a]https://shiny.rstudio.com/articles/#reactivity
> [b]https://shiny.rstudio.com/articles/reactivity-overview.html
> [c]https://shiny.rstudio.com/articles/understanding-reactivity.html

19.4 Publishing Shiny Apps

While the previous sections discussed building and running Shiny apps on your own computer, the entire point of an interactive application is to be able to share it with others. To do that, you will need a website that is able to *host* the application so that others can navigate to it using their web browsers. However, you can't just use GitHub Pages to host the application because—in addition to the UI—you need an R interpreter session to run the server that the UI can connect to (and GitHub does not provide R interpreters). For this reason, sharing a Shiny app with the world is a bit more involved than simply pushing the code to GitHub.

While there are a few different solutions for hosting Shiny apps, the simplest is hosting through **shinyapps.io**[8]. *shinyapps.io* is a platform provided by RStudio that is used for hosting and running Shiny apps. Anyone can deploy and host five small(ish) applications to the platform for free, though deploying large applications costs money.

To host your app on *shinyapps.io*, you will need to create a free account.[9] You can sign up with GitHub (recommended) or a Google account. After you sign up, follow the site's instructions:

- Select an account name, keeping in mind it will be part of the URL people use to access your application.

- Install the required rsconnect package (it may have been included with your RStudio download).

- Set your authorization token ("password") for uploading your app. To do this, click the green "Copy to Clipboard" button, and then paste that selected command into the Console in RStudio. You should need to do this just once per machine.

_____

[8]**shinyapps.io** web hosting for Shiny apps: https://www.shinyapps.io
[9]**shinyapps.io signup**: https://www.shinyapps.io/admin/#/signup

Don't worry about the listed "Step 3 - Deploy"; you should instead publish directly through RStudio!

After you have set up an account, you can publish your application by running your app through RStudio (i.e., by clicking the "Run App" button), and then clicking the *"Publish"* button in the upper-right corner of the app viewer (see Figure 19.7).

After a minute of processing and uploading, your app should become available online to use at the URL:

```
# The URL for a Shiny app hosted with shinyapps.io
https://USERNAME.shinyapps.io/APP_NAME/
```

While it sounds as simple as clicking a button, publishing to *shinyapps.io* is unfortunately one of the "pain points" in working with Shiny. Things can unexpectedly go wrong, and it's even more difficult to determine the problem than with local Shiny apps! Here are some useful tips for successfully publishing an application:

1. Always test and debug your app *locally* (e.g., on your own computer, by running the app through RStudio). It's easier to find and fix errors locally; make sure the app works on your machine before you even try to put it online.

2. You can view the error logs for your deployed app by either using the "Logs" tab in the application view or calling the **showLogs()** function (part of the `rsconnect` package). These logs will show `print()` statements and often list the errors that explain the problem that occurred when deploying your app.

3. Use correct folder structures and *relative paths*. All of your app files should reside in a single folder (usually named after the project). Make sure any `.csv` or `.R` files referenced are inside the app folder, and that you use relative paths to refer to them in your code. Do not ever include any `setwd()` statements in your code; only set the working directory through RStudio (because *shinyapps.io* will have its own working directory).

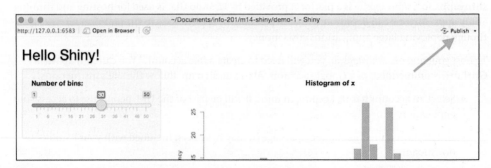

Figure 19.7 Click the Publish button in the upper-right corner of a Shiny app to publish it to shinyapps.io.

4. Make sure that any external packages you use are referenced with the `library()` function in your `app.R` file. The most common problem we've seen involves external packages not being available. See the documentation[10] for an example and suggested solutions.

For more options and details, see the *shinyapps.io* user guide.[11]

19.5   Shiny in Action: Visualizing Fatal Police Shootings

This section demonstrates building a Shiny application to visualize a data set of people who were fatally shot by the police in the United States in the first half of 2018 (January through June). The data set was compiled by the *Washington Post*,[12] which made the data available on GitHub.[13] Latitude and longitude have been added to the data based on the city and state; the code for this data preparation is available alongside the full application code in the book code repository.[14]

As of the time of writing, there were 506 fatalities captured in the data set during the time period, each one of which has 17 pieces of information about the incident, such as the name, age, and race of the victim (a subset of the data is shown in Figure 19.8). The purpose of the Shiny application is to understand the geographic distribution of *where* people have been killed by the police, and to provide summary information about the incidents, such as the total number of people killed broken down by race or gender. The final product (shown in Figure 19.10) allows users to select a variable in the data set—such as `race` or `gender`—through which to analyze the data. This choice will dictate the color encoding in the map as well as the level of aggregation in a summary table.

A main component of this application will be an interactive map displaying the location of each shooting. The color of each point will expresses additional information about that individual (such

	date	manner_of_death	age	gender	race	city	state
1	2018–01–02	shot	30	Female	White, non–Hispanic	Camp Wood	TX
2	2018–01–02	shot	49	Male	White, non–Hispanic	Ozark	AR
3	2018–01–02	shot	66	Male	White, non–Hispanic	Joplin	MO
4	2018–01–03	shot	28	Male	Black, non–Hispanic	Oakland	CA
5	2018–01–04	shot	27	Male	White, non–Hispanic	Boise	ID
6	2018–01–04	shot	31	Male	White, non–Hispanic	Crandon	WI
7	2018–01–05	shot	46	Male	White, non–Hispanic	Springfield	MO
8	2018–01–05	shot	28	Male	Black, non–Hispanic	Whitehall	OH
9	2018–01–05	shot	35	Male	Asian	Long Beach	CA

Figure 19.8   A subset of the police shootings data set, originally compiled by the *Washington Post*.

---

[10] *shinyapps.io* **build errors on deployment:** http://docs.rstudio.com/shinyapps.io/Troubleshooting.html#build-errors-on-deployment

[11] **shinyapps.io user guide:** http://docs.rstudio.com/shinyapps.io/index.html

[12] **"Fatal Force"**, *Washington Post*: https://www.washingtonpost.com/graphics/2018/national/police-shootings-2018/

[13] **Fatal Shootings GitHub page:** https://github.com/washingtonpost/data-police-shootings

[14] **Shiny in Action:** https://github.com/programming-for-data-science/in-action/tree/master/shiny

as race or gender). While the column used to dictate the color will eventually be dynamically selected by the user, you can start by creating a map with the column "hard-coded." For example, you can use Leaflet (discussed in Section 17.3) to generate a map displaying the location of each shooting with points colored by race of the victim (shown in Figure 19.9):

```r
# Create a map of fatal police shootings using leaflet

# Load the leaflet function for mapping
library("leaflet")

# Load the prepared data
shootings <- read.csv("police-shootings.csv", stringsAsFactors = FALSE)

# Construct a color palette (scale) based on the `race` column
# Using double-bracket notation will make it easier to adapt for use with Shiny
palette_fn <- colorFactor(palette = "Dark2", domain = shootings[["race"]])

# Create a map of the shootings using leaflet
# The `addCircleMarkers()` function will make circles with radii based on zoom
leaflet(data = shootings) %>%
  addProviderTiles("Stamen.TonerLite") %>% # add Stamen Map Tiles
  addCircleMarkers( # add markers for each shooting
    lat = ~lat,
    lng = ~long,
    label = ~paste0(name, ", ", age), # add a hover label: victim's name and age
    color = ~palette_fn(shootings[["race"]]), # color points by race
    fillOpacity = .7,
    radius = 4,
    stroke = FALSE
  ) %>%
  addLegend( # include a legend on the plot
    position = "bottomright",
    title = "race",
    pal = palette_fn, # the palette to label
    values = shootings[["race"]], # again, using double-bracket notation
    opacity = 1 # legend is opaque
  )
```

> **Tip:** Because server inputs in a Shiny application are strings, it's helpful to use R's **double-bracket notation** to select data of interest (e.g., df[[input$some_key]]), rather than relying on dplyr functions such as select().

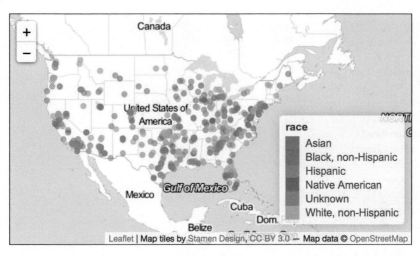

Figure 19.9   A map of each person killed by the police in 2018, created using `leaflet`.

> **Tip:** A great way to develop a Shiny application is to first build a static version of your content, then swap out static values (variable names) for dynamic ones (information stored in the `input` variable). Starting with a working version of your content will make debugging the application much easier.

While this map allows you to get an overall sense of the geographic distribution of the fatalities, supporting it with specific quantitative data—such as the total number of people killed by race—can provide more precise information. Such summary information can be calculated using the `dplyr` functions `group_by()` and `count()`. Note the use of double-bracket notation to pass in the column values directly (rather than referencing the column by name), which will allow you to more easily make the column name dynamic in the Shiny application.

```
# Calculate the number of fatalities by race
# Use double-bracket notation to support dynamic column choice in Shiny
table <- shootings %>%
  group_by(shootings[["race"]]) %>% # pass the column values directly
  count() %>%
  arrange(-n) # sort the table in decreasing order by number of fatalities

colnames(table) <- c("race", "Number of Victims") # Format column names
```

With these data representations established, you can begin implementing the Shiny application. For every Shiny app, you will need to create a UI and a server. It's often useful to start with the UI to help provide structure to the application (and it's easier to test that it works). To create the UI that will render these elements, you can use a structure similar to that described in Section 19.2.4 and declare a `fluidPage()` layout that has a `sidebarPanel()` to keep the control widgets (a "dropdown box" that lets the user select which column to analyze) along the side, and a `mainPanel()` in which to show the primary content (the leaflet map and the data table):

```
# Define the UI for the application that renders the map and table
my_ui <- fluidPage(
  # Application title
  titlePanel("Fatal Police Shootings"),

  # Sidebar with a selectInput for the variable for analysis
  sidebarLayout(
    sidebarPanel(
      selectInput(
        inputId = "analysis_var",
        label = "Level of Analysis",
        choices = c("gender", "race", "body_camera", "threat_level")
      )
    ),

    # Display the map and table in the main panel
    mainPanel(
      leafletOutput(outputId = "shooting_map"), # reactive output from leaflet
      tableOutput(outputId = "grouped_table")
    )
  )
)
```

You can check that the UI looks correct by providing an empty server function and calling the shinyApp( ) function to run the application. While the map and data won't show up (they haven't been defined), you can at least check the layout of your work.

```
# A temporarily empty server function
server <- function(input, output) {

}

# Start running the application
shinyApp(ui = my_ui, server = server)
```

Once the UI is complete, you can fill in the server. Since the UI renders two reactive outputs (a leafletOutput( ) and a tableOutput( )), your server needs to provide corresponding render functions. These functions can return versions of the "hard-coded" map and data table defined previously, but using information taken from the UI's input to select the appropriate column—in other words, replacing the "race" column with the column named by input$analysis_var.

Notice the use of input$analysis_var to dynamically set the color of each point, as well as the aggregation column for the data table.

```r
# Define the server that renders a map and a table
my_server <- function(input, output) {

  # Define a map to render in the UI
  output$shooting_map <- renderLeaflet({

    # Construct a color palette (scale) based on chosen analysis variable
    palette_fn <- colorFactor(
      palette = "Dark2",
      domain = shootings[[input$analysis_var]]
    )

    # Create and return the map
    leaflet(data = shootings) %>%
      addProviderTiles("Stamen.TonerLite") %>% # add Stamen Map Tiles
      addCircleMarkers( # add markers for each shooting
        lat = ~lat,
        lng = ~long,
        label = ~paste0(name, ", ", age), # add a label: name and age
        color = ~palette_fn(shootings[[input$analysis_var]]), # set color w/ input
        fillOpacity = .7,
        radius = 4,
        stroke = FALSE
      ) %>%
      addLegend( # include a legend on the plot
        "bottomright",
        title = input$analysis_var,
        pal = palette_fn, # the palette to label
        values = shootings[[input$analysis_var]], # again, double-bracket notation
        opacity = 1 # legend is opaque
      )
  })

  # Define a table to render in the UI
  output$grouped_table <- renderTable({
    table <- shootings %>%
      group_by(shootings[[input$analysis_var]]) %>%
      count() %>%
      arrange(-n)

    colnames(table) <- c(input$analysis_var, "Number of Victims") # format columns
    table # return the table
  })
}
```

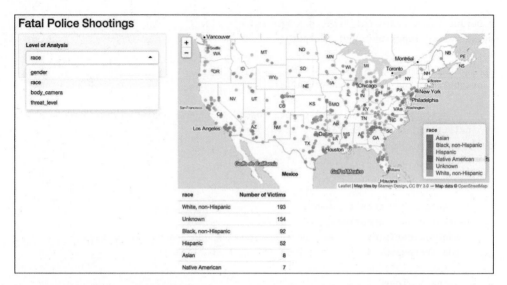

Figure 19.10   A Shiny application exploring fatal police shootings in 2018. A dropdown menu allows users to select the feature that dictates the color on the map, as well as the level of aggregation for the summary table.

As this example shows, in a little less than 80 lines of well-commented code, you can build an interactive application for exploring fatal police shootings. The final application is shown in Figure 19.10, and the full code appears below.

```
# An interactive exploration of police shootings in 2018
# Data compiled by the Washington Post

# Load libraries
library(shiny)
library(dplyr)
library(leaflet)

# Load the prepared data
shootings <- read.csv("police-shootings.csv", stringsAsFactors = FALSE)

# Define UI for application that renders the map and table
my_ui <- fluidPage(
  # Application title
  titlePanel("Fatal Police Shootings"),

  # Sidebar with a selectInput for the variable for analysis
  sidebarLayout(
    sidebarPanel(
      selectInput(
```

```r
      inputId = "analysis_var",
      label = "Level of Analysis",
      choices = c("gender", "race", "body_camera", "threat_level")
    )
  ),

  # Display the map and table in the main panel
  mainPanel(
    leafletOutput("shooting_map"), # reactive output provided by leaflet
    tableOutput("grouped_table")
  )
)
)
)

# Define server that renders a map and a table
my_server <- function(input, output) {

  # Define a map to render in the UI
  output$shooting_map <- renderLeaflet({

    # Construct a color palette (scale) based on chosen analysis variable
    palette_fn <- colorFactor(
      palette = "Dark2",
      domain = shootings[[input$analysis_var]]
    )

    # Create and return the map
    leaflet(data = shootings) %>%
      addProviderTiles("Stamen.TonerLite") %>% # add Stamen Map Tiles
      addCircleMarkers( # add markers for each shooting
        lat = ~lat,
        lng = ~long,
        label = ~paste0(name, ", ", age), # add a label: name and age
        color = ~palette_fn(shootings[[input$analysis_var]]), # set color w/ input
        fillOpacity = .7,
        radius = 4,
        stroke = FALSE
      ) %>%
      addLegend( # include a legend on the plot
        "bottomright",
        title = "race",
        pal = palette_fn, # the palette to label
        values = shootings[[input$analysis_var]], # double-bracket notation
        opacity = 1 # legend is opaque
      )
  })
```

```
# Define a table to render in the UI
output$grouped_table <- renderTable({
  table <- shootings %>%
    group_by(shootings[[input$analysis_var]]) %>%
    count() %>%
    arrange(-n)

  colnames(table) <- c(input$analysis_var, "Number of Victims") # format column names
  table # return the table
})
}

# Start running the application
shinyApp(ui = my_ui, server = my_server)
```

By creating interactive user interfaces for exploring your data, you can empower others to discover relationships in the data, regardless of their technical skills. This will help bolster their understanding of your data set and eliminate requests for you to perform different analyses (others can do it themselves!).

> **Tip:** Shiny is a very complex framework and system, so RStudio provides a large number of resources to help you learn to use it. In addition to providing a cheatsheet available through the RStudio menu (Help > Cheatsheets), RStudio has compiled a detailed and effective set of video and written tutorials.[a]
>
> ――――――――――――――
> [a]http://shiny.rstudio.com/tutorial/

For practice building Shiny applications, see the set of accompanying book exercises.[15]

――――――――――――――

[15]**Shiny exercises:** https://github.com/programming-for-data-science/chapter-19-exercises

Working Collaboratively

To be a successful member of a data science team, you will need to be able to effectively collaborate with others. While this is true for nearly any practice, an additional challenge for collaborative data science is working on shared code for the same project. Many of the techniques for supporting collaborative coding involve writing clear, well-documented code (as demonstrated throughout this book!) that can be read, understood, and modified by others. But you will also need to be able to effectively *integrate* your code with code written by others, avoiding any "copy-and-pasting" work for collaboration. The best way to do this is to use a version control system. Indeed, one of the biggest benefits of git is its ability to support **collaboration** (working with other people). In this chapter, you will expand your version control skills to maintain different versions of the same code base using git's **branching model**, and familiarize yourself with two different models for collaborative development.

20.1 Tracking Different Versions of Code with Branches

To work effectively with others, you need to understand how git supports nonlinear development on a project through branches. A **branch** in git is a way of labeling a *sequence of commits*. You can create labeled commit sequences (branches) that exist side by side within the same project, allowing you to effectively have different "lines" of development occurring in parallel and diverging from each other. That is, you can use git to track multiple different, diverging versions of your code, allowing you to work on multiple versions at the same time.

Chapter 3 describes how to use git when you are working on a single branch (called master) using a *linear sequence* of commits. As an example, Figure 20.1 illustrates a series of commits for a sample project history. Each one of these commits—identified by its **hash** (e.g., e6cfd89 in short form)—follows sequentially from the previous commit. Each commit builds directly on the other; you would move back and forth through the history in a straight line. This linear sequence represents a workflow using a *single line of development*. Having a single line of development is a great start for a work process, as it allows you to track changes and revert to earlier versions of your work.

In addition to supporting single development lines, git supports a *nonlinear model* in which you "branch off" from a particular line of development to create new concurrent change histories. You can think of these as "alternate timelines," which are used for developing different features or fixing bugs. For example, suppose you want to develop a new visualization for your project, but you're

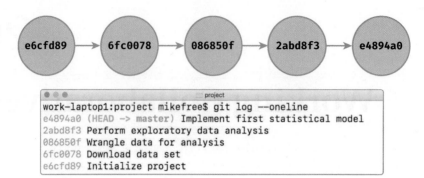

Figure 20.1 A diagram of a linear sequence of commits alongside a log of the commit history as shown in the terminal. This project has a single history of commits (i.e., *branch*), each represented by a six-character *commit hash*. The HEAD—most recent commit—is on the `master` branch.

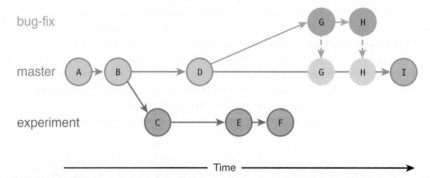

Figure 20.2 A sequence of commits spread across multiple branches, producing "alternate time-lines." Commits switch between being added to each branch (timeline). The commits on the `bug-fix` branch (labeled G and H) are *merged* into the `master` branch, becoming part of that history.

unsure if it will look good and be incorporated. You don't want to pollute the primary line of development (the "main work") with experimental code, so instead you *branch off* from the line of development to work on this code at the same time as the rest of the core work. You are able to commit iterative changes to both the experimental visualization branch and the main development line, as shown in Figure 20.2. If you eventually decide that the code from the experimental branch is worth keeping, you can easily *merge* it back into the main development line as if it were created there from the start!

20.1.1 Working with Different Branches

All `git` repositories have at least one branch (line of development) where commits are made. By default, this branch is called `master`. You can view a list of current branches in the repo with the **git branch** command:

```
# See a list of current branches in the repo
git branch
```

The line printed with the asterisk (*) is the "current branch" you're on. You can use the same `git branch` command to create a *new* branch:

```
# Create a new branch called BRANCH_NAME
git branch BRANCH_NAME
```

This will create a new branch called BRANCH_NAME (replace BRANCH_NAME with whatever name you want; usually not in all caps). For example, you could create a branch called `experiment`:

```
# Create a new branch called `experiment`
git branch experiment
```

If you run `git branch` again, you will see that this *hasn't actually changed what branch you're on*. In fact, all you have done is create a new branch that *starts* at the current commit!

> **Going Further**: Creating a new branch is similar to creating a new pointer to a node in the *linked list* data structure from computer science.

To switch to a different branch, you use the `git checkout` command (the same one described in Section 3.5.2).

```
# Switch to the BRANCH_NAME branch
git checkout BRANCH_NAME
```

For example, you can switch to the `experiment` branch with the following command:

```
# Switch to the `experiment` branch
git checkout experiment
```

Checking out a branch doesn't actually create a new commit! All it does is change the HEAD so that it now refers to the latest commit of the target branch (the alternate timeline). **HEAD** is just an alias for "the most recent commit on the current branch." It lets you talk about the most recent commit generically, rather than needing to use a particular *commit hash*.

You can confirm that the branch has changed by running the `git branch` command and looking for the asterisk (*), as shown in Figure 20.3.

Alternatively (and more commonly), you can create *and* checkout a branch in a single step using the -b option with `git checkout`:

```
# Create and switch to a branch called BRANCH_NAME
git checkout -b BRANCH_NAME
```

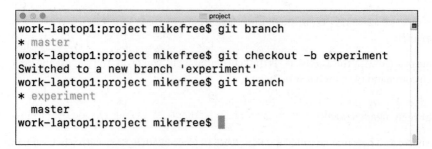

Figure 20.3   Using `git` commands on the command line to display the current branch (`git branch`), and create and checkout a new branch called `experiment` (`git checkout -b experiment`).

For example, to create and switch to a new branch called `experiment`, you would use the following command:

```
# Create and switch to a new branch called `experiment`
git checkout -b experiment
```

This effectively does a `git branch BRANCH_NAME` followed by a `git checkout BRANCH_NAME`. This is the recommended way of creating new branches.

Once you have checked out a particular branch, any new commits from that point on will occur in the "alternate timeline," without disturbing any other line of development. New commits will be "attached" to the HEAD (the most recent commit on the *current* branch), while all other branches (e.g., `master`) will stay the same. If you use `git checkout` again, you can switch back to the other branch. This process is illustrated in Figure 20.4.

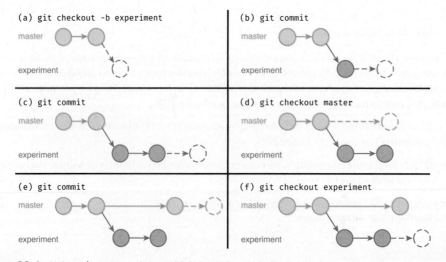

Figure 20.4   Using `git` to commit to multiple branches. A hollow circle is used to represent where the next commit will be added to the history. Switching branches, as in figures (a), (d), and (f), will change the location of the HEAD (the commit that points to the hollow circle), while making new commits, as in figures (b), (c), and (e), will add new commits to the current branch.

Importantly, checking out a branch will "reset" the files and code in the repo to whatever they looked like when you made the last commit on that branch; the code from the other branches' versions is stored in the repo's `.git` database. You can switch back and forth between branches and watch your code change!

For example, Figure 20.5 demonstrates the following steps:

1. `git status`: Check the status of your project. This confirms that the repo is on the `master` branch.

2. `git checkout -b experiment`: Create and checkout a new branch, `experiment`. This code will *branch off* of the `master` branch.

3. Make an update to the file in a text editor (still on the `experiment` branch).

4. `git commit -am "Update README"`: This will `add` and `commit` the changes (as a single command)! This commit is made *only* to the `experiment` branch; it exists in that timeline.

5. `git checkout master`: Switch back to the `master` branch. The file switches to show the latest version of the `master` branch.

6. `git checkout experiment`: Switch back to the `experiment` branch. The file switches to show the latest version of the `experiment` branch.

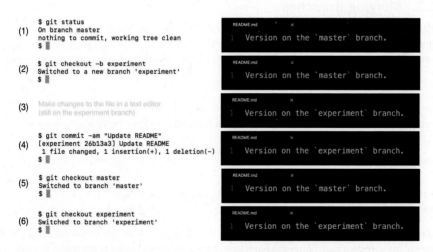

Figure 20.5 Switching branches allows you to work on multiple versions of the code simultaneously.

> **Caution:** You can only check out a branch if the *current working directory* has no uncommitted changes. This means you will need to `commit` any changes to the current branch before you `checkout` another branch. If you want to "save" your changes but don't want to commit to them, you can use git's ability to temporarily **stash**[a] changes.
>
> ───────────────────
> [a]https://git-scm.com/book/en/v2/Git-Tools-Stashing-and-Cleaning

Finally, you can delete a branch using `git branch -d BRANCH_NAME`. Note that this command will give you a warning if you might lose work; be sure to read the output message!

Taken together, these commands will allow you to develop different aspects of your project *in parallel*. The next section discusses how to bring these lines of development together.

> **Tip:** You can also use the `git checkout BRANCH_NAME FILE_NAME` command to checkout an individual file from a particular branch. This will load the file directly into the current working directory as a file change, *replacing* the current version of the file (`git` will not merge the two versions of the file together)! This is identical to checking out a file from a past commit (as described in Chapter 3), just using a branch name instead of a commit hash.

20.1.2 Merging Branches

If you have changes (commits) spread across multiple branches, eventually you will want to combine those changes back into a single branch. This process is called **merging**: you "merge" the changes from one branch into another. You do this with the (surprise!) **git merge** command:

```
# Merge OTHER_BRANCH into the current branch
git merge OTHER_BRANCH
```

For example, you can merge the `experiment` branch into the `master` branch as follows:

```
# Make sure you are on the `master` branch
git checkout master

# Merge the `experiment` branch into the current (`master`) branch
git merge experiment
```

The `merge` command will (in effect) walk through each line of code in the two versions of the files, looking for any differences. Changes to each line of code in the *incoming* branch will then be applied to the equivalent line in the current branch, so that the current version of the files contains all of the incoming changes. For example, if the `experiment` branch included a commit that added a new code statement to a file at line 5, changed the code statement at line 9, and deleted the code statement at line 13, then `git` would add the new line 5 to the file (pushing everything else down), change the code statement that was at line 9, and delete the code statement that was at line 13. `git` will automatically "stitch" together the two versions of the files so that the current version contains *all* of the changes.

> **Tip:** When merging, think about where you want the code to "end up"—that is the branch you want to checkout and merge into!

In effect, merging will take the commits from another branch and insert them into the history of the current branch. This is illustrated in Figure 20.6.

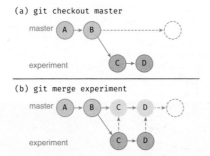

Figure 20.6 Merging an experiment branch into the master branch. The committed changes from the experiment branch (labeled C and D) are inserted into the master branch's history, while also remaining present in the experiment branch.

Note that the git merge command will merge OTHER_BRANCH into the branch you are currently on. For example, if you want to take the changes from your experiment branch and merge them into your master branch, you will need to first checkout your master branch, and merge in the changes from the experiment branch.

> **Caution:** If something goes wrong, don't panic and close your command shell! Instead, take a breath and look up how to fix the problem you've encountered (e.g., how to exit *vim*). As always, if you're unsure why something isn't working with git, use git status to check the current status and to determine which steps to do next.

If the two branches have not edited the same line of code, git will stitch the files together seamlessly and you can move forward with your development. Otherwise, you will have to resolve any *conflict* that occurs as part of your merge.

20.1.3 Merge Conflicts

If you perform a merge between two branches that have different commits that edit the *same lines of code* the result will be a **merge conflict** (so called because the changes are in "conflict"), as demonstrated in Figure 20.7.

git is just a simple computer program, and has no way of knowing which version of the conflicting code it should keep—is the master version or the experiment version better? Since git can't

(a)
```
README.md              ×
1    Change on the `master` branch.
```

(b)
```
$ git commit -am "Change to master version"
[master 644f035] Change to master version
 1 file changed, 1 insertion(+), 1 deletion(-)
$
```

(c)
```
$ git checkout experiment
Switched to branch 'experiment'
$
```

(d)
```
README.md              ×
1    Other change to the `experiment` branch.
```

(e)
```
$ git commit -am "Change to experiment version"
[experiment 512d274] Change to experiment version
 1 file changed, 1 insertion(+), 1 deletion(-)
$
```

(f)
```
$ git checkout master
Switched to branch 'master'
$
```

(g)
```
$ git merge experiment
Auto-merging README.md
CONFLICT (content): Merge conflict in README.md
Automatic merge failed; fix conflicts and then commit the result.
$
$ git status
On branch master
You have unmerged paths.
  (fix conflicts and run "git commit")
  (use "git merge --abort" to abort the merge)

Unmerged paths:
  (use "git add <file>..." to mark resolution)

        both modified:   README.md

no changes added to commit (use "git add" and/or "git commit -a")
$
```

Figure 20.7   A merge conflict as shown in the command shell.

determine which version of the code to keep, it stops the merge in the middle and forces you to choose what code is correct *manually*.

To resolve the merge conflict, you will need to edit the files (code) to pick which version to keep. git adds special characters (e.g., <<<<<<<<) to the files to indicate where it encountered a conflict (and thus where you need to make a decision about which code to keep), as shown in Figure 20.8.

```
Use me ▪                                                                    our changes
<<<<<<< HEAD
Change on the `master` branch.
=======
Other change to the `experiment` branch.
>>>>>>> experiment
Use me ▪                                                                  their changes
```

Figure 20.8 A merge conflict as shown in Atom. You can select the version of the code you wish to keep by clicking one of the Use me buttons, or edit the code in the file directly.

To resolve a merge conflict, you need to take the following steps:

1. Use git status to see which files have merge conflicts. Note that multiple files may have conflicts, and each file may have more than one conflict.

2. Choose which version of the code to keep. You do this by editing the files (e.g., in RStudio or Atom). You can make these edits manually, though some IDEs (including Atom) provide buttons that let you directly choose a version of the code to keep (e.g., the "Use me" button in Figure 20.8).

   Note that you can choose to keep the "original" HEAD version from the current branch, the "incoming" version from the other branch, or some combination thereof. Alternatively, you can replace the conflicting code with something new entirely! Think about what you want the "correct" version of the final code to be, and make it so. Remember to remove the <<<<<<< and ======= and >>>>>>> characters; these are not legal code in any language.

   > **Tip:** When resolving a merge conflict, pretend that a cat walked across your keyboard and added a bunch of extra junk to your code. Your task is to fix your work and restore it to a clean, working state. *Be sure to test your code to confirm that it continues to work after making these changes!*

3. Once you are confident that the conflicts are all resolved and everything works as it should, follow the instructions shown by git status to add and commit the change you made to the code to resolve the conflict:

```
# Check current status: have you edited all conflicting files?
git status

# Add and commit all updated files
git add .
git commit -m "Resolve merge conflict"
```

This will complete the merge! Use `git status` to check that everything is clean again.

> **Tip:** If you want to "cancel" a merge with a conflict (e.g., you initiated a merge, but you don't want to go through with it because of various conflicts), you can cancel the merge process with the **git merge --abort** command.

> **Remember:** *Merge conflicts are expected.* You didn't do something wrong if one occurs! Don't worry about getting merge conflicts or try to avoid them: just resolve the conflict, fix the "bug" that has appeared, and move on with your life.

20.1.4   Merging from GitHub

When you push to and pull from GitHub, what you're actually doing is merging your commits with the ones on GitHub! Because GitHub won't know which version of your files to keep, you will need to resolve all merge conflicts on your machine. This plays out in two ways:

1. You will *not* be able to push to GitHub if merging your commits *into* GitHub's repo might cause a merge conflict. git will instead report an error, telling you that you need to pull changes first and make sure that your version is up to date. "Up to date" in this case means that you have downloaded and merged all the commits on your local machine, so there is no chance of divergent changes causing a merge conflict when you merge by pushing.

2. Whenever you pull changes from GitHub, there may be a merge conflict. These are resolved in the exact same way as when merging local branches; that is, you need to edit the files to resolve the conflict, then add and commit the updated versions.

Thus, when working with GitHub (and especially with multiple people), you will need to perform the following steps to upload your changes:

1. pull (download) any changes you don't have

2. *Resolve* any merge conflicts that occur

3. push (upload) your merged set of changes

Of course, because GitHub repositories are repos just like the ones on your local machine, they can have branches as well. You gain access to any *remote* branches when you clone a repo; you can see a list of them with `git branch -a` (using the "all" option).

If you create a new branch on your local machine, it is possible to push that branch to GitHub, creating a mirroring branch on the remote repo (which usually has the alias name `origin`). You do this by specifying the branch in the `git push` command:

```
# Push the current branch to the BRANCH_NAME branch on the `origin`
# remote (GitHub)
git push origin BRANCH_NAME
```

where `BRANCH_NAME` is the name of the branch you are currently on (and thus want to push to GitHub). For example, you could push the `experiment` branch to GitHub with the following command:

```
# Make sure you are on the `experiment` branch
git checkout experiment
```

```
# Push the current branch to the `experiment` branch on GitHub
git push origin experiment
```

You often want to create an association between your local branch with the remote one on GitHub. You can establish this relationship by including the `-u` option in your push:

```
# Push to the BRANCH_NAME branch on origin, enabling remote tracking
# The -u creates an association between the local and remote branches
git push -u origin BRANCH_NAME
```

This causes your local branch to "track" the one on GitHub. Then when you run a command such as `git status`, it will tell you whether one repo has more commits than the other. Tracking will be remembered once set up, so you only need to use the `-u` option *once*. It is best to do this the first time you push a local branch to GitHub.

20.2   Developing Projects Using Feature Branches

The main benefit of branches is that they allow you (and others) to simultaneously work on different aspects of the code without disturbing the main code base. Such development is best organized by separating your work across different **feature branches**—branches that are each dedicated to a different **feature** (capability or part) of the project. For example, you might have one branch called `new-chart` that focuses on adding a complex visualization, or another branch called `experimental-analysis` that tries a bold new approach to processing the data. Importantly, each branch is based on a feature of the project, not a particular person: a single developer could be working on multiple feature branches, and multiple developers could collaborate on a single feature branch (more on this later).

The goal when organizing projects into feature branches is that the `master` branch should always contain "production-level" code: valid, completely working code that you could deploy or publish (read: give to your boss or teacher) at a whim. All feature branches branch off of `master`, and are allowed to contain temporary or even broken code (since they are still in development). This way there is always a "working" (if incomplete) copy of the code (`master`), and development can be kept isolated and considered independent of the whole. Note that this organization is similar to how the earlier example uses an experiment branch.

Using feature branches works like this:

1. You decide to add a new feature to the project: a snazzy visualization. You create a new feature branch off of master to isolate this work:

```
# Make sure you are on the `master` branch
git checkout master

# Create and switch to a new feature branch (called `new-chart`)
git checkout -b new-chart
```

2. You then do your coding work while on this branch. Once you have completed some work, you would make a commit to add that progress:

```
# Add and commit changes to the current (`new-chart`) branch
git add .
git commit -m "Add progress on new vis feature"
```

3. Unfortunately, you may then realize that there is a bug in the master branch. To address this issue, you would switch back to the master branch, then create a *new branch* to fix the bug:

```
# Switch from your `new-chart` branch back to `master`
git checkout master

# Create and switch to a new branch `bug-fix` to fix the bug
git checkout -b bug-fix
```

(You would fix a bug on a separate branch if it was complex or involved multiple commits, in order to work on the fix separate from your regular work).

4. After fixing the bug on the bug-fix *branch*, you would add and commit those changes, then checkout the master branch to merge the fix back into master:

```
# Add and commit changes that fix the bug (on the `bug-fix` branch)
git add .
git commit -m "Fix the bug"

# Switch to the `master` branch
git checkout master

# Merge the changes from `bug-fix` into the current (`master`) branch
git merge bug-fix
```

5. Now that you have fixed the bug (and merged the changes into master), you can get back to developing the visualization (on the new-chart branch). When it is complete, you will add and commit those changes, then checkout the master branch to merge the visualization code back into master:

```
# Switch back to the `new-chart` branch from the `master` branch
git checkout new-chart

# Work on the new chart...

# After doing some work, add and commit the changes
git add .
git commit -m "Finish new visualization"

# Switch back to the `master` branch
git checkout master

# Merge in changes from the `new-chart` branch
git merge new-chart
```

The use of feature branches helps isolate progress on different elements of a project, reducing the need for repeated merging (and the resultant conflicts) of half-finished features and creating an organized project history. Note that feature branches can be used as part of either the *centralized workflow* (see Section 20.3) or the *forking workflow* (see Section 20.4).

20.3   Collaboration Using the Centralized Workflow

This section describes a model for working with multiple collaborators on the same project, coordinating and sharing work through GitHub. In particular, it focuses on the **centralized workflow**,[1] in which all collaborators use a single repository on GitHub. This workflow can be extended to support the use of feature branches (in which each feature is developed on a different branch) as described in Section 20.2—the only additional change is that multiple people can work on each feature! Using the centralized workflow involves configuring a shared repository on GitHub, and managing changes across multiple contributors.

20.3.1   Creating a Centralized Repository

The centralized workflow uses a single repository stored on GitHub—that is, every single member of the collaboration team will push and pull to the same GitHub repo. However, since each repository needs to be created under a particular account, a *single member* of the team will need to create that repository (e.g., by clicking the *"New"* button on the "Repositories" tab on the GitHub web portal).

To make sure everyone is able to push to the repository, whoever creates the repo will need to add the other team members as collaborators.[2] They can do this under the *"Settings"* tab of the repo's web portal page, as shown in Figure 20.9. (The creator will want to give all team members "write" access so they can push changes to the repo.)

---

[1] **Atlassian: Centralized Workflow:** https://www.atlassian.com/git/tutorials/comparing-workflows#centralized-workflow

[2] **GitHub: Inviting collaborators to a personal repository:** https://help.github.com/articles/inviting-collaborators-to-a-personal-repository/

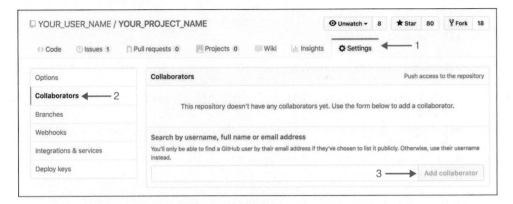

Figure 20.9   Adding a collaborator to a GitHub repository via the web portal.

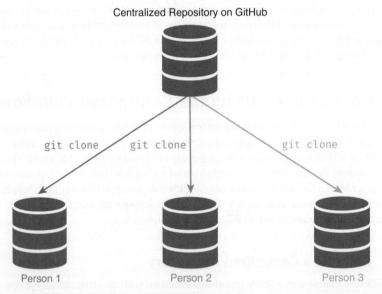

Figure 20.10 With the centralized workflow, each collaborator clones the same repository from GitHub. All users must have *write permissions* to this repository in order to push their changes to it.

Once everyone has been added to the GitHub repository, *each team member* will need to `clone` the repository to their local machines to work on the code individually, as shown in Figure 20.10. Collaborators can then push any changes they make to the central repository, and `pull` any changes made by others.

When you are contributing to the same repository along with multiple other people, it's important to ensure that you are working on the most up-to-date version of the code. This means that you will regularly have to *pull changes* from GitHub that your team members may have committed. As a result, developing code with the centralized workflow follows these steps:

1. To begin your work session, `pull` in the latest changes from GitHub. For example:

```
# Pull latest changes from `origin` (GitHub's) `master` branch
# You could specify a different branch as appropriate
git pull origin master
```

2. Do your work, making changes to the code. Remember to add and `commit` your work each time you make notable progress!

3. Once you are satisfied with your changes and want to share them with your team, you'll need to upload the changes back to GitHub. But note that if someone pushes a commit to GitHub *before you push your own changes*, you will need to integrate those changes into your code (and test them!) before doing your own push up to GitHub. Thus you'll want to first `pull` down any changes that have been made in the interim (there may not be any) so that you are up to date and ready to push:

```
# Pull latest changes from `origin` (GitHub's) `master` branch
# You could specify a different branch as appropriate
git pull origin master

# In case of a merge conflict, fix the changes
# Once fixed, add and commit the changes (using default commit message)
git add .
git commit --no-edit

# Push changes to `origin` (GitHub's) `master` branch
# You could specify a different branch as appropriate
git push origin master
```

Remember that when you `pull` in changes, `git` is really merging the remote branch with your local one, which may result in a merge conflict you need to resolve; be sure to fix the conflict and then mark it as resolved. (The `--no-edit` argument used with `git commit` tells `git` to use the default commit message, instead of specifying your own with the `-m` option.)

While this strategy of working on a single `master` branch may suffice for small teams and projects, you can spend less time merging commits from different team members if your team instead uses a dedicated feature branch for each feature they work on.

20.3.2   Using Feature Branches in the Centralized Workflow

The centralized workflow supports the use of feature branches for development (often referred to as the *feature branch workflow*). This is similar to the procedure for working with feature branches described previously. The only additional complexity is that you must `push` and `pull` multiple branches to GitHub so that multiple people can work on the same feature.

> **Remember:** In the feature branch workflow, each branch is for a different feature, *not* a different developer! This means that a developer can work on multiple different features, and a feature can be worked on by multiple developers.

As an example of this workflow, consider the collaboration on a feature occurring between two developers, Ada and Bebe:

1. Ada decides to add a new feature to the code, a snazzy visualization. She creates a new feature branch off of `master`:

    ```
    # Double-check that the current branch is the `master` branch
    git checkout master

    # Create and switch to a new feature branch (called `new-chart`)
    git checkout -b new-chart
    ```

2. Ada does some work on this feature, and then commits that work when she's satisfied with it:

    ```
    # Add and commit changes to the current (`new-chart`) branch
    git add .
    git commit -m "Add progress on new vis feature"
    ```

3. Happy with her work, Ada decide to takes a break. She pushes her feature branch to GitHub to back it up (and so her team can also contribute to it):

    ```
    # Push to a new branch on `origin` (GitHub) called `new-chart`,
    # enabling tracking
    git push -u origin new-chart
    ```

4. After talking to Ada, Bebe decides to help finish up the feature. She checks out the feature branch and makes some changes, then pushes them back to GitHub:

    ```
    # Use `git fetch` to "download" commits from GitHub, without merging
    # This makes the remote branch available locally
    git fetch origin

    # Switch to local copy of the `new-chart` branch
    git checkout new-chart

    # Work on the feature is done outside of terminal...

    # Add, commit, and push the changes back to `origin`
    # (to the existing `new-chart` branch, which this branch tracks)
    git add .
    git commit -m "Add more progress on feature"
    git push
    ```

    The **git fetch** command will "download" commits and branches from GitHub (but without merging them); it is used to get access to branches that were created after the repo was originally cloned. Note that `git pull` is actually a shortcut for a `git fetch` followed by a `git merge`!

5. Ada then downloads Bebe's changes to her (Ada's) machine:

```
# Download and merge changes from the `new-chart` branch on GitHub
# to the current branch
git pull origin new-chart
```

6. Ada decides the feature is finished, and merges it back into `master`. But first, she makes sure she has the latest version of the `master` code:

```
# Switch to the `master` branch, and download any changes
git checkout master
git pull

# Merge the feature branch into the master branch (locally)
git merge new-chart
# Fix any merge conflicts!
# Add and commit these fixes (if necessary)

# Push the updated `master` code back to GitHub
git push
```

7. Now that the feature has been successfully added to the project, Ada can delete the feature branch (using `git branch -d new-chart`). She can delete GitHub's version of the branch through the web portal interface (recommended), or by using `git push origin -d new-chart`.

This kind of workflow is very common and effective for supporting collaboration. Moreover, as projects grow in size, you may need to start being more organized about how and when you create feature branches. For example, the **Git Flow**[3] model organizes feature branches around product releases, and is a popular starting point for large collaborative projects.

20.4 Collaboration Using the Forking Workflow

The **forking workflow** takes a fundamentally different approach to collaboration from the centralized workflow. Rather than having a single shared remote repository, each developer has their own repository on GitHub that is **forked** from the original repository, as shown in Figure 20.11. As discussed in Chapter 3, developers can create their own copy of a repository on GitHub by forking it. This allows the individual to make changes (and contribute) to the repository, without necessarily needing permission to modify the "original" repo. This is particularly valuable when contributing to open source software projects (such as R packages like `dplyr`) to which you may not have ownership.

In this model, each person contributes code to their own personal copy of the repository. The changes between these different repos are merged together through a GitHub process called a **pull**

---

[3]**Git Flow: A successful Git branching model:** http://nvie.com/posts/a-successful-git-branching-model/

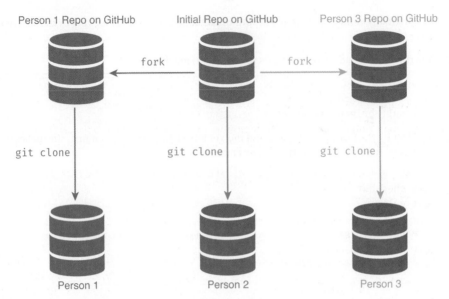

Figure 20.11 In the forking workflow, collaborators create their own version of the repository on GitHub by forking it. Each member then clones *their fork* and pushes changes to *that* remote repository. Changes across forks are integrated using a pull request.

request.[4] A pull request (colloquially called a "PR") is a request for the changes in one version of the code (i.e., a fork or branch) to be pulled (merged) into another. With pull requests, one developer can send a request to another developer, essentially saying *"I forked your repository and made some changes: can you integrate **my** changes into **your** repo?"* The second developer can perform a **code review**: reviewing the proposed changes and making comments or asking for corrections to anything that appears problematic. Once the changes are made (committed and pushed to the "source" branch on GitHub), the pull request can be **accepted** and the changes merged into the "target" branch. Because pull requests can be applied across (forked) repositories that share history, a developer can fork an existing professional project, make changes to that fork, and then send a pull request back to the original developer asking that developer to merge in changes.

> **Caution**: You should only use pull requests to integrate changes on *remote* branches (i.e., two different forks of a repo). To integrate commits from different branches of the same repository, you should merge changes on your local machine (*not* using GitHub's pull request feature).

To issue a pull request, you will need to make changes to your fork of a repository and push those to GitHub. For example, you could walk through the following steps:

---

[4]**GitHub: About pull requests**: https://help.github.com/articles/about-pull-requests/

1. Fork a repository to create your own version *on GitHub*. For example, you could fork the repository for the `dplyr` package[5] if you wanted to make additions to it, or fix a bug that you've identified.

   You will need to `clone` your fork of the repository to your own machine. Be careful that you clone the correct repo (look at the username for the repo in the GitHub web portal—where it says YOUR_USER_NAME in Figure 20.12).

2. After you've cloned your fork of the repository to your own machine, you can make any changes desired. When you're finished, `add` and `commit` those changes, then `push` them up to GitHub.

   You can use feature branches to make these changes, including pushing the feature branches to GitHub as described earlier.

3. Once you've pushed your changes, navigate to the web portal page for your fork of the repository on GitHub and click the *"New Pull Request"* button as shown in Figure 20.12.

4. On the next page, you will need to specify which branches of the repositories you wish to merge. The **base** branch is the one you want to merge *into* (often the `master` branch of the original repository), and the **head** branch (labeled "compare") is the branch with the new changes you want to be merged in (often the `master` branch of your fork of the repository), as shown in Figure 20.13.

5. After clicking the *"Create Pull Request"* button (in Figure 20.13), you will write a title and a description for your pull request (as shown in Figure 20.14). After describing your proposed changes, click the *"Create pull request"* button to issue the pull request.

Figure 20.12   Create a new pull request by clicking the *"New Pull Request"* button on *your fork* of a repository.

---

[5]**dplyr** Package GitHub Repository: https://github.com/tidyverse/dplyr

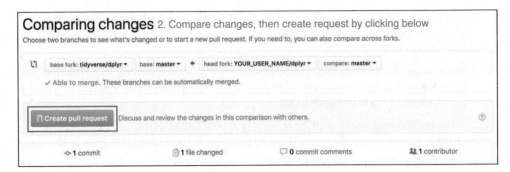

Figure 20.13 Compare changes between the two forks of a repository on GitHub before issuing a pull request.

Figure 20.14 Write a title and description for your pull request, then issue the request by clicking the *"Create Pull Request"* button.

Remember: The pull request is a request to merge two branches, not to merge a specific set of commits. This means that you can *push more commits* to the head ("merge-from") branch, and they will automatically be included in the pull request—the request is always up to date with whatever commits are on the (remote) branch.

If the code reviewer requests changes, you make those changes to your local repo and just push the changes as normal. They will be integrated into the existing pull request automatically without you needing to issue a new request!

You can view all pull requests (including those that have been accepted) through the *"Pull Requests"* tab at the top of the repository's web portal. This view will allow you to see comments that have been left by the reviewer.

If someone (e.g., another developer on your team) sends you a pull request, you can accept that pull request[6] through GitHub's web portal. If the branches can be merged without a conflict, you can do this simply by clicking the *"Merge pull request"* button. However, if GitHub detects that a conflict may occur, you will need to pull down the branches and merge them locally.[7]

Note that when you merge a pull request via the GitHub website, the merge is done in the repository on GitHub's servers. Your copy of the repository on your local machine will not yet have those changes, so you will need to use `git pull` to download the updates to the appropriate branch.

In the end, the ability to effectively collaborate with others on programming and data projects is one of the biggest benefits of using `git` and GitHub. While such collaboration may involves some coordination and additional commands, the techniques described in this chapter will enable you to work with others—both within your team and throughout the open source community—on larger and more important projects.

> **Tip:** Branches and collaboration are among the most confusing parts of `git`, so there is no shortage of resources that aim to help clarify this interaction. *Git and GitHub in Plain English*[a] is an example tutorial focused on collaboration with branches, while *Learn Git Branching*[b] is an interactive tutorial focused on branching itself. Additional interactive visualizations of branching with `git` can be found here.[c]
>
> ---
> [a]https://red-badger.com/blog/2016/11/29/gitgithub-in-plain-english
> [b]http://learngitbranching.js.org
> [c]https://onlywei.github.io/explain-git-with-d3/

For practice working with collaborative version control methods, see the set of accompanying book exercises.[8]

---

[6]**GitHub: Merging a pull request:** https://help.github.com/articles/merging-a-pull-request/
[7]**GitHub: Checking out pull requests locally:** https://help.github.com/articles/checking-out-pull-requests-locally
[8]**`git` collaboration exercises:** https://github.com/programming-for-data-science/chapter-20-exercises

21

Moving Forward

In this text, you have learned the foundational programming skills necessary for entering the data science field. The ability to *write code to work with data* empowers you to explore and communicate information in transparent, reusable, and collaborative ways. As many data scientists will attest, the most time-consuming part of a project is organizing and exploring the data—something that you are now more than capable of doing. These skills on their own are quite valuable for gaining insight from quantitative information, but there is always more to learn. If you are eager to expand your skills, there are a few areas that serve as obvious next steps in data science.

21.1 Statistical Learning

The term **statistical learning** encompasses the statistical and computational techniques used to transform data into information. This book has laid the groundwork for performing these techniques in R, but has not explored specific functions or packages. The aims of statistical learning can typically be reduced to two categories: assessing relationships between variables and making predictions for unobserved values.

21.1.1 Assessing Relationships

The programming skills covered in this book allow you to make comparisons across groups using summary statistics and visualization. However, it does not discuss statistical assessments for measuring the size or significance of these variations. A multitude of statistical techniques are available in R for assessing the strength of relationships between variables. This includes questions such as *Are salaries consistent across genders?* and *Is investment in education associated with lower healthcare costs for a city?* While this text has taught you to perform exploratory data analysis techniques, it did not describe methods for measuring the strength of association that exists between variables. To draw conclusions about **causality**, and control for complex relationships in your data, you will need to understand the **statistical methods** available for answering these questions. Here are a few resources that may help you in this area:

- *R for Everyone*[1] introduces statistical modeling and evaluation in R, including linear and non-linear methods.

[1] Lander, J. P. (2017). *R for everyone: Advanced analytics and graphics* (2nd ed.). Boston, MA: Addison-Wesley.

- *An Introduction to Statistical Learning*[2] is a more general introduction to statistical learning problems, which includes an implementation in R (though the focus is more conceptual than programming oriented).

- *OpenIntro Stats*[3] is an open source[4] set of texts that focus on the basics of probability and statistics.

21.1.2 Making Predictions

The other major domain of data science is making predictions for unobserved values. This includes questions like *Which students are most likely to pass a course?* and *How is a congressperson likely to vote on a piece of legislation?* Broadly speaking, statistical methods are better suited for *assessing relationships*, while **machine learning** techniques are optimized for *making predictions*. These techniques involve the application of specific algorithms to make predictions based on patterns identified in data (for a wonderful visual introduction to machine learning, see this online interactive tutorial[5]). While a vast amount of domain knowledge is needed to properly select and interpret machine learning algorithms, many of them can be implemented in R in a *single line of code* using external packages.

21.2 Other Programming Languages

R is an excellent language for programmatically working with data (if it wasn't, we wouldn't have written a book about it!). Depending on which skills you want to expand, and what techniques your team is using, it may be worth investing in learning another programming language. Luckily, after learning one language, it's much easier to learn another—you have already practiced the skills of installing software, reading documentation, debugging code, and writing programs. To advance your data science skills, you could invest in learning the following languages:

- **Python** is another popular language for doing data science. Like R, it is open source, and has a large community of people contributing to its statistical, machine learning, and visualization packages. Because R and Python largely enable you to solve the same problems in data science, the motivations to learn Python would include collaboration (to work with people who *only* use Python), curiosity (about how a similar language solves the same problems), and analysis (if a specific sophisticated analysis is only available in a Python package). A great book for learning to program for data science in Python is the *Python Data Science Handbook*.[6]

- **Web development technologies** including *HTML, CSS*, and *JavaScript* represent a complementary skill set for data scientists. If you are passionate about building visual interfaces on the web for interacting with data, you will likely become frustrated by the

---

[2]James, G., Witten, D., Hastie, T., & Tibshirani, R. (2013). *An introduction to statistical learning* (Vol. 112). Springer. http://www-bcf.usc.edu/~gareth/ISL/

[3]Diez, D. M., Barr, C. D., & Cetinkaya-Rundel, M. (2012). OpenIntro statistics. CreateSpace. https://www.openintro.org/stat/textbook.php

[4]**OpenIntro Statistics**: https://www.openintro.org

[5]**A Visual Introduction to Machine Learning**: http://www.r2d3.us/visual-intro-to-machine-learning-part-1/

[6]VanderPlas, J. (2016). *Python Python data science handbook: Essential tools for working with data*. O'Reilly Media, Inc.

limitations of the Shiny framework. Building interactive websites from scratch requires a notable time investment, but it gives you complete control over the style and behavior of your webpages. If you are seriously interested in building custom visualizations, look into using the **d3.js**[7] JavaScript library, which you can also read about in *Visual storytelling with D3*.[8]

21.3   Ethical Responsibilities

The power of data science to change the world around us—for better and for worse—has become evident over the past decade. Data science has helped move forward research in a variety of socially impactful domains such as public health and education. At the same time it has been used to develop systems that systematically disenfranchise groups of people (both intentionally and unintentionally). Algorithms that *appear* to be unbiased have had profoundly negative impacts. For example, an analysis[9] by ProPublica revealed the racist nature of a piece of software that predicts criminal activity (you can see all of the code on GitHub[10]).

Such consequences of unchecked assumptions in data science can be difficult to detect and have outsized effects on people, so tread carefully as you move forward with your newly acquired skills. Remember: *you* are responsible for the impact of the programs that you write. The analytical and programming skills covered in this text empower you to identify and communicate about the injustices in the world. As a data scientist, you have a **moral responsibility** to *do no harm* with your skills (or better yet, to work to *undo* harms that have occurred in the past and are occurring today). As you begin to work in data science, you must *always* consider how people will be differentially impacted by your work. Think carefully about *who* is represented in—and excluded from—your data, what *assumptions* are built into your analysis, and how any decisions made using your data could differentially benefit different communities—particularly those communities that are often overlooked.

Thank you for reading our book! We hope that it provided inspiration and guidance in your pursuit of data science, and that you use these skills for good.

---

[7] **d3.js** https://d3js.org

[8] King, R. S. (2014). *Visual storytelling with D3: An introduction to data visualization in JavaScript.* Addison-Wesley.

[9] Angwin, J. L. (2016, May 23). Machine bias. ProPublica. https://www.propublica.org/article/machine-bias-risk-assessments-in-criminal-sentencing

[10] **Machine Bias Analysis, Complete Code**: https://github.com/propublica/compas-analysis

Index

Symbols

, (comma)

    data frame syntax, 122

    function syntax, 69

    key-value pair syntax, 191

" (double quotes), character data syntax, **61**

' (single quotes), character data syntax, **61**

.. (double dot), moving up directory, **14**

. (single dot), referencing current folder, **14**

| (pipe)

    directing output, 20

    pipe table, 48

! (exclamation point), Markdown image syntax, **47**

(pound/hashtag symbol)

    comment syntax, 10, 58

$ (dollar notation)

    accessing data frames, 122

    accessing list elements, 97–98

%>% (pipe operator), `dplyr` package, **141–142**

( ) (parentheses)

    function syntax, 70

    Markdown hyperlink syntax, 46

* (asterisk wildcard)

    loading entire table from database, 173

    using wildcards with files, 17–18

? (question mark), query parameter syntax, **184**

[ ] (single-bracket notation)

    accessing data frames, 122–123

    comparing single- and double-bracket notation, 101

    Markdown hyperlink syntax, 46

    retrieving value from vector, 88

[[ ]] (double-bracket notation)

    list syntax, 98–99, 101

    selecting data of interest for application, 312

{} (braces)

    code chunk syntax, 279

    key-value pair syntax, 191

    render function syntax, 308

<- (assignment operator), 59, 92

>> directing output, 20

> directing output, 20

~ (tilde), home directory shorthand, 10, 15

A

Absolute path

    for CSV data, 125

    finding R and RScript, 57

    for images, 48

    specifying paths, 14–15

    URLs and, 47

Access tokens (API keys)

    example finding Cuban food in Seattle, 196–197

    registering with web services, 186–188

add (git). *See also* **Staging Area**

    add and commit changes, 38–39, 322, 327–328, 333, 337

    adding files to repository, 32–33

    unadd, 35

aes( ) function, for aesthetic mappings, 237

Aesthetics

    adding titles and labels to charts, 246

    aesthetic mappings, 234, 237–238

    data visualization, 229–230

Aggregation

    proportional representation of data and, 212–213

    in Shiny example, 315–316

    statistical transformation of data, 255

    using summarize( ), 138–139

Analysis. *See* **Data analysis**

Annotation

    capabilities of version control systems, 28

    ggplot2 package, 246–248

Anonymous variables, 71, 140

anscombe data set, in R, 208

Anscombe's Quartet, 208

API keys (Access tokens)

    example finding Cuban food in Seattle, 196–197

    registering with web services, 186–188

APIs (application programming interfaces).
  See also **Web APIs**

    defined, 181

    in plotly package, 258

Application servers, developing, 306–309

Applications

    Shiny app example applying to fatal police shootings, 311–318

    structure in Shiny framework, 295–299

app.R file, 295–296

Apps, publishing Shiny, 309–311

Area encoding, visualizing hierarchical data, 218

Arguments

    commands and, 13

    creating data frames, 120

    creating lists, 96

    debugging functions, 78

    function inputs, 69–70

    function parts, 76

    named arguments, 72–73

    syntax of, 16

    vectorized functions and, 87

arrange( )

    dplyr core functions, 131, 137–138

    summarizing information using dplyr functions, 313

Arrays, JSON support, 191–192

AS keyword, renaming columns, 173

Assignment operator (<-)

    assigning values to variables, 59

    modifying vectors, 92

Atom

    preview rendering support, 49–50

    selecting text editor, 6–7

    writing code, 3

Authentication, API authentication service, 187

B

Bar charts

    facets and, 245

    position adjustments, 240

    proportional representation of data, 211–213

    visualizing data with single variable, 210–211

Bash shell. *See also* **Git Bash**

    commands, 13

    executing code, 4

    ls command, 13

Bins, breaking data into different variables, 142

BitBucket, comparing with GitHub, 29

Blockquotes, markdown options, 48

Blocks, markdown formatting syntax, 47

Body, function parts, 76–77

Bokeh package, 261

Bold, text formatting, 45–46

Books, resources for learning R, 65

Boolean. *See* Logical (boolean)

Box plots, 210

Bracket notation

double. *See* [[ ]] (double-bracket notation)

retrieving value from vector using bracket notation, 88

single. *See* [ ] (single-bracket notation)

Branches

git branching model, 319–320

merging, 324–325

merging from GitHub, 328–329

resolving merge conflicts, 327–328

tracking code versions with, 319–320

using in feature branch workflows, 333–335

using in forking workflows, 335–339

working with, 320–324

working with feature branches, 329–331

C

c( ) function, creating vectors, 81–82

Case sensitivity, variable names, 58

Categorical data. *See* Nominal (categorical) data

Causality, assessing statistical relationships, 341

cd, change directory command, 12–13

Centralized workflow

creating centralized repository, 331–333

feature branches in, 333–335

overview of, 331

working with feature branches, 333–335

Character data type

lists and, 95

overview of, 61

vectorized functions and, 87

Charts, 229. *See also* by individual types of graphs

Cheatsheets

for dplyr, 148

for ggplot2, 255

for GitHub, 43

for markdown, 48

for R functions, 71

for RStudio, 56, 280, 318

checkout (git)

switching between branches, 321–324

working with feature branches, 329–330

working with feature branches in centralized workflow, 335

Checkpoints. *See* Commit

Choropleth maps

drawing and examples, 248–251

overview of, 248

Chunks

breaking data into different variables, 142

inline code and, 280

options, 279–280

.Rmd files and, 277–278

Circle packing, visualizing hierarchical data, 218–219

clone (git)

collaboration using forking workflow, 336

creating centralized repository, 332

forks, 337

merging branches and, 328

repos, 36–39, 43

understanding/using git commands, 43

Code

chunks, 142, 277–280

executing, 4–5

inline code, 280

managing, 3–4

running, 54–57

syntax-colored code blocks, 48

tracking versions with branches, 319–320

Visual Studio Code (VS Code), 7, 49

writing, 3

Collaboration

centralized workflow for, 331

creating centralized repository, 331–333

interactive web applications and. *See* Shiny framework

merging branches, 324–325, 328–329

overview of, 273–274, 319

reports. *See* R Markdown

resolving merge conflicts, 327–328

tracking code versions, 319–320

working with branches, 320–324

working with feature branches, 329–331, 333–335

working with forking workflows, 335–339

collect( ), manipulating table data, 177–178

Colon operator (a:b)

creating vectors, 82

specifying range of vector index, 90

Color

adding to Leaflet map, 270

color palettes, 223–225, 242

effective for data visualization, 222–226

ggplot2 color scales, 242–243

ColorBrewer tool
color palettes, 242
examples, 289
overview of, 223–225
`colorFactor( )`, Leaflet maps, 270
Columns
changing to/from rows using `tidyr`, 157–159
`dplyr arrange( )` operation, 137–138
`dplyr filter( )` operation, 135
`dplyr mutate( )` operation, 136
Columns (fields), in relational databases, 168
Comma-separated value data. *See* CSV (comma-separated value) data
Command line
accessing, 9–10
changing directories, 12–13
cloning repository, 37
commit history, 320
directing/redirecting output, 20
executing code, 4
handling errors, 18–19
interacting with databases, 31
learning new commands, 16–17
listing files, 13
managing files, 15–16
navigating files, 11–12
networking commands, 20–23
overview of, 9
running R code, 56–57
set up tools, 4–5
specifying paths, 14–15
wildcards, 17–18
working with, 4
Command prompt. *See* **Command line**
Command Prompt (Windows)
accessing, 9–10
executing code, 4
working with, 5
Command shell (terminal). *See* **Command line**
Commands. *See also* **by individual types**
issuing, 13
list of advanced, 18
list of basic, 15
Comments
R language, 58
syntax for code comments, 10
`commit (git)`
add and commit changes, 33, 38–39, 327–328, 337
creating centralized repository, 333
git core concepts, 28
history, 40

message etiquette, 34–35
reverting to earlier versions, 40–42
tracking code versions, 319–320
understanding/using `git` commands, 43
working with branches, 320–324
working with feature branches, 330–331, 334
Communities
resources for learning R, 66–67
sources of data, 109
Comparison operators, logical values and, 62
Compiled languages, 53
Complex data type, 63, 99
Comprehensive R Archive (CRAN), 6
Computer, set up, 3–4
Concurrency, capabilities of version control systems, 28
Conditional statements, 79–80
`config`, configuring `git` for first-time use, 30
Console, RStudio, 55
Content
building Shiny application, 313
content elements in designing UIs, 299
extracting from HTTP request, 200
static content in Shiny framework, 300–301
`content( )`, extracting content from HTTP request, 200
Continuous color scales, 225–226
Continuous data
choosing effective colors for data visualization, 223
selecting visual layouts, 209–210
visualization with multiple variables, 213–216
visualizing with single variable, 210
Control widgets
developing application servers, 307
in Shiny framework, 295
user interactions in Shiny apps, 301–303
`coord_` functions
`coord_flip( )` example, 244
types of coordinate systems for geometric objects, 243–244
Coordinate systems
`coord_flip( )` example, 244
creating choropleth maps, 249–250
creating dot distribution maps, 252
Grammar of Graphics, 232
types for geometric objects, 243–244
`cor( )`, correlation function in R, 161
`count( )`, summarizing information, 313
Courses, resources for learning R, 65–66
CRAN (Comprehensive R Archive), 6
CSS language, 342
CSV (comma-separated value) data
factor variables, 126–129

loading data sets from .csv file, 167

read.csv( ), 161

viewing working directory, 125–126

working with, 124–125

ctrl+c, stopping or canceling program or command, 19

D

d3.js JavaScript library, 343

Data

acquiring domain knowledge, 112–113

analyzing. *See* Data analysis

answering questions, 116–118

dplyr example analyzing flight data, 148–153

dplyr grammar for manipulating, 131–132

encoding, 220–222, 229, 237

finding, 108–109

flattening JSON data, 196–197

generating, 107–108

interactive presentation, 293

interpreting, 112

measuring, 110–111

overview of, 107

ratio data, 111

reusable functions in managing, 70

schemas, 113–116

structures, 111–112, 122

transforming into information, 341

understanding data schemas, 113–116

visualization of. *See* Data visualization

working with CSV data, 124–125

wrangling, 106

Data analysis

generating data, 108

reusable functions, 70

tidyr package. *See* tidyr package

Data frames

accessing, 122–123

analyzing by group, 142–144

creating, 120–121

describing structure of, 121–122

factor variables, 126–129

joining, 144–148

overview of, 119–120

viewing working directory, 125–126

working with CSV data, 124–125

data( ) function, viewing available data sets, 124–125

Data-ink ratio, aesthetics of graphics, 229

Data schemas, 113–116

Data structures

overview of, 111–112

two-dimensional, 122

Data types

factors, 120

lists and, 95

R language, 60–63

selecting visual layouts, 209–210

vectorized functions and, 87

vectorized operations and, 83

Data visualization

aesthetics, 229–230

choosing effective colors, 222–226

choosing effective graphical encodings, 220–222

expressive displays, 227–229

ggplot2. *See* ggplot2 package

of hierarchical data, 217–220

leveraging preattentive attributes, 226–227

with multiple variables, 213–217

overview of, 205–207

purpose of, 207–209

reusable functions, 70

selecting visual layouts, 209–210

with single variable, 210–213

tidyr package. *See* tidyr package

Data visualization, interactive

example exploring changes to Seattle, 266–272

leaflet package, 263–266

overview of, 257–258

plotly package, 258–261

rbokeh package, 261–263

Databases

accessing from R, 175–179

designing relational, 144

overview of relational, 167–169

setting up relational, 169–171

SQL statements, 171–175

***DataCamp*, resources for learning R, 66**

dbConnect( ), accessing SQLite, 176–177

dbListTables( ), listing database tables, 177

dbplyr package, 176–179

dbplyr package, accessing databases, 174

Debugging functions, 78. *See also* **Error handling**

Directories

accessing command line and, 10

changing from command line, 12–13

printing working directory, 11

tree structure of, 12

turning into a repository, 31

viewing working directory, 125–126

Displays, expressive, 227–229

Distributions, of x and y values (statistics), 208–209

Documentation

of commands, 16

getting help via, 64

resources for learning R, 66

Shiny layouts, 304

Documents

creating, 275

knitting, 278

Domain, interpreting data by, 112–113

Dot distribution maps, 248, 251–252

Double-bracket notation. *See* **[[]] (double-bracket notation)**

dplyr package

analyzing data frames, 142–144

analyzing flight data, 148–153

arrange( ), 137–138

converting dplyr functions into SQL statements, 178

core functions, 131–132

example mapping evictions in San Francisco, 252

example report on life expectancy, 289

filter( ), 135–136

grammar for data manipulation, 131–132

group_by( ), 244

joining data frames, 144–148

mutate( ), 136–137

orienting data frames for plotting, 239

overview of, 131

performing sequential operations, 139–141

pipe operator (%>%), 141–142

select( ), 133–134

summarize( ), 138–139

Dynamic inputs, Shiny framework, 301–303

Dynamic outputs, Shiny framework, 303–304

Dynamically typed languages, 60

E

Encoding data

aesthetic graphics, 229

aesthetic mappings, 237

choosing effective graphical encodings, 220–222

Endpoints, web APIs, 183–185

Environment pane, RStudio, 55

Error handling

command line, 18–19

debugging functions, 78

reading error messages, 63

Ethical responsibilities, 343

Excel, working with CSV data, 124

exit

disconnecting from remote computer, 22

stopping or canceling program or running command, 19

Expressions, multiple operators in, 61

Extensions, file, 6, 48–49

F

facet_ functions, 244–245

Facets

ggplot2 package, 244–245

Grammar of Graphics, 232

Factors

creating data frames, 120

variables, 126–129

Feature branches

in centralized workflow, 333–335

working with, 329–331

Fields (columns), in relational databases, 168

figure( ), creating Bokeh plots, 262–263

Files

adding to repository, 32–33

changing directories, 12–13

creating .Rmd files, 276–278

extensions, 6, 48–49

ignoring, 42–44

listing, 13

managing, 15–16

navigating, 11–12

specifying paths, 14–15

fill( ), aesthetic layouts, 238–240

filter( )

dplyr core functions, 131, 135–136

example report on life expectancy, 289

manipulating table data, 177–178

Filtering

joins, 148

vectors, 90–91, 93

flatten( )

example finding Cuban food in Seattle, 200, 202

JSON data, 196–197

for loops, 87

Foreign keys, in relational databases, 168–169

fork, repos on GitHub, 36–38

Forking workflow

  feature branches in, 331, 333–335

  working with, 335–339

Formats

  table, 157

  text, 46

Formulas, 245

Frameworks

  defined, 293

  Shiny framework. *See* Shiny framework

fromJSON( ), converting JSON string to list, 193–194, 200

full_join( ), 148

function keyword, 76

Functions

  for aesthetic mappings (aes( )), 237–238

  applying to lists, 102–103

  built-in, 71–72

  c( ) function, 81–82

  conditional statements, 79–80

  converting dplyr functions into SQL statements, 178

  coord_ functions, 243–244

  correlation function (cor( )), 161

  creating lists, 96

  debugging, 78. *See also* Error handling

  developing application servers, 307–309

  geometry. *See* geom_ functions

  inspecting data frames, 121–122

  loading, 73–75

  named arguments, 72–73

  nested statements within, 140–141

  overview of, 69–70

  referencing database table, 177

  in Shiny layouts, 305

  syntax, 70–71

  tidyr functions for changing columns to/from rows, 157–159

  vectorized, 86–88

  viewing available data sets (data( )), 124–125

  writing, 75–77

Functions, dplyr

  arrange( ), 137–138

  core functions, 131–132

  filter( ), 135–136

  group_by( ), 142–144

  left_join( ), 145–147

  mutate( ), 136–137

  overview of, 132

  select( ), 133–134

  summarize( ), 138–139

  summarizing information using, 313

G

gather( )

  applying to educational statistics, 161–163

  combining with spread( ), 159

  tidyr function for changing columns to rows, 157–158

geom_ functions

  adding titles and labels to charts, 247–248

  aesthetic mappings and, 237–238

  creating choropleth maps, 249–250

  creating dot distribution maps, 252

  example mapping evictions in San Francisco, 253–256

  rendering plots, 284

  specifying geometric objects, 234

  specifying geometries, 235–237

  statistical transformation of data, 237

Geometries

  ggplot2 layers, 232

  position adjustments, 238–240

  specifying geometric objects, 234–235

  specifying with ggplot2 package, 235–237

GET

  example finding Cuban food in Seattle, 197–198, 202

  HTTP verbs, 188–189

  sending GET requests, 189–190

getwd( ), viewing working directory, 125

ggmap package

  example finding Cuban food in Seattle, 200–203

  example mapping evictions in San Francisco, 253

  map tiles, 252

ggplot( )

  creating plots, 232, 234

  example mapping evictions in San Francisco, 256

ggplot2 package

  aesthetic mappings, 237–238

  basic plotting, 232–235

  choropleth maps, 248–251

  coordinate systems, 243–244

  dot distribution maps, 252

  example finding Cuban food in Seattle, 200

  example mapping evictions in San Francisco, 252–256

  facets, 244–245

  Grammar of Graphics, 231–232

  labels and annotations, 246–248

  map types, 248

  position adjustments, 238–240

  rendering plots, 284

  specifying geometries, 235–237

static plot of iris data set, 257–258

statistical transformation of data, 255

styling with scales, 240–242

tidyr example, 160–161

ggplotly( ), 259

ggrepel package, preventing labels from overlapping, 247–248

git

accessing project history, 40–42

adding files, 32–33

branching model. *See* Branches

checking repository status, 31–33

committing changes, 33–35

core concepts, 27–28

creating repository, 30–31

ignoring files, 42–44

installing, 5

leveraging using GitHub, 6

local git process, 35

managing code with, 3–4

overview of, 27–28

project setup and configuration, 30

tracking changes, 32

tutorials, 43–44

version control, 4

Git Bash. *See also* **Bash shell**

accessing command line, 9–10

commands used by, 13

executing code using Bash shell, 4–5

ls command, 13

tab-completion support, 15

Git Flow model, 335

GitHub

accessing project history, 40–42

creating centralized repository, 331–333

creating GitHub account, 6

forking/cloning repos on GitHub, 36–38

ignoring files, 42–44

managing code with, 3

overview of, 29

pushing/pulling repos on GitHub, 38–40

README file, 48–49

sharing reports as website, 285–286

storing projects on, 36

tutorials, 43–44

.gitignore, ignoring files, **42–44**

GitLab, comparing with GitHub, **29**

Google Docs, version control systems compared with, **28**

Google, getting help via, **63**

Google Sheets, working with CSV data, **124**

Government publications, sources of data, **108**

Grammar of Data Manipulation (Wickham), **131**

Grammar of Graphics, **231–232**

Graphics. *See also* by individual types of graphs; Data visualization

aesthetics, 229–230

choosing effective graphical encodings, 220–222

expressive displays, 227–229

with ggplot2. *See* ggplot2 package

Grammar of Graphics, 231–232

leveraging preattentive attributes, 226–227

selecting visual layouts, 209–210

visualizing hierarchical data, 217–220

group_by( )

analyzing data frames by group, 142–144

facets and, 244

statistical transformation of data, 255

summarizing information using, 313

GROUP_BY clause, **SQL SELECT, 174**

H

Heatmaps. *See also* **Choropleth maps**

data visualization with multiple variables, 215, 217

example mapping evictions in San Francisco, 256

Help

R language, 63–64

RStudio, 55

Hidden files, 42–44

Hierarchical data, visualization of, **217–220**

Histograms

data visualization with multiple variables, 216

expressive displays, 229

visualizing data with single variable, 210

Hosts, Shiny apps, 309–310

HSL Calculator, 223

HSL (hue-saturation-lightness) color model, 222–223

HTML (Hypertext Markup Language)

HTML Tags Glossary, 300–301

markup languages, 45

sharing reports as website, 284–286

web development language, 342

HTTP (HyperText Transfer Protocol)

header, 196–197

overview of, 181–182

verbs, 188–189

HTTP requests

example finding Cuban food in Seattle, 196–200

response header and body, 190

web services and, 181

HTTP verbs, Web APIs, 188–189

httr package

parsing JSON data, 192–193

sending GET requests, 189–190

Hue

choosing effective colors for data visualization, 222

multi-hue color scales, 225

Hue-saturation-lightness (HSL) color model, 222–223

Hyperlinks, markdown, 46–47

I

Icons, types of interfaces, 9

IDE (integrated development environment), 54

if_else, conditional statements, 79–80

Images, markdown, 47–48

Indices

for getting subsets of vectors, 88–89

multiple indices, 89–90

init (git), turning a directory into a git repository, 31

Inline code, in R Markdown, 280

INNER JOIN clause, SQL SELECT, 174

inner_join(), 147–148

Inputs

dynamic inputs with Shiny framework, 301–303

functions and, 69

Shiny framework, 293–294

Integer data type, 63

Integrated development environment (IDE), 54

Interactivity

interactive data visualization. *See* Data visualization, interactive

interactive web applications. *See* Shiny framework

Interface

command line as, 9

defined, 181

user. *See* UIs (user interfaces)

web APIs. *See* Web APIs

Interpreted languages, 53

Interval data, measuring data, 111

iris data set, interactive plots in, 257–258

Italics, text formatting, 45–46

J

JavaScript, 342–343

join()

dplyr core functions, 131

joining data frames, 144–148

JOIN clause, SQL SELECT, 174–175

Journalism, sources of data, 109

JSON (JavaScript Object Notation)

flattening JSON data, 195–197

list of lists structure in, 97

parsing JSON data, 193–195

processing JSON data, 191–193

jsonlite package, 192–193

K

kable(), knitr package, 283–284, 291

Key-value pairs

JSON (JavaScript Object Notation), 191

query parameters and, 184

tidyr data tables, 157

knitr package

creating R Markdown documents, 275

kable(), 283–284, 291

Knitting documents, 278

L

Labels

adding to plots, 246–248

aesthetics of graphics, 230

labs(), adding titles and labels to charts, 246

lapply(), applying functions to lists, 102–103

Layers, ggplot2 package, 232

layout(), 260–261, 268

Layouts

coordinate systems, 243–244

designing UIs, 299

example exploring changes to Seattle, 268

facets, 244–245

labels and annotations, 246–248

plotly package, 260–261

position adjustments, 238–240

selecting visual, 209–210

Shiny framework, 304–306

styling with scales, 240–242

Lazy evaluation, in dplyr package, 178

leaflet()

creating Leaflet map, 264

example exploring changes to Seattle, 269

leaflet package
    creating interactive plots, 264–266
    example exploring changes to Seattle, 269–271
    installing and loading, 263
    Shiny app example applying to fatal police shootings, 312–313
Learn Git Branching, 339
LEFT JOIN clause, SQL SELECT, 174
left_join()
    example of join operation, 145–146
    join types, 146–147
Legends
    adding to Leaflet map, 270–271
    aesthetics of graphics, 230
length() function, determining number of elements in a vector, 82
Libraries. *See* Packages
library(), referencing external packages, 311
Lightness, choosing effective colors for data visualization, 223
Linux
    command-line tools on, 5
    installing git, 5
list() function, creating lists, 96
Lists
    accessing elements of, 97–99
    applying functions to, 102–103
    converting JSON string to list, 193–194
    creating, 96–97
    creating data frames, 120–121
    double-bracket notation, 101
    JSON structures compared with, 192–193
    listing files from command line, 13
    modifying, 100
    overview of, 95
    rendering Markdown lists, 282–283
log, viewing commit history, 40
Logical (boolean)
    data type, 61–63
    debugging functions, 78
    operators, 62–63
    vector filtering by values, 90–91
Loops, vectorized functions and, 87
ls
    list folder contents, 13
    using with remote computer, 22

M

-m option, adding messages to commit command, 34
Mac OSs. *See also* Terminal (Mac)
    accessing command line, 9–10
    command-line tools on, 4
    installing git, 5
Machine learning, making predictions, 342
Mackinlay's Expressiveness Criteria, 227–229
man, looking up commands in manual, 16–17
Map tiles
    adding to Leaflet map, 264
    ggmap package, 252
Maps
    aesthetic mappings, 237–238
    choropleth maps, 248–251
    dot distribution maps, 251–252
    example mapping evictions in San Francisco, 252–256
    interactive, 263
    types of, 248
Markdown
    hyperlinks, 46–47
    images, 47–48
    overview of, 45
    rendering, 48–50
    rendering lists, 282–283
    rendering strings, 281
    rendering tables, 283–284
    static content elements of UIs, 300–301
    tables, 48
    text formatting and blocks, 46
Markdown Reader, 49
Markers, adding to Leaflet map, 264
Markup languages, 45
Mathematical operators
    applying to vectors, 83
    assigning values to variables, 59
    using on numeric data types, 60
    vectorized functions and, 86–87
Matrix, two-dimensional data structures in R, 122
.md file extension, for markdown files, 48
Menus, types of interfaces, 9
merge (git)
    combining branches, 324–325
    forking/cloning repository on GitHub, 337–338
    resolving merge conflicts, 327–328
    working with feature branches, 330, 334–335
Merging, git core concepts, 29
message etiquette, commit, 34–35
Meta-data, 114–116, 277
Microsoft Excel, 124
Microsoft Windows. *See* Windows OSs

mkdir, documentation of commands, 16–17

Moral responsibility, 343

mutate()

dplyr core functions, 131, 136–137

example finding Cuban food in Seattle, 202

example report on life expectancy, 289–290

Mutating joins, 148

MySQL, 171

N

NA value

compared with NULL, 100

logical values and, 89

modifying vectors and, 92

Named arguments, **R** functions, 72–73

Named lists, creating data frames, 120

names() function, creating lists and, 96

Negative index, vector indices, 89

Nested objects, JSON support, 192

Nested statements, within other functions, 140–141

Nested structures, visualizing hierarchical data, 217–220

Networking commands, 20–23

News, sources of data, 109

Nominal (categorical) data

choosing effective colors for data visualization, 223

data visualization with multiple variables, 215

measuring data, 110

proportional representation of data and, 212

selecting visual layouts and, 209–210

visualizing single variable, 210

Non-standard evaluation (NSE), **dplyr**, 133

NULL value, modifying lists and, 100

Numbers, working with CSV data, 124

Numeric data type, 60–61, 95

O

OAuth, API authentication service, 187

Observations, data structures, 111–112

ON clause, SQL **SELECT**, 174

Online communities, sources of data, 109

Open source, **R** language as, 53

OpenStreetMap, 264

Operationalization, using data to answer questions, 116–118

Optional arguments, functions and, 72

Options (flags), argument syntax, 16

OPTIONS, HTTP verbs, 188

ORDER_BY clause, SQL **SELECT**, 174

Ordinal data

measuring data, 110–111

selecting visual layouts and, 209–210

Orientation, **tidyr** data tables, 157

Out-of-bounds indices, vector indices, 89

OUTER JOIN clause, SQL **SELECT**, 174

Outliers, visualizing data with single variable, 210

Output

directing/redirecting, 20

dynamic, 303–304

functions and, 69

reactive, 295

Shiny framework, 293–294

P

Packages

Bokeh, 261

dbplyr, 176–179

dplyr. *See* dplyr package

ggmap. *See* ggmap package

ggplot2. *See* ggplot2 package

ggrepel, 247–248

httr, 189–190, 192–193

jsonlite, 192–193

knitr. *See* knitr package

leaflet. *See* leaflet package

plotly, 258–261

of R functions, 73–75

rbokeh, 261–263

RColorBrewer, 224–225

referencing external, 311

rmarkdown, 275

RStudio, 55

tidyr. *See* tidyr package

tidyverse, 132, 142

Panning, interactive data visualization, 257

Parameters

function inputs, 69–70

query parameters, 184–186, 202

Passing arguments

debugging functions, 78

to functions, 70

PATCH, HTTP verbs, 188

Paths

finding, 57

on remote computers, 22

specifying from command line, 14–15

viewing working directory, 125

Pie charts, 211–213, 221

pipe operator (%>%), **dplyr** package, 141–142

pipe table, 48

plot_ly( )

creating plots, 260

example exploring changes to Seattle, 268

plotly package

creating interactive plots, 259–261

example exploring changes to Seattle, 268

loading, 258

Plots

ggplot2 package. *See* ggplot2 package

plotly package. *See* plotly package

plotting, 232–235

rendering in R Markdown, 284

RStudio, 55

Pointers, types of interfaces, 9

Popups, adding interactivity to Leaflet map, 266

Positional arguments

functions and, 72–73

ggplot2 geometries, 238–240

PostgreSQL, 170–171, 176

Powershell, Windows Management Framework, 5

Preattentive processing, in data visualization, 226–227

Predictions, 342

Preview Markdown rendering, 49

Primary keys, in relational databases, 168–169

print( ), analyzing flight data, 152

Probability, 342. *See also* Statistics

Problem domain, interpreting data by domain, 112–113

Programming/programming languages

compiled languages, 53

data wrangling, 106

dynamically vs. statically typed languages, 60

interpreted languages, 53

learning, 342–343

markup languages, 45

R language. *See* R language

S language, 53

SQL. *See* SQL (Structured Query Language)

statically typed, 60

statistical languages, 53

Proportional representation, visualizing data with single variable, 211–212

publishing apps, Shiny framework, 309–311

pull (git)

creating centralized repository, 333

merging from GitHub, 328

repos on GitHub, 38–40

understanding/using git commands, 43

working with feature branches, 335

Pull request, GitHub, 335–339

push (git)

creating centralized repository, 333

merging from GitHub, 328–329

repos on GitHub, 38–40

understanding/using git commands, 43

working with feature branches, 333–335

pwd, print working directory, 11, 22

Python, 342

Q

qmplot( ), creating background maps, 253–254

Query parameters

example finding Cuban food in Seattle, 202

in Web URIs, 184–186

quit (q), stopping or canceling program or running command, 19

R

R for Everyone, 341

R language

accessing databases, 175–179

accessing Web APIs, 189–190

anscombe data set in, 208

arguments, 72–73

built-in functions, 71–72

code chunks and, 279–280

comments, 58

data types, 60–63

downloading, 6–8

as dynamically typed language, 60

function packages, 73–75

function syntax, 70–71

functions in Shiny layouts, 305

help resources, 63–64

interactive data visualization. *See* Data visualization, interactive

learning, 64–67

overview of, 4

programming with, 53–54

running R code from command line, 56–57

running R code using RStudio, 54–56

two-dimensional data structures, 122

variable definition, 58–60

web application framework. *See* Shiny framework

R Markdown

code chunks and, 279–280

creating .Rmd files, 276–278

example report on life expectancy, 287–292

inline code and, 280

knitting documents, 278

rendering lists, 282–283

rendering plots, 284

rendering strings, 281–282

rendering tables, 283–284

setting up reports, 275

sharing reports, 284–286

static content elements of UIs, 300–301

Ratio data, measuring, 111

rbokeh package

creating interactive plots, 262–263

installing and loading, 261–262

RColorBrewer package, 224–225

RDMS (relational database management system), 169.
See also **Relational databases**

Reactive output

dynamic outputs with Shiny framework, 303–304

render functions and, 308

in Shiny framework, 295

Reactivity, in Shiny framework, 295

read.csv( )

creating choropleth maps, 250

example mapping evictions in San Francisco, 253

in R, 161

README file, GitHub, 48–49

Records

data structures, 111–112

keeping, 107–108

Recycling operation, vectors, 84–85

Redirects, output, 20

Relational databases

accessing, 175–179

designing, 144

overview of, 167–169

setting up, 169–171

SQL statements, 171–175

Relational operators

logical values and, 62

vector filtering with, 91

Relationships

assessing in statistical learning, 341–342

between x and y values (statistics), 208–209

Relative path

images, 48

specifying paths, 14

URLs, 47

viewing working directory, 125–126

Remote repository

`git` core concepts, 29

repositories as remotes, 36

Remote computers, accessing, 20–21

Render function

developing application servers, 307–309

in Shiny framework, 295–296

Rendering markdown, 48–50

Reports, 275. *See also* **R Markdown**

Repository (repo)

checking status, 31–33

creating, 30–31

creating centralized repository, 331–333

forking/cloning on GitHub, 36–38, 336–337

`git` core concepts, 28

linking online to local, 36

pushing/pulling on GitHub, 38–40

viewing current branch, 320–321

REpresentational State Transfer. *See* **REST (REpresentational State Transfer)**

Required arguments, functions and, 72

Research, sources of data, 109

reset, destroying commit history, 42

Response body, HTTP requests, 190

Response header, HTTP requests, 190

REST (REpresentational State Transfer)

responding to HTTP requests, 189

web APIs, 182

web services and, 181

Return value

c( ) function, 81–82

function parts, 77

writing functions, 75–76

Reversibility

capabilities of version control systems, 28

reverting to earlier versions, 40–42

revert, reverting to earlier versions, 40–42

RIGHT JOIN clause, SQL SELECT, 174

right_join( ), 145–147

rmarkdown package, creating R Markdown documents, 275

.Rmd files, creating, 276–278

round( ) function, vectorized functions and, 86–87

Rows

arrange( ) operation, 137–138

changing from columns to/from, 157–159

filter( ) operation, 135

Rows (records), in relational databases, 168

`RScript`, running scripts from command line, 57

RStudio

    changing working directory, 125

    cheatsheet, 56, 280, 318

    creating list elements, 97

    creating .Rmd files, 276–278

    debugging functions, 78

    downloading, 8

    getting help via RStudio community, 64

    `ggplot2` graphics in RStudio window, 233

    knitting documents, 278

    running R code, 54–56

    running Shiny apps, 297–298

    writing code with, 3

`rworldmap`, example report on life expectancy, 289, 291

S

`sapply()`, applying functions to lists, 103

Saturation, choosing effective colors for data visualization, 222

Scalable vector graphics (SVGs), 266

Scalar, example adding, 85–86

Scale, `ggplot2`

    color scales, 242–243

    styling with, 240–241

Scatterplot matrix, 213

Scatterplots

    Anscombe's Quartet, 209

    data visualization with multiple variables, 213–217

    `ggplot2` example, 233

Scientific research, sources of data, 109

Scripts

    programming with R language, 53–54

    running from command line, 57

    running using RStudio, 54

`select()`

    `dplyr` core functions, 131, 133–134

    example report on life expectancy, 289–290

    manipulating table data, 177–178

SELECT statement

    ON clause, 174

    JOIN clause, 174–175

    ORDER_BY and GROUP_BY clauses, 174

    SQL statements, 171–174

    WHERE clause, 173–174

Sensors, generating data, 107

`seq()` function, creating vectors and, 82–83

Sequences, performing sequential operations, 139–141

Servers

    application structure in Shiny framework, 296

    building Shiny application, 313–318

    defined, 294

    developing application servers, 306–309

    division of responsibility in Shiny apps, 298–299

Shapefiles, creating choropleth maps, 248–249

Shapes, adding to Leaflet map, 264

Sharing. *See* **Collaboration**

Shiny framework

    application structure, 295–299

    core concepts, 294–295

    designing user interfaces, 299

    developing application servers, 306–309

    dynamic inputs, 301–303

    dynamic outputs, 303–304

    example applying to fatal police shootings, 311–318

    layouts, 304–306

    overview of, 293–294

    publishing Shiny apps, 309–311

    static content, 300–301

`shinyApp()`, 296–297, 299

`shinyapp.io`, hosting Shiny apps, 309–310

Sidebar, in Shiny example, 316

Single-bracket notation. *See* **`[]` (single-bracket notation)**

Slideshows, 275

`snake_case`

    variable names, 58

    writing functions, 76

Snapshots. *See* **Commit**

`source()`, loading and running API keys, 188

`spread()`

    applying to educational statistics, 164–165

    changing rows to columns, 158–159

Spreadsheets, working with CSV data, 124

SQL (Structured Query Language)

    converting `dplyr` functions into SQL equivalents, 178

    JOIN clause, 174–175

    ORDER_BY and GROUP_BY clauses, 174

    resources for learning, 171

    SELECT statement, 171–173

    WHERE clause, 173–174

SQLite

    accessing from R, 176–177

    SELECT statement in, 172

    types of RDMSs, 169–170

    WHERE clause, 173–174

`ssh`, accessing remote computers, 21–22

Stacked bar charts, 211–213, 239

StackOverflow, getting help via, 64

Staging area, adding files, 33. *See also* **add** (**git**)

Statements

    conditional, 79–80

    SQL, 171–175

Static content

    building Shiny application, 313

    Shiny framework, 300–301

Statically typed language, 60

Statistical learning

    assessing relationships, 341–342

    making predictions, 342

    overview of, 341

Statistics

    Anscombe's Quartet, 208–209

    applying `tidyr` to educational statistics, 160–165

    statistical transformation of data, 237, 255

status (**git**)

    checking project status, 323

    checking repository status, 31–33

    pushing branches to GitHub, 329

    resolving merge conflicts, 327–328

    understanding/using `git` commands, 43

Strings

    character data types, 61

    rendering in R Markdown, 281–282

Style, vs. syntax, 59

Sublime Text, selecting text editor, 7

Subplots, facets and, 244

Subset, of vector, 88–89

summarize( ), **dplyr** core functions, 131, 138–139

Sunburst diagrams, 218, 220

Surveys, generating data, 107

SVGs (scalable vector graphics), 266

Syntax

    debugging functions, 78

    vs. style, 59

Syntax-colored code blocks, markdown options, 48

T

Tab-completion, command shells supporting, 15

Tables

    building Shiny application, 314–318

    creating data frames, 120

    data structures, 111–112

    JOIN clause, 174

    markdown, 48

    referencing database table, 177

    in relational databases, 168

    rendering, 283–284

    `tidyr`, 157

Tagged elements, in lists, 95–96

tbl( ), referencing database table, 177

Terminal (command shell). *See* Command line

Terminal (Linux), 5

Terminal (Mac)

    accessing, 9–10

    connecting to remote server, 21

    executing code, 4

    `ls` command, 13

    manuals (man pages), 17

    running R code, 56–57

    setting up, 4

    tab-completion support, 15

Text blocks, markdown, 46

Text editor, 6–7

Text formatting, 46

theme( ), creating choropleth maps, 251

Tibble data frame, 142–143

tidyr package

    applying to educational statistics, 160–165

    changing from columns to/from rows, 157–159

    example mapping evictions in San Francisco, 252

    orienting data frames for plotting, 239

    overview of, 155–157

    reshaping data sets, 165

The tidyverse style guide

    defining variables, 58

    `dplyr` package, 132

    tibble data frame, 142–143

    writing functions, 76

Treemaps, 211–213, 218–220

Tutorials, for learning **R**, 65–66

U

UIs (user interfaces)

    application structure in Shiny framework, 295–296

    building Shiny application, 313–318

    defined, 294

    designing, 299

    division of responsibility in Shiny apps, 298–299

Unit of analysis, grouping for redefining, 144

Unordered lists, rendering Markdown lists, 282–283

URIs (Uniform Resource Identifiers)

    example finding Cuban food in Seattle, 202

    HTTP requests and, 182–184

    hyperlink syntax, 46–47

URLs (Uniform Resource Locators), 182, 286

User interfaces. See UIs (user interfaces)

Users, accessing command line, 10

V

Values

    creating vectors, 81–82

    modifying vectors, 92–93

    tidyr cells representing, 155

    vectors as one-dimensional collections of, 81

Variables

    anonymous, 71, 140

    breaking data into, 142

    creating intermediary variables for use in analysis, 139

    data visualization with multiple, 213–217

    data visualization with single, 210–213

    defining, 58–60

    factor variables, 126–129

    storing Shiny layouts in, 305

    tidyr columns representing, 155

VCS (version control system), 28

Vectorized functions, 86–88

Vectors

    creating, 81–83

    creating data frames, 120

    example adding, 85–86

    filtering, 90–91

    lists and, 95

    modifying, 92–93

    multiple indices, 89–90

    overview of, 81

    performing operations on, 83–84

    recycling operation, 84–85

    subsets of, 88–89

    vectorized functions, 86–88

Verbs

    dplyr package, 131

    HTTP verbs, 188–189

Version control

    accessing project history, 40–42

    adding files, 32–33

    checking repository status, 31–33

    command line in, 9

    committing changes, 33–35

    creating repository, 30–31

    forking/cloning repos and, 36–38

    git for, 4, 27–29

    GitHub for, 29

    ignoring files, 42–44

    local git process, 35

    overview of, 27

    project setup and configuration, 30

    pushing/pulling repos and, 38–40

    storing projects on GitHub, 36

    tracking changes, 32, 319–320

Version control system (VCS), 28

Videos, resources for learning R, 65

Violin plots

    data visualization with multiple variables, 215

    data visualization with single variable, 210

Visual channels, aesthetic mappings and, 237

Visual storytelling with D3, 343

Visualization. See Data visualization

VS Code (Visual Studio Code)

    preview rendering support, 49

    selecting text editor, 7

W

Web APIs

    access tokens (API keys), 186–188, 196–197

    accessing from R, 189–190

    example locating Cuban food in Seattle, 197–203

    flattening JSON data, 195–197

    HTTP verbs, 188–189

    overview of, 181–182

    parsing JSON data, 193–195

    processing JSON data, 191–193

    query parameters, 184–186

    RESTful requests, 182

    URIs and, 182–184

Web applications

    defined, 293

    interactive. See Shiny framework

Web browsers, Shiny framework as interface, 293–294

Web servers, 182. See also Servers

Web services. See also Web APIs

    overview of, 181

    registering with, 186–188

Webpage, URL for, 286

Websites

    creating using R Markdown, 275

publishing Shiny apps, 309–311

sharing R Markdown reports, 284–286

WHERE clause, SELECT statement, 173–174

Widgets. *See* **Control widgets**

Wildcards, command line, 17–18

Windows, icons, menus, and pointers (WIMP), 9

Windows Management Framework, 5

Windows OSs

accessing command line, 9–10

command-line tools, 4–5

installing git, 5

Windows, types of interfaces, 9

Workflows

centralized, 331

creating centralized repository, 331–333

tracking code versions with branches, 319–320

working with feature branch workflows, 333–335

working with forking workflows, 335–339

X

Xcode command line developer tools, 5

Z

Zooming, interactive data visualization, 257

Credits

Cover: Garry Killian/Shutterstock

Chapter 2, Figures 2.1a, 2.2, 2.4, 2.5, 2.6, 2.7: Screenshot of Mac © 2018 Apple Inc.

Chapter 2, Figure 2.1b: Screenshot of Git Bash © Software Freedom Conservancy, Inc.

Chapter 2, Table 2.1: Lagmonster

Chapter 3: "A version control system ... reversibility, concurrency, and annotation". Raymond, E. S. (2009). Understanding version-control systems. http://www.catb.org/esr/writings/versioncontrol/version-control.html

Chapter 3, Figures 3.1, 3.2, 3.3, 3.8: Screenshot of Mac © 2018 Apple Inc.

Chapter 3: "If you forget the -m option, git ... happens to everyone". Stack Overflow: Helping One Million Developers Exit Vim, David Robinson, Stack Exchange Inc.

Chapter 3, Figure 3.3: Freepik Company S.L.

Chapter 3, Figures 3.5, 3.6: Screenshot of GitHub's web portal © 2018 GitHub Inc.

Chapter 4, Figures 4.1, 4.2: Screenshot of Markdown © 2002–2018 The Daring Fireball Company LLC

Chapter 5, Figures 5.1, 5.4, 5.5: Screenshot of Rstudio © 2018 Rstudio

Chapter 5, Figures 5.2, 5.3a: Screenshot of Mac © 2018 Apple Inc.

Chapter 5, Figure 5.3b: Screenshot of Git Bash © Software Freedom Conservancy, Inc.

Chapter 6, Table 6.1: Built-in Functions, Quick-R, Robert I. Kabacoff, Ph.D.

Chapter 6, Figure 6.1: Screenshot of Rstudio © 2018 Rstudio

Chapter 7, Figure 7.1: Screenshot of Rstudio © 2018 Rstudio

Chapter 8, Figure 8.1: Screenshot of Rstudio © 2018 Rstudio

Chapter 9, Table 9.1: Stevens, S. S. (1946). On the theory of scales of measurement. Science, 103(2684), 677–680. https://doi.org/10.1126/science.103.2684.677

Chapter 9, Figure 9.1: Screenshot of Rstudio © 2018 Rstudio

Chapter 9, Figure 9.2: City of Seattle, Land Use Permits: https://data.seattle.gov/Permitting/Land-Use-Permits/uyyd-8gak

Chapter 9, Figure 9.3: Screenshot of Treemaps © IHME

Chapter 10, Figures 10.1, 10.2, 10.3: Screenshot of Rstudio © 2018 Rstudio

Chapter 11, Figures 11.1, 11.3, 11.4, 11.5, 11.6, 11.8, 11.9, 11.10, 11.11, 11.13, 11.14: Screenshot of Rstudio © 2018 Rstudio

Chapter 11, Figures 11.2, 11.15: Screenshot of ggplot2 © Hadley Wickham

Chapter 11, Figure 11.12: SQL Joins, http://www.sql-join.com/sql-join-types/

Chapter 12, Figures 12.1, 12.2, 12.3, 12.5: Screenshot of Rstudio © 2018 Rstudio

Chapter 12, Figures 12.4, 12.6, 12.7: Screenshot of ggplot2 © Hadley Wickham

Chapter 13, Figures 13.3, 13.4, 13.5: Screenshot of SQLite Browser © DB Browser

Chapter 13, Figure 13.6: Screenshot of Rstudio © 2018 Rstudio

Chapter 14, Figures 14.3, 14.5, 14.6, 14.8, 14.9, 14.11: Screenshot of Rstudio © GitHub Inc.

Chapter 14, Figure 14.4: Screenshot of Mac © 2018 Apple Inc.

Chapter 14, Figure 14.7: Screenshot of JSON © JSON

Chapter 14, Figure 14.10: Screenshot of Yelp Fusion © 2004–2018 Yelp Inc.

Photo izusek/gettyimages

Register Your Product at informit.com/regiser

Access additional benefits and **save 35%** on your next purchase

- Automatically receive a coupon for 35% off your next purchase, vald for 30 days. Look for your code in your InformIT cart or the Manage Codes section of your account page.

- Download available product updates.

- Access bonus material if available.*

- Check the box to hear from us and receive exclusive offers on new editions and related products.

Registration benefits vary by product. Benefits will be listed on your account page under Registered Products.

InformIT.com—The Trusted Technology Learning Source

InformIT is the online home of information technology brands at Pearson, the world's foremost education company. At InformIT.com, you can:

- Shop our books, eBooks, software, and video training
- Take advantage of our special offers and promotions (informit.com/promotions)
- Sign up for special offers and content newsletter (informit.com/newsletters)
- Access thousands of free chapters and video lessons

Connect with InformIT—Visit informit.com/community

the trusted technology learning source

Addison-Wesley • Adobe Press • Cisco Press • Microsoft Press • Pearson IT Certification • Prentice Hall • Que • Sams • Peachpit Press

Ⓟ Pearson